Refashioning the Renaissance

Manchester University Press

To buy or to find out more about the books currently available in this series, please go to: https://manchesteruniversitypress.co.uk/series/studies-in-design-and-material-culture/

STUDIES IN DESIGN & MATERIAL CULTURE

Refashioning the Renaissance

Everyday dress in Europe, 1500–1650

Edited by
Paula Hohti

Manchester University Press

This book is part of the Refashioning the Renaissance project that has received funding from the European Research Council (ERC) under the European Union's Horizon 2020 Research and Innovation Programme (Grant Agreement No 726195).

Published by Manchester University Press
Oxford Road, Manchester, M13 9PL
www.manchesteruniversitypress.co.uk

British Library Cataloguing-in-Publication Data
A catalogue record for this book is available from the British Library

ISBN 978 1 5261 6465 0 hardback

First published 2025

Typeset
by Cheshire Typesetting Ltd, Cuddington, Cheshire
Printed in Great Britain
by Bell & Bain Ltd, Glasgow

For mum and dad

Contents

Figures

Chapters

Experiments in focus

Tables

Notes on contributors

Paula Hohti is a professor and a fashion and material culture historian at Aalto University (Finland). Her research focuses on the Italian Renaissance, with a special focus on dress and decorative arts. Hohti has been a principal investigator within the international The Material Renaissance and Fashioning the Early Modern projects, and a Marie Curie Research Fellow at the Centre for Textile Research in Copenhagen. In 2016, she received a €2 million ERC consolidator grant to develop material hands-on experimentation and historical reconstruction as a methodology in dress and fashion history.

John Styles is Professor Emeritus in History at the University of Hertfordshire (UK). Before taking up his professorship at Hertfordshire, he was Head of Graduate Studies at the Victoria and Albert Museum in London, where he remains Honorary Senior Research Fellow. He specialises in the history of early modern Britain and its empire, especially the study of textiles, material life, manufacturing and design. His books include *Gender, Taste and Material Culture in Britain and North America, 1700–1830* (co-edited with Amanda Vickery, Yale University Press, 2006), *The Dress of the People: Everyday Fashion in Eighteenth-Century England* (Yale University Press, 2007), and *Threads of Feeling: The London Foundling Hospital's Textile Tokens, 1740–1770* (Foundling Museum, 2010). He curated the exhibition 'Threads of Feeling' at the London Foundling Museum in 2010–11 and at the de Witt Wallace Decorative Arts Museum at Colonial Williamsburg, Virginia, USA in 2013–14. He is currently writing a book on fashion, textiles and the origins of the Industrial Revolution.

Andrea Caracausi is Professor of Economic History at the University of Padua (Italy). He specialises in the social and economic history of Italy and the Mediterranean world, with particular focus on textile manufacturing, guilds, women's work and child labour. He received a BA in History (University of Padua) and a PhD in Economic and Social History (Bocconi University, Milan). Recent publications include 'Fashion, capitalism and ribbon-making in early modern Europe', in Thomas Max Safley (ed.), *Labor before the Industrial Revolution: Work, Technology and Their Ecologies in an Age of Early Capitalism* (Routledge, 2019).

Sophie Pitman is the Pleasant Rowland Textile Specialist and Research Director of the Helen Louise Allen Textile Collection at the University of Wisconsin Madison (USA). A cultural historian specialising in the early modern period, she is particularly interested in clothing, textiles, material culture and reconstruction methodology. Formerly she was a Leverhulme Early Career Fellow (UCL) and held postdoctoral positions on the Making and Knowing Project (Columbia University) and the 'Refashioning the Renaissance' project (Aalto University), where she led the experimental strand of research including the reconstruction of a mockado doublet. Pitman holds a PhD in History from the University of Cambridge. She has published widely on clothing and textiles, including investigations into sumptuary law, luxury and identity, and early modern materials and craft practice. Her current research explores clothing, weather and the environment. She recently curated the exhibition *Remaking the Renaissance* (February-May 2024, Lynn Mecklenburg Textile Gallery) which featured many of the Refashioning experimental reconstructions.

Michele Nicole Robinson prepared her chapter for this book as a postdoctoral researcher on the project 'Refashioning the Renaissance, Popular Groups and the Material and Cultural Significance of Clothing in Europe 1550–1650', which was funded by the European Research Council. Her work explored the role of print culture in the dissemination of ideas about fashion in early modern Italy.

Natasha Awais-Dean is a Visiting Research Fellow in the Department of History, King's College London (UK). She holds degrees from the University of Cambridge, the Royal College of Art and a doctorate from Queen Mary, University of London. Natasha has curatorial experience from working at three national museums – the British Museum, the Victoria and Albert Museum and the National Portrait Gallery, London – where she developed a love for working with objects. Her research interests include the material culture of early modern Europe, with specialism in jewellery and metalwork. Her publication *Bejewelled: Men and Jewellery in Tudor and Jacobean England* (British Museum Press, 2017, rev. edn 2022) reinterprets collections within the British Museum, revealing that jewellery was owned, worn and valued by a range of men in sixteenth- and early seventeenth-century

English society. Natasha is a Trustee of the Society of Jewellery Historians, where she has previously held editorial roles (Features Editor of *Jewellery History Today* and Co-editor of *Jewellery Studies*).

Victoria Bartels is a cultural historian of early modern Italy. She holds a PhD in Early Modern European History from the University of Cambridge, where her doctoral research examined the cultural role of arms and armour in Renaissance Florence. She worked as a Research Fellow on the project 'Refashioning the Renaissance: Popular Groups and the Material and Cultural Significance of Clothing in Europe 1550–1650', funded by the European Research Council. She also previously held a Kress Fellowship at the Medici Archive Project in Florence. She currently works as a post-doctoral researcher at Aalto University and a Lecturer at Syracuse University in Florence.

Stefania Montemezzo is an Ahmanson Fellow at I Tatti, the Harvard Center for Italian Renaissance Studies, in Florence, Italy. She specialises in Renaissance Italy and the Mediterranean, focusing on late medieval trade and Venetian mercantile practices. Her other main area of interest is the material culture of popular groups in Renaissance Italy. She has held fellowships from the Universities of Bologna and Padua Aalto University and Villa I Tatti-Harvard and is co-author of the database 'Refashioning the Renaissance', which collects thousands of fashion items from Italian inventories. She was a visiting scholar at Cambridge University, St Andrews University and Gustave Eiffel University (Paris). In 2017 she received the 'Ugo Tucci' award from the Istituto Veneto di Lettere Scienze e Arti (Venice) for her doctoral dissertation.

Astrid Wendel-Hansen is a postdoctoral researcher at the Department of History, Friedrich Schiller University, Jena (Germany). She is interested in early modern northern Europe, particularly Sweden and its relationship with its Baltic territories. Her research focuses on relations of inequality, material culture and consumption as well as the daily economic and provisioning activities of middling and poor people. Her current research focuses on thefts and the economy of early modern Sweden. She defended her PhD in 2020 at the University of Uppsala with a dissertation titled *Dress Matters. Clothes and Social Order in Tallinn, 1600–1700.* Her other publications include 'The fabric of a corporate society: Sumptuary laws, social order and propriety in early modern Tallinn', in Johanna Ilmakunnas and Jon Stobart (eds), *A Taste for Luxury in Early Modern Europe* (Bloomsbury Academic, 2017) and an edited volume *To Take Us Lands Away: Essays in Honour of Margaret R. Hunt* (Acta Universitatis Upsaliensis, 2022).

Elizabeth Currie is a lecturer and author specialising in the history of fashion and textiles. She has published articles in the *Bulletin of the John Rylands Library*, *Fashion Theory*, *Renaissance Studies* and the *Journal of Medieval and Early Modern Studies* on tailoring, clothing retail and production, health, gender and the body. Her books include *Fashion and Masculinity in Renaissance Florence* (Bloomsbury Academic, 2016) and the edited volume *A Cultural History of Dress and Fashion: The Renaissance* (Bloomsbury Academic, 2017). She is currently researching the clothing of non-elite groups in Italian Baroque art.

Jordan Mitchell-King is a PhD researcher at De Montfort University (UK). Her research is centred on the relationship between the materiality of textiles and the cultures surrounding their use. Her current work focuses on eighteenth-century women's undress clothing in Britain, exploring the complex ways in which elite women experienced and understood informal forms of dress in the period. Jordan is also a maker and often experiments with research through the making and wearing of reproduced historical clothing. She studied at the University of York for her undergraduate degree in History and at the Royal College of Art/Victoria and Albert Museum for her MA in History of Design.

Anne-Kristine Sindvald Larsen is a cultural historian from Denmark. She received her master's degree in history from the University of Copenhagen in 2016. She was a PhD candidate in the ERC-funded 'Refashioning the Renaissance' project and the Department of Art and Media at Aalto University, Finland. In her PhD dissertation she focused on the culture and meanings of clothing of Danish artisans and their wives living in the town of Elsinore from 1550 to 1650.

Piia Lempiäinen holds a MA in Finnish and Nordic History from the University of Helsinki. Her own research focused on medieval and early modern clothing terminology and class, and over the years she has deepened her knowledge on various historical textile crafts. She worked in the ERC funded 'Refashioning the Renaissance' project for four years as a project co-ordinator, where among other responsibilities she was part of developing the project's participatory research framework through co-ordinating the citizen science project on early modern stockings.

Valerio Zanetti is the Sir John Elliott Junior Research Fellow in History at Oriel College, University of Oxford. He completed a History PhD at the University of Cambridge in 2020 and has since held fellowships with the Society for Renaissance Studies, the Warburg Institute, I Tatti – The Harvard University Center for Italian Renaissance Studies, the Herzog August Bibliothek and the British School at Rome. His research and publications focus on the history of the body across early modern Europe, exploring topics ranging from women's athletic culture to horse-riding and hunting, embodied nationality and educational games, dress and fashion.

Maarit Kalmakurki is a scenographer, researcher and educator. She is a lecturer at the Department of Culture at Xamk – South-Eastern Finland University of Applied Sciences, where she also leads the digital design part in a Creative Europe funded project *Time for (S)heroes* (2024–28). Maarit's doctoral thesis (Aalto University, 2021) investigated digital character costume design in computer-animated feature films. This pioneering work has been recognised by the industry and media, such as *The Walt Disney Animation* and *The New York Times*. She has conducted practice-led research by digitally reconstructing historical clothing in *Refashioning the Renaissance* and in an international, multidisciplinary research project *Visualising Lost Theatres*. Maarit has received the Herbert D. Greggs Honor Award from the USITT editorial staff and the TD&T Publications Committee for her article studying performance costumes in Stockholm during the Gustavian era.

Elena Kanagy-Loux descends from the Amish and grew up between the USA and Japan, immersed in both traditional Mennonite craft and the DIY fashion scene in Tokyo. After earning her BFA in Textile Design from FIT, she won a grant to study lace-making across a dozen European countries for four months in 2015. Upon returning to NYC, she co-founded Brooklyn Lace Guild and began teaching bobbin-lace classes. In 2018 she completed her MA in Costume Studies at New York University and now works as the Collections Specialist at the Antonio Ratti Textile Center at the Metropolitan Museum of Art.

Acknowledgements

The research for this book began in 2016, when I received an ERC Consolidator grant for my five-year research project 'Refashioning the Renaissance: Popular Groups, Fashion and the Material and Cultural Significance of Clothing in Europe, 1550–1650' at Aalto University, Helsinki. The generous funding from the European Research Council enabled me, alongside a team of five talented international dress historians and numerous collaborators from academia, museums, conservators and craft experts, to conduct extensive archival work and experiments. Together, we aimed to create new perspectives in early modern everyday and fashion history. My heartfelt appreciation goes out to the remarkable 'Refashioning' team: Sophie Pitman, who led the experimental research; Michele Robinson, who meticulously co-ordinated the research in the printed sources and archival work on Italian dress fashions; Stefania Montemezzo, who conducted the core archival work and transcriptions for the database; Victoria Bartels, who carried out extensive supplementary archival work; and Anne-Kristine Sindvald Larsen, who studied numerous documents in the Scandinavian archives and translated them into creative research. Additionally, I extend my warm thanks to our co-ordinator Piia Lempiäinen for project administration and co-ordination of the citizen science experiment, Lena Kingelin for helping us to bring the project to a successful completion, and our research assistants Mattia Viale and Umberto Signori for their efforts with transcriptions. It was the energy, expertise and insights of this incredible team that enabled us not only to dive deep into the symbolic and material meanings of Renaissance fashion but also to organise numerous successful research activities, workshops, hands-on experiments, exhibitions and conferences.

Our research, experiments and activities would not have been possible, however, without extensive collaboration. I extend my heartfelt gratitude to the members of our advisory board, John Styles, Evelyn Welch, Ulinka Rublack, Maria Hayward, Flora Dennis, Tessa Storey and Susan North, for their insightful guidance and participation in the project. Additionally, I am also profoundly thankful to all the skilled experts in crafts and costume-making whose expertise made our historical reconstructions possible, with special thanks to Jenny Tiramani and the London School of Historical Dress for the amazing work on the doublet, the thirty-five volunteer knitters from Finland who participated in our citizen science knitting project and Elena Kanagy-Loux for her bobbin-lace reconstruction.

Throughout the project, we learned new historical craft skills from many of our colleagues and engaged in discussions with peers. Special thanks are extended to Jo Kirby, Pamela Smith, Art Proaño Gabor and Natalie Ortega-Saez for sharing their expertise on historical colours and recipes; Miriam Pugliese for allowing us to learn how silk fibres were traditionally made; Melanie Brown, Claire Thornton and Jane Malcolm-Davis for teaching how early modern garments were stitched together, Maj Ringaard and Christel Brandenbourgh for their advice concerning early modern knitting and citizen science; Valerio Zanetti for being a model for our doublet fittings; Krista Wright for carrying out fibre analysis for our project, and Tim McCall, Flora Dennis, Tessa Storey and Elizabeth Currie for sharing their expert knowledge at our experimental workshops. Additionally, collaboration with creative design students, photographers, designers, filmmakers and editors was crucial for the dissemination of our results. My sincere gratitude to Maarit Kalmakurki, Eero Brandt, Jodie Cox, Ana deMatos and others for their exceptional contributions.

During the project, we visited numerous craft workshops and museums storerooms to study early modern textiles. I am very grateful for the many curators, conservators and researchers who generously allowed us to study textile objects with them, including those at the Metropolitan Museum of Art and Antonio Ratti Textile Centre, New York, Museum of Copenhagen, Museum of London, Bata shoe museum, Fondazione Lisio, Museo del tessuto Prato, Tessitura Bevilaqua, Maison de Canuts Lyon, Trelleborgen Viking museum, Museo del merletto Burano, Museo del tessuto Venice and Turku Cathedral Museum.

This book is a product of the fruitful collaboration during the project. I express my gratitude to all the authors of this book, as well as Emma Brennan from Manchester University Press for her invaluable support.

Lastly, I extend my deepest thanks to my loving family for their continuing support and encouragement.

Introduction: Refashioning the Renaissance

Paula Hohti

In 1581, the Italian writer Giovan Andrea Corsuccio complained in his *Il vermicello* that 'today anyone, vile as he may be, dresses in silk, ... so that even the charlatans, if they have no velvet cap or doublet, are not able to draw a crowd of listener'.[1] Corsuccio was troubled by the fact that men of inferior status wore excessive garments and too expensive materials, such as silk, traditionally considered the badge of a gentleman.

Similar concerns about extravagant appearance and 'confused mingle-mangle of apparell' among the lower social orders were also raised in other parts of Renaissance Europe, such as England and Italy.[2] For instance, in Spain, towards the end of the sixteenth century, a group of curious observers claimed that 'it is well-known the excess of the way of dressing among Spanish people, because on a festive day the craftsman and his wife do not differ from the nobility'.[3] The high-ranking Spanish citizens strongly disapproved with the idea that it would no longer be possible to identify people and their social rank on the basis of the wearer's dress, as clothing had been traditionally seen as one of the most fundamental and visible means of recognising and distinguishing class.[4]

While Renaissance and early modern authors and commentators might have exaggerated to emphasise particular points, complaints about sumptuous dress among common artisanal populations were increasingly voiced during a period when stylish garments, fashionable accessories and desirable materials in a wide range of qualities became more accessible than ever before, not just for the elites but also for those of modest means and inferior status. The impact of this changing fashion context on every-day dress and on the development of fashion among the ordinary artisanal population is the subject of this book.

The changing context of fashion

The sixteenth century represented an important turning point in European fashion. The emergence of new crafts, industries and technical innovations, and the wide circulation of new fashion innovations, popularised by courts and promoted by the flourishing cloth and clothing trade, introduced new concepts into the traditional, local ways of dressing, making a break with traditional ways of dressing and the significance attributed to clothing.[5] This not only changed the way clothing was made, decorated and worn, but it also introduced a wide range of new products, consumption patterns and cultural values into the systems of dress. Heavy, brocaded velvets and damasks provided no longer the most exclusive powerful tools that made distinctions of rank visible in society. Instead, the display of new fashion manufactures that were designed to be worn or carried, such as light silk fabrics, gloves, fans, handkerchiefs, hats, trims, silk bands and buttons became essential parts of a fashionable outfit and indicators of the wearer's rank.[6] Less magnificent silks, like taffeta and tabby, became acceptable even in the formal wear of the ruling men and women, and could be combined with elaborate surface decorations, such as applied braid and slashing.[7] The multiple ways in which fashionable dress could be put together, mixing and matching detachable sleeves, bands of decoration, silk ribbons and small-scale personal items such as gloves, shoes, handkerchiefs and fans, is visible in surviving sixteenth-century visual images and surviving garments all over Europe (Figures 0.1a–c).

Dress historians have connected the new emphasis placed on light silks, accessories, trims and surface decoration, in part, to the broadening markets for more marketable goods that were suited to the gathering pace of elite dress fashions.[8] However, the cheaper price and the smaller size of new products that were on offer in the local shops and second-hand markets made fashion accessible also to new social groups. Shop records and other archival evidence demonstrate that ready-made items, from woven ribbons and lace veils to velvet hats and gilt netting, were available at varying prices and qualities in local fairs and shops to a wide range of consumers. At the same time, cheaper imitations of the desired goods, such as false pearls, foiled gems or stamped mock velvets, appeared on the market, turning extraordinary textiles, garments and jewellery into something familiar.[9]

This changing fashion context provided a new dynamic ability for urban men and women way below the nobility, including local artisans and shopkeepers from barbers and bakers to shoemakers, innkeepers and book dealers, to experiment with appearance. Foreign travellers occasionally noted the extravagance of the ordinary people's dress. Pietro da Casola, a Milanese cleric who visited Venice on his way to the Holy Land

Figure 0.1a–c Accessories and surface decorations in sixteenth and seventeenth century Spanish and Italian dress. Museo Nacional del Prado, Madrid, Museo Nazionale di Palazzo Reale, Pisa, and Kunsthistorisches Museum Vienna.

at the turn of the sixteenth century, for example, remarked that in Venice 'those [women] who are able as well as those who are not dress very sumptuously'.[10]

The increasing ability of ordinary people in the Renaissance period to acquire a wide range of material goods through pawning, second-hand markets, renting, borrowing, gift giving, theft, lotteries, auctions, inheritances or advanced credit systems has been noted by Renaissance art and material culture historians who have referred to examples of popular classes in their studies of consumption, material culture and dress.[11] This evidence suggests that ordinary Europeans enjoyed a greater access to consumption of new clothing and fashion manufactures than ever before. Yet, as Margaret Rosenthal has noted, exactly how and to what extent lower social orders were connected to fashion and new consumption practices, and what characteristics defined their appearance, is still an ongoing area of investigation.[12] Could individuals from lower social classes, such as tailors, bakers, barbers, shoemakers and butchers, participate in Renaissance fashion and culture and engage with the latest trends?

Non-elite engagement with fashion

Early modern probate inventories, sumptuary law statutes, guild documents, auction records and account books suggest that the wardrobe of ordinary artisanal individuals and families underwent a transformation in Europe in the second half of the sixteenth century. Preliminary research based on probate inventories from Italy, for example, indicates that the number and kind of garments and dress accessories among local urban craftspeople, shopkeepers and tradespeople increased significantly in the second half of the sixteenth century. The plain and durable woollen and linen garments suited to work, typical for artisans, were complemented with a notable range of garments and dress accessories that embraced new fashions. These ranged from colourful silk skirts and aprons to muffs, gloves, silk scarves and hats decorated with gold badges and feathers. Male artisans had a taste in particular for ruffs that grew from a narrow frill at neck and wrists to a broad 'cartwheel' style that required a wire support by the 1580s, and women favoured gold frontals and rosettes, silver and gold nets and thin silk veils. In addition, a range of elaborate bright stockings and detachable sleeves that were made from coloured silks, particularly red, white, green, or yellow taffeta, sarcenet or satin, were included in artisan wardrobes.[13]

This evidence suggests that many types of fashion novelties were worn by the general population as well as the more affluent elites, and, when contextualised with pictorial evidence, their cultural significance as functional or fashionable wear at lower levels of society becomes evident. For example, in Vincenzo Campi's *Fruitseller*, painted in the 1580s (Figure 0.2), the young woman wears a plain linen garment decorated with a matching partlet and sleeve cuffs and red sleeve ribbons. A yellow band is embroidered across her green apron, perhaps made in imitation of gold.

Figure 0.2 Campi, *Fruitseller*, 1560s. Young women wearing a linen gown with matching partlet and sleeve cuffs and red sleeve ribbons. A yellow band is embroidered across her green apron. Pinacoteca di Brera, Milan.

Historians have shown, furthermore, that ordinary people's engagement with fashion in the Renaissance period was more than just a matter of 'getting and spending'.[14] Instead, it extended to the way ordinary people began – through their garments and images of clothes – to adopt and express attitudes towards life, explore their connections to others and interact with the surrounding world and culture.[15]

To comprehend the significance of this profound change, we can examine surviving visual evidence. One notable example is the 'democratisation' of portraits in the sixteenth century, which demonstrate how urban citizens below the nobility began to place importance on clothes, accessories, hairstyles and their comportment to assert their social value and how they desired to be perceived. In Moroni's portrait from the 1560s, for example, the tailor appears in a self-assured pose, wearing a fine cream doublet and red hose (Figure 0.3). Although he holds a pair of scissors, symbolising his craft and manual labour, his posture and elegant clothes establish a visual connection between the tailor and portraits of the contemporary elite.

The significance of clothing below the nobility, moreover, extended beyond self-fashioning and visual display. The 'Book of Clothes', studied in detail by Maria Hayward, Ulinka Rublack and Jenny Tiramani, shows that, in Germany, Matthäus Schwarz, an accountant and a son of a wine merchant, commissioned over 130 watercolour illustrations between 1520 and 1560, in which he posed in different styles of outfits, in front and back (Figure 0.4).[16] What makes this collection of illustrations remarkable, as Rublack notes, is not only the accountant's fascination with manipulating clothing and adapting changing fashions to create the desired impression but also how he experienced his clothing in relation to his body. It appears that the accountant was worried about putting on weight. In 1529, at the age of twenty-nine, he had himself depicted in the nude and inscribed a note next to the image, stating, 'this was my proper figure from behind, for I had become fat and round' (Figure 0.5).[17] This example, Rublack argues, is one demonstration of how deeply clothing became embedded with the perceptions people across social classes had about their dress, bodies and the social and cultural meanings associated with fashion.[18]

These findings highlight the issue that has been largely overlooked in dress history: that fashion transformation in the sixteenth and seventeenth centuries was interconnected across social classes. All this evidence demonstrates, then, that there is a demand for an interdisciplinary study of Renaissance and early modern fashion that investigates how fashion evolved in dialogue with various social groups and economic contexts.

This book explores how fashion emerged and developed in Europe in 1500–1650 among the middling classes, including craftspeople, shopkeepers and local traders residing in commercial centres, trading towns and the surrounding countryside across various regions in southern and

Figure 0.3 Moroni, *Portrait of a Tailor*, 1565–70. The sitter is dressed in fine pinked cream colour doublet and red breeches. The National Gallery, London.

northern Europe such as Italy, England, Scandinavia and Estonia.[19] Their social group consisted of individuals such as bakers, barbers, shoemakers, innkeepers and others who earned their livelihood by creating or selling goods. While the collective terms used in this book to refer to the group of artisans and small local shopkeepers include 'artisans', 'artisanal groups', the 'artisan classes', 'non-elite' and the 'middling sort', it is important to recognise the diversity within this group. While

Figure 0.4 Schwarz, 'Book of Clothes', fol. 113r, 1538. The Herzog Anton Ulrich Museum, Brunswick.

most were ordinary members of minor craft and trade guilds, the group encompassed a wide range of individuals, from prosperous pewterers and goldsmiths to destitute members of the textile and building crafts.[20] Dress played a crucial role in making social distinctions visible between these diverse categories.[21]

Approaching everyday artisan fashion

This book is the outcome of research conducted within the 'Refashioning the Renaissance' project, a five-year collaborative dress-history initiative funded by the European Research Council (ERC, Aalto University 2017–23).[22] The project was set up to identify new sources and develop methodologies that allow us to investigate fundamental questions relating to the transformation of fashion in early modern Europe, with a special focus on popular taste, dissemination, transformation and adaption of fashion, on imitation and meaning, and on changing cultural attitudes to dress among popular groups in sixteenth- and seventeenth-century Europe.

Figure 0.5 Schwarz, 'Book of clothes', fol. 80r, 1526. The Herzog Anton Ulrich Museum, Brunswick.

One of the key challenges of studying non-elite dress is the question of how we can gain access to the cultural practices, artefacts and dress fashions at popular levels of society. Museums and archives primarily document the culture of the rich and powerful, leaving scant information about the garments and lifestyles of the ordinary population. Consequently, understanding how dress fashions evolved at popular levels of society, as well as how to study them effectively, presents significant hurdles.

This challenge is particularly evident regarding the physical texture and visual and sensory properties of clothing and accessories, along with the skills and artisanship involved, as written accounts and visual depictions provide limited information about materials and construction techniques, and surviving garments worn by the ordinary population are scarce, often fragmented and in poor condition.[23] Addressing these gaps of evidence requires innovative methodologies and a combination of diverse approaches.

The research presented in this book aims to overcome the limitations in sources by incorporating a novel methodology that combines traditional historical empirical evidence with a range of explorations of early modern materials and crafting techniques.

Methods and sources

The core historical findings are based on archival sources, written records and visual representations to investigate the clothing worn by lower social classes. At the heart of our research is an extensive archival dataset, created during the 'Refashioning the Renaissance' project. This dataset comprises nearly a hundred thousand items of textiles, clothing and jewellery documented in post-mortem inventories of ordinary artisanal families between 1550 and 1650, currently accessible online in an open-access format (Figure 0.6).[24] Referred to as the 'Refashioning the Renaissance database' in this book, it contains 80,076 records from Siena, Florence and Venice, as well as 12,207 records from early modern Denmark.[25] The dataset is cross-referenced with a range of other archival evidence from guild records, petitions, trade accounts, contracts, sumptuary law statutes and printed and painted depictions of dress. These ranges of traditional historical records

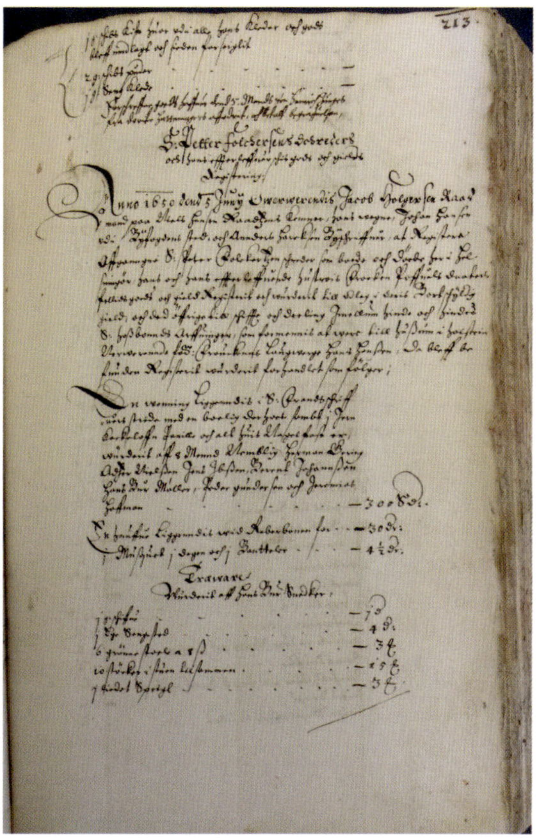

Figure 0.6 Inventory of the Danish tailor Peter Folchersen, 5 June, 1650, fol. 213r. National archives, Copenhagen.

offer a solid foundation and reliable evidence for exploring essential elements of early modern popular fashion. They shed light on the garments, accessories and textile materials that circulated among lower social groups in this period, and reveal what economic, symbolic and cultural meanings were associated with dress and fashion in their own time.

The conventional historical research method is complemented in this book with practical hands-on experiments. Designed and organised during the 'Refashioning' project in collaboration with museum professionals, textile and craft experts, costume makers, historical re-enactors and scientists, these enabled us to re-create objects that no longer survive, showing how they were made, what they might have looked and felt like, and what levels of skill and sophistication were involved in making them.

The efforts carried out during the 'Refashioning' project included a range of experiments aimed at getting closer to the materials and garments worn by the artisanal population of the past. For instance, in collaboration with costume experts at the London School of Historical Dress, we brought back to life a seventeenth-century male doublet made from stamped mock velvet, recorded in the inventory of the modest Florentine waterseller Francesco Ristori, who passed away in Florence in 1631 (Figure 0.7). Additionally, we initiated a knitting project to re-create early modern knitted stockings, including a seventeenth-century hand-knitted silk stocking of high fashion conserved at Turku Cathedral Museum, Finland (Experiment in focus II). A group of dedicated volunteer knitters used tiny 1 mm knitting needles and silk yarn produced by hand at the silk farm Nido di seta in Calabria to bring to light the patterns and fine artisanship of the delicate stocking (Figure 0.8).

These explorative historical reconstructions, informed by detailed archival and visual research, scientific testing and object analysis carried out by the 'Refashioning' team, were made by hand by skilled craft experts using historically appropriate materials. Even though reconstructions are, as Jenny Tiramani has pointed out, 'acts of interpretation' and never precise copies of the authentic, engaging with reconstructed garments through touching, looking and even wearing them offered new insights into the fashionable aspirations and innovative methods used by the artisan classes to participate in early modern culture of fashion.[26]

Additionally, several small hands-on experiments were conducted during the project to explore, for example, how early modern garments were stitched together and shaped by tailors, in what ways silk thread was traditionally made from cocoons, how the colours of clothing mentioned in our archival sources were created using natural dyes and historical colour recipes, how imitations of precious pearls and amber could be made at home following recipes from cheap printed advice manuals and how textiles were cleaned and cared for (Figures 0.9a–b). These material investigations allowed us to explore, for instance, a sense of the visual and sensory effects offered by counterfeit materials or imitative objects,

Figure 0.7 Reconstruction of a seventeenth-century male artisan's doublet made by the 'Refashioning the Renaissance' project and London School of Historical Dress.

Figure 0.8 Two hand-knitted replications of an extant seventeenth-century silk stocking, reconstructed by the 'Refashioning the Renaissance' project. The stockings are knitted using hand-reeled Bombyx mori silk and dyed with fustic and logwood.

such as false amber or fake leopard fur, as well as to re-create some of the essential colours, materials and shapes of garments and objects of adornment that were mentioned in artisans' clothing inventories.

Experimental recreative methods and object-based research have emerged as a significant development in the field of dress history in recent years. Building on methods of experimental archaeologists, costume makers, re-enactors and textile conservators, historians such as Sarah Bendall, Serena Dyer, Hilary Davidson, Pamela Smith, John Styles and Ulinka Rublack have demonstrated the transformative capacity of replication to turn static narratives of the past into a dynamic process to comprehend the experiences of wearing and making of early modern textiles.[27] Despite the promising outcomes, experimental approaches in the study of cultural history of dress remain relatively unfamiliar, with possibilities yet to be fully explored and methodologies still needing to be established.[28]

This book presents, alongside more traditional essays, an exploration of the most important material experiments and historical reconstructions conducted as part of the 'Refashioning the Renaissance' project.[29] Through short Experiment in focus texts, readers can examine and understand how experimental work and material-based approaches can help us

Figure 0.9a Red dye experiments at historical colour workshop, held by the 'Refashioning the Renaissance' project at Aalto University, September 2019.

to provide access to the visual, material and sensory properties of early modern of fashion and gain new insights into the skills of makers and the significance of materials now to lost to us. Additionally, the experiments in focus demonstrate how a range of 're-methods' can provide historians with new ways of sharing findings about the making, wearing and historical importance of textiles and clothing.

Cultural studies of dress and fashion have been traditionally separated from the study of the real physical and material objects, due to the preference for interpreting semiotic symbols and signs of dress through visual or written representations rather than the physical and material properties of clothing and textiles.[30] As a result, we have largely lost touch with the

Figure 0.9b Cheap sixteenth-century printed 'book of secrets' by maestro Giovanni da Lucca, *Opera nuova nella quale troverai molti bellissimi secreti* ... (1540?). Wellcome Institute, London, 4630/B.

materiality of historical objects and fashion and the material experiences linked to their creation and use.

By focusing on the materials and materiality fashion, along with its symbolic, social and cultural significance, this book seeks to offer new insights not only into everyday artisan fashion but also to the lived experiences associated with the early modern materials and garments that influenced fashion trends. Understanding the visual and physical attributes of garments during the centuries covered in this book, ranging from the early sixteenth century to about 1650, is crucial, as the period witnessed a number of significant shifts in the ways in which textiles and clothing were produced, adapted and worn. Materials, skilled artisanship and the sensory qualities of garments and accessories constituted defining elements of fashion and an important part of its meanings.[31]

Previous studies by notable dress, textile and consumer historians such as John Styles, Giorgio Riello, Beverly Lemire and Maxine Berg have provided a valuable framework for understanding the dynamics of economic, social and cultural change in premodern Europe, and the role that textiles and dress played in shaping such changes.[32] However, social groups below the wealthy elites have been largely overlooked in the elite-dominated dress studies of the Renaissance and early modern periods, because of either the perception that their fashions were characterised by emulation and passive copying of the elites or the assumption that individuals on the societal margins had limited contact with the rapid changes in European fashion and remained culturally isolated.[33] As a result, a significant portion of the consumer population has been marginalised in the history of fashion in urban Renaissance and early modern life, raising doubts among historians about the impact of the Renaissance on the lives of ordinary Italians.[34]

By integrating traditional historical methods and sources with exploration of materials and techniques of early modern textiles and garments, this book seeks to enhance our understanding of the fashion of the lower social orders and challenge the inherent biases often present in conventional historical sources, which tend to emphasise the perspectives of the privileged elites. In doing so, this book aspires to present a fresh and socially more diverse perspective of early modern dress, aiming to inspire new inquiries and broaden the horizons of how narratives in cultural historical studies of dress can be constructed.

'Refashioning the Renaissance'

This book presents the key findings of the 'Refashioning the Renaissance' project. In collaboration with economic, cultural and social historians, dress and material culture historians, museum specialists and craft experts, it explores what materials and objects constituted the key elements of everyday dress and agents of fashion change among the lower social groups in Europe in the sixteenth and seventeenth centuries. By focusing on the creation, adaptation, innovation, uses and meanings of a wide range of fashion items among the ordinary artisanal population, along with the materials of fashion, ranging from affordable light textiles and imitations to expensive jewellery and valuable fabrics, the central questions explored in the chapters and experiments in focus include: What types of novelties and low-cost fashion manufactures were available to emerging consumer groups and how was their use regulated by sumptuary laws? What materials and objects shaped consumer preferences at the lower social levels? How was fashionable appearance physically, materially and visually constructed by ordinary artisans and what processes were involved in dressing oneself?

The book is organised into three sections, each dedicated to different aspects of everyday artisan fashion. The first part concentrates on innovations and imitations in fashion, showing the range of new commercial fashion products available to consumers of modest means from the early sixteenth century onwards. In Chapter 1, John Styles frames the book chronologically and methodologically from an economic-history perspective. His chapter shows that, between the fifteenth and the eighteenth centuries, western Europe witnessed a surge of novelty in textiles and a wave of innovations, embracing both fabrics and equipment used for textile production. Styles examines the shift from heavily napped woollen broadcloths and silk velvets, which dominated the European high-prestige textile market, to lighter and more affordable silk, wool and mixed-fibre fabrics, and explores the impact of these innovations and how they extended to non-elite consumers. Styles shows that the rise in popularity of the new affordable and visually appealing textile goods, particularly cheaper silks, had a profound impact not only on the vibrant consumer economy of early modern Europe but also on ordinary people's engagement with fashion. The transformations in the textile market expanded the range of attractive clothing options accessible to ordinary women and men and shaped their broader fashion preferences.

This shift in the European textile market was accompanied by a corresponding transformation in the variety and availability of textile trimmings and clothing accessories, such as knitted stockings, kerchiefs, lace trims and ribbons. While ribbons and tapes, for example, had ancient origins and knitted goods were present in medieval Europe, the period after 1500, as Styles emphasises, witnessed a remarkable expansion and elaboration of these items.

Andrea Caracausi, in Chapter 2, explores the impact of this transformation on fashion. His chapter reveals how the market for ribbons expanded during the late Renaissance and beyond, offering a plethora of choices for decorating garments, ranging from expensive luxury braids to affordable plain ribbons made from cheap waste silk. This expansion in the ribbon market stimulated demand and made ribbons accessible to people from all economic and social backgrounds. Because of their small size, ribbons provided an affordable means to incorporate precious materials in dress and maintain a connection with prevailing fashionable trends.

Sophie Pitman's contribution in Chapter 3 shifts the focus to imitations and the manipulation of materials, exploring how imitations of expensive and rare items like fur, silk velvet and gold were created by skilled craftspeople. Pitman's chapter shows the expanding range of artificial novelties available, ranging from small buttons made of precious metal and silk threads wrapped around a wooden bead to imitation fur and fake pearls. Such imitations allowed men and women from diverse backgrounds to convey an impression of luxury without high cost. Yet Pitman challenges the prevailing assumptions about their value, showing that imitations were not necessarily seen just as fakes, cheap copies or inferior substitutes of the originals, reserved only for those with limited economic means. Instead, many types of replicas, worn by individuals across all social strata, were often appreciated for their artisanship and regarded at times even more attractive and suitable options than the genuine articles.

The three experiments in focus in this part – imitation fur, knitted stockings and stamped mock velvet doublet – focus on some of the key materials and techniques that played an instrumental role in driving the innovative fashion trends discussed in the chapters. They underscore the significant role played by affordable and fashionable semi-durable or 'semi-luxuries' from the early sixteenth century onward as a means to engage with current fashions.[35]

The second part of the book focuses on adornment and display, investigating the availability for artisans of costly dress accessories such as jewellery, precious gems and protective arms. Traditionally, these items have been regarded by scholars as expensive status symbols exclusively reserved for the affluent and influential members of society. Michele Nicole Robinson and Natasha Awais-Dean in Chapters 4 and 5 challenge these assumptions by demonstrating that artisans, both men and women, commonly possessed and adorned themselves with expensive jewellery, including pearls and gold and silver jewellery, despite their cost and regulations imposed by sumptuary laws. However, their chapters also highlight that while all jewels and jewellery were generally expensive, we need to be cautious and avoid making broad generalisations about their material value. Items such as pearls, gold strings, medallions, rings, silver buttons and hat badges were available in various grades and lower price ranges for ordinary consumers through pedlars, local fairs, markets, auctions, pawnshops and goldsmiths' workshops. Misshapen and tiny lower-grade pearls, for example, were sold by the ounce rather than being individually priced like larger pearls, making them much more affordable than more costly and desirable Asian pearls.

Victoria Bartels, in Chapter 6, focuses on male artisans, arms and armoury, showing that the possession and display of arms and armoury – once reserved for princes, lords, knights or upper-class citizens – was not limited to the elite echelons of society in the sixteenth and seventeenth centuries. Men from all social classes, including artisans and even humble farmers, legally and illegally owned, and carried weapons. While arms served vital important protective functions, swords and daggers worn on belts were also considered fashionable accessories and essential components of respectable male dress by artisans as well as high-ranking citizens.

The three experiments in focus in this section explore the embodied experience linked with wearing a Renaissance male doublet and examine the visualisation of the garment's complex construction and hidden layers through digital reconstruction. These are followed by an exploration of how imitation pearls and amber could be made following recipes of that time. The experiments underscore the importance of considering sensory aspects such as scent, warmth, physical sensations and bodily postures, as well as the concealed layers of garments which may not be visible to the eye, alongside visual aesthetics, when assessing the experience and value of artisan fashions. For example, in the case of false amber beads frequently used in rosaries, as argued by Michele Nicole Robinson, it might have been amber's transformative effect that was sought rather than the material or visual look itself. Ensuring that counterfeit amber beads emitted a similar scent and warmth to authentic amber when touched was important, as these supported religious practices and protected bodily health.

The final part of the book examines everyday artisan fashion in relation to status and reputation. In Chapters 7 and 8, Stefania Montemezzo and Astrid Wendel-Hansen challenge the prevailing notion of the Renaissance and early modern periods, which suggests that only wealthy and powerful citizens could demonstrate success and express interest in fine garments through expensive textiles. Their research reveals that costly materials such as fur, patterned silk velvets, damasks, brocades and red fabrics dyed with expensive insect dyes – traditionally associated with elite luxury and symbols of status in sumptuary laws – were owned and embraced by a broad range of individuals and from different social groups. Despite the clear desire for novelty and the ability to adapt to evolving fashion trends demonstrated by ordinary men and women, these prestigious textiles – often seen as important stores of wealth that were circulated in inheritances – did not disappear from artisanal wardrobes. The coexistence of traditional luxury fabrics and emerging textile innovations, such as new lightweight fabrics and mixed silks, some of which imitated the exotic silks of Asia and were crafted in foreign styles, complicates the notion of linear change in fashion.

Chapters 9 by Elizabeth Currie and Jordan Mitchell-King and 10 by Anne-Kristine Sindvald Larsen highlight the importance of a respectable appearance among artisan, peasant and trading communities, not only on festive days or special occasions but also in their everyday professional lives. Their research suggests that individuals from modest Italian countrywomen to humble artisans from small Danish trading towns recognised the importance of maintaining a clean, fashionable and modestly prosperous appearance for their social and marital position, public influence, and professional image as trustworthy and reliable businessmen or businesswomen. As Currie, Mitchell-King and Sindvald Larsen show, presenting a favourable image through a simple linen apron or other unassuming garments decorated with embroidery, ribbons, lace trims and accessories, while ensuring clothes remained clean and maintained a pleasant scent, held similar importance for artisans and peasant women as it did for their social superiors. Some artisans, like Jacob Jensen Nordmand, a skilled Danish art turner, even expressed their social ambitions and aspirations by commissioning portraits that immortalised their professional and prosperous image (see Figure 10.5).

The three experiments in focus in the final section, focusing on stain removal recipes, historical colours and dyes, and bobbin-lace making, explore the processes involved in caring for, dyeing and embellishing linens and other textiles at home. These explorations reveal that instructions found in printed collections of recipes, advice manuals, pattern books and books of secrets – increasingly circulated in Europe in cheap printed media – offered individuals of the lower social strata new possibilities for creating vibrant and desirable colours and decorations within their household environment. They highlight the importance of widespread new fashion knowledge in driving the expansion and evolution of everyday artisan fashion.

Conclusion

Through an exploration of written records, visual representations, material objects, hands-on experiments and historical reconstruction, it becomes abundantly clear that ordinary European men and women not only possessed access to contemporary fashions but also placed great importance on their dress and outward appearance. While practicality of garments was often a primary concern in the early modern period, especially in the north of Europe which experienced extreme cold due to the 'little ice age', the belongings of artisans, shopkeepers and small local traders show a clear awareness and understanding of prevailing clothing trends. Their clothing and accessories reflect an ability to embrace novelty and adapt to the evolving dynamics of fashion.

The key components of fashion among the popular groups, spanning various regions from Italy and England to Denmark and Estonia, included cost-effective textile and fashion innovations of the period, such as new light silks, imitation fabrics and ribbons as well as small accessories such as bobbin-lace-trimmed handkerchiefs, hats, muffs and knitted stockings. At the same time, however, their clothing cupboards incorporated significant and expensive textile objects and objects of adornment, such as silk velvets, fur, gold, pearls and arms, associated traditionally in sumptuary laws and visual images with prestige, power and high social status.

Materials played a significant role in driving transformations and shaping fashions at the lower social levels. The ways in which artisans applied imitation fabrics in stylish garments, or blended materials of high intrinsic value with cheaper textiles in order to appear fashionable, reveal the creative and personalised ways in which early modern European artisans engaged with fashion. By incorporating trendy elements such as ornamental silk ribbons, trimmings of lace or imitation fur or fake gems into the most visible areas of their dress, even individuals of modest means, such as blacksmiths, gardeners or rural women were able to engage actively in fashion at an affordable cost.

This creative and innovative approach to fashion emerged during a time when the traditional hierarchy of clothing, based on financial value of textiles, was being challenged. The value of garments and accessories depended increasingly as much, if not more, on the novelty and artisanship involved rather than solely on the intrinsic value of the material itself. This emphasises the complex nature of how dress functioned in the early modern period.

The material culture of Renaissance fashion among the middling classes was characterised by liveliness, vibrancy and creativity. Simply labelling everyday artisan clothing and fashion as inferior and plain, defined by inexpensive alternatives and uninformed imitation of elite fashion, overlooks the intricate dynamics at play in sixteenth- and seventeenth-century Europe.

Notes

The research leading to these results was done as part of the 'Refashioning the Renaissance' project that has received funding from the European Research Council (ERC) under the European Union's Horizon 2020 research and innovation programme (grant agreement No. 726195).

1 Luca Molà, *The Silk Industry of Renaissance Venice* (Baltimore: Johns Hopkins University Press, 2000), 91. The original is found in Giovan Andrea Corsuccio da Sascobaro's treatise, *Il vermicello della seta* (Rimini, 1581), 11–12.
2 The 'confused mingle-mangle of apparel' was referred to in Philip Stubbes's *Anatomie of the abuses in England* (1583). See Carlo M. Belfanti, 'Clothing and social inequality in early modern Europe: Introductory remarks', *Continuity and Change*, 15, 3 (2000), 359–65, 360.
3 Quoted in Belfanti, 'Clothing and social inequality', 360.
4 For clothing as a visible statement of rank and status, see, for example, the Introduction in Carole C. Frick, *Dressing Renaissance Florence: Families, Fortunes, and Fine Clothing* (Baltimore: Johns Hopkins University Press, 2002). Visual appearances were defined in sumptuary laws: see Catherine Kovesi Killerby, 'Practical problems in the enforcement of Italian sumptuary law, 1200–1500', in Trevor Dean and Kate J. P. Lowe (eds), *Crime, Society, and the Law in Renaissance Italy* (Cambridge: Cambridge University Press, 1994); and for a global perspective, Giorgio Riello and Ulinka Rublack (eds), *The Right to Dress: Sumptuary Laws in a Global Perspective, c. 1200–1800* (Cambridge: Cambridge University Press, 2019).
5 Carlo M. Belfanti, 'Was fashion a European invention?', *Journal of Global History*, 3 (2008), 419–43. For an overview of these changes, see Evelyn Welch, 'Introduction', in *Fashioning the Early Modern: Dress, Textiles, and Innovation in Europe, 1500–1800* (Oxford: Oxford University Press, 2017), 4–6.
6 For Renaissance and early modern dress fashions, see Frick, *Dressing Renaissance Florence*; Ulinka Rublack, *Dressing Up: Cultural Identity in Renaissance Europe* (Oxford: Oxford University Press, 2010); Roberta N. Orsi Landini, *Moda a Firenze: Lo stile di Eleonora di Toledo e la sua influenza* (Florence: Edizioni Polistampa, 2007); Elizabeth Currie (ed.), *A Cultural History of Dress and Fashion in the Renaissance* (London: Bloomsbury, 2017); Maria Hayward, *Rich Apparel: Clothing and the Law in Henry VIII's England* (London: Routledge, 2009); Timothy McCall, *Making the Renaissance Man: Masculinity in the Courts of Renaissance Italy* (Chicago: University

of Chicago Press, 2023). For the growing range of accessories and objects of personal adornment, see Timothy McCall, *Brilliant Bodies: Fashioning Courtly Men in Early Renaissance Italy* (University Park: Penn State University Press, 2022); Evelyn Welch, 'Art on the edge: Hair and hands in Renaissance Italy', *Renaissance Studies*, 23, 3 (2008), 241–68; and Bella Mirabella (ed.), *Ornamentalism: The Art of Renaissance Accessories* (Ann Arbor: University of Michigan Press, 2011). For these developments, and their connections to the manufacturing sector, see John Styles and Andrea Caracausi, Chapters 1 and 2 below.

7 Currie, Elizabeth, 'Clothing and a Florentine style, 1550–1620', *Renaissance Studies*, 23, 1 (2009), 51.

8 Currie, 'Clothing and a Florentine style', 51.

9 Rublack, *Dressing Up*, 6.

10 Quoted by Patricia Allerston, 'Clothing and early modern Venetian society', *Continuity and Change*, 15, 3 (2000), 367.

11 For example, Allerston, 'Clothing and early modern Venetian society', 367–90; and Paula Hohti, 'The art of artisan fashions: Moroni's tailor and the changing culture of clothing in sixteenth-century Italy', in Rembrandt Duits (ed.), *The Art of the Poor: The Aesthetic Material Culture of the Lower Classes in Europe 1300–1600* (London: Bloomsbury Academic, 2020), 109–16; Hohti, 'Dress, dissemination, and innovation: Artisan "fashions" in sixteenth- and early seventeenth-century Italy', in Evelyn Welch (ed.), *Fashioning the Early Modern: Dress, Textiles, and Innovation in Europe, 1500–1800* (Oxford: Oxford University Press, 2017), 148–58. See also Belfanti, 'Clothing and social inequality', 259–65.

12 Margaret F. Rosenthal, 'Cultures of clothing in later medieval and early modern Europe', *Journal of Medieval and Early Modern Studies*, 39, 3 (Fall 2009), 461.

13 For comprehensive data, see www.refashioningrenaissance.eu/database. See also Paula Hohti, 'Dress, dissemination, and innovation', 148–58.

14 John Styles, *The Dress of the People: Everyday Fashion in Eighteenth-Century England* (New Haven and London: Yale University Press, 2007), 1.

15 For dress codes and arguments about clothes and their values, see Rublack, *Dressing Up*.

16 Ulinka Rublack, Maria Hayward and Jenny Tiramani (eds), *The First Book of Fashion: The Book of Clothes of Matthaus and Veit Konrad Schwarz of Augsburg* (London: Bloomsbury, 2015).

17 Rublack et al., *The First Book of Fashion*, 128–9, 288–9.

18 Rublack, *Dressing Up*.

19 Although most of the evidence is from the period 1500–1650, occasional references, when appropriate, are made to later decades by authors of this book.

20 Margaret A. Pappano and Nicole R. Rice, 'Medieval and early modern artisan culture', *Journal of Medieval and Early Modern Studies*, 43, 3 (2013), 473–85, at 475, quoting Heather Swanson, *Medieval Artisans: An Urban Class in Late Medieval England* (Oxford: Basil Blackwell, 1989), 2. For the difficulty of defining the artisan class, see Paula Hohti, '"Conspicuous" consumption and popular consumers: Material culture and social status in sixteenth-century Siena', *Renaissance Studies*, 24, 5 (2010), 654–56; and Pappano and Rice, 'Medieval and early modern artisan culture', 473–85. The social structures and guild organisation varied according to regions: for England and Italy, see Hohti, *Artisans, Objects and Everyday Life*, 31–3; for Estonia, see Chapter 8 below by Astrid Wendel-Hansen; and for Denmark, Anne-Kristine Sindvald Larsen, 'Clothes, culture and crafts: Dress and fashion among artisans and small shopkeepers in the Danish town of Elsinore, 1550–1650' (PhD dissertation, Aalto University, Aalto ARTS Books, 2023), 25–8; and for the rural setting, Elizabeth Currie, Chapter 9 below.

21 See Hohti, Artisans, *Objects and Everyday Life*, chapter 3, especially p. 121, and note 4 above.

22 www.refashioningrenaissance.eu.
23 As noted in Evelyn Welch, 'Introduction', *Fashioning the Early Modern*, 30.
24 www.refashioningrenaissance.eu/database. The dataset has been created collectively by Stefania Montemezzo, Paula Hohti, Mattia Viale, Umberto Signori, with the online database design by Hohti, Piia Lempiäinen and Jane Malcolm-Davis, in collaboration with Jodie Cox from Wildside, UK. Note that in this period the year traditionally began on 25 March; the years in the online database or the references of this book have not been adjusted to the modern calendar's new year on 1 January.
25 For further discussion of the sources included, and the institutions that compiled the inventories, see Chapter 7 below by Stefania Montemezzo.
26 On the question of authenticity in reconstruction, Jenny Tiramani, 'Reconstructing a Schwartz outfit', in Rublack, Hayward and Tiramani (eds), *The First Book of Fashion*, 374.
27 Sarah A. Bendall, *Shaping Femininity: Foundation Garments, the Body and Women in Early Modern England* (London: Bloomsbury, 2021), 10–16. Reconstruction is a relatively new approach in material culture and cultural studies of dress. For a useful summary and for the 'making turn' dress and textile history, see Hilary Davidson, 'The embodied turn: making and remaking dress as an academic practice', *Fashion Theory*, 23, 3 (2019), 329–62, esp. 338; Peter McNeil and Melissa Bellanta, 'Fashion, embodiment and the "making turn"', *Fashion Theory: The Journal of Dress, Body & Culture*, 23, 3 (2019, 325–8; Pamela Smith, Amy R. W. Mayers and Harold J. Cook (eds), *Ways of Making and Knowing: The Material Culture of Empirical Knowledge* (Chicago: University of Chicago Press/Bard Graduate Centre, 2017), 8; Rublack, Hayward, and Tiramani (eds), *The First Book of Fashion*; Jenny Tiramani, 'Reconstructing a Schwartz outfit', 374; Ulinka Rublack, 'Matter in the material Renaissance', *Past and Present*, 219 (2013), 41–85; John Styles, 'The spinning project', http://spinning-wheel.org/author/john-styles. For a an example of object-based research in dress research, see Rebecca Unsworth, 'Hands deep in history: Pockets in men and women's dress in western Europe, c. 1480–1630', *Costume*, 51, 2 (2017), 148–70; and for 're-methods', Sven Dupré, Anna Harris and Julia Kursell (eds), *Reconstruction, Replication and Re-enactment in the Humanities and Social Sciences* (Amsterdam: Amsterdam University Press, 2020) and the critical edition published by the 'Making and Knowing' project at https://edition640.makingand-knowing.org, accessed 22 June 2024. Cultural studies of dress have drawn influence and inspiration from experimental archaeology. For experimental archaeology in textiles, for example, see Jane Malcolm-Davis, 'An early modern mystery: A pilot study of knitting, napping and capping', *Archaeological Textile Review*, 58 (2016), 57–74, and Center for Textile Research website for guidelines for using experimental archaeology as a scientific method, at https://ctr.hum.ku.dk.
28 The need for better theorization of reconstruction as a methodology has been emphasised, for example, in Davidson, 'The embodied turn', 8. The ongoing AHRC funded 'Making Historical Dress' network by Sarah Bendall and Serena Dyer aims to establish a methodological framework for combining recreative methods in the work of academics, curators and costumers. For the project, including a bibliography on recreative methods, see https://makinghistoricaldress.dmu.ac.uk, accessed 28 January 2024.
29 All experiments, including video recordings of the experiments, are available at our website at www.refashioningrenaissance.eu. For a full methodological analysis, see Sophie Pitman and Paula Hohti (eds), *Remaking Dress History: Applying Reconstruction Methods to Early Modern Textiles and Clothing*, Special Issue, *Textile History*, forthcoming 2024; and www.refashioningrenaissance.eu.
30 Cultural historians, borrowing from social anthropologists and semiotics such as Barthes and Saussure, have been successful in showing what clothing items meant in their own time, but this has been divorced from the actual clothing and

the materiality of fashion. For an analysis of the developments, see Christopher Breward, 'Cultures, identities, histories: Fashioning a cultural approach to dress', *Fashion Theory*, 2, 4 (1998), 301–13; and for a study of dress as part of cultural arguments about display and identity and their visual presentation, see Rublack, *Dressing Up*, esp. 25. For Barthes, who created his theory of fashion mostly between the years 1957 and 1963, see Roland Barthes, *The Fashion System*, trans. Matthew Ward and Richard Howard (Berkeley: University of California Press, 1990).

31 For the importance of how things were made and what visual skills were involved in the making processes in understanding fashion, see Ulinka Rublack, 'Renaissance dress, cultures of making, and the period eye', *Journal of Decorative Arts, Design History, and Material Culture*, 23, 1 (2016), 6–34. Evelyn Welch also points out that tangible, tactile aspects of how things were made or how they felt may have been equally crucial to the success or failure of new textile innovations in her 'Introduction', in Welch (ed.), *Fashioning the Early Modern*, 6. See also Beverly Gordon, 'The hand of the maker: The importance of understanding textiles from the "inside out"', in *Silk Roads, Other Roads* (Proceedings of the 8th Biennial Symposium of the Textile Society of America, Northampton, MA, September 2002), 189–98.

32 Maxine Berg, *Luxury and Pleasure in Eighteenth-Century Britain* (Oxford: Oxford University Press, 2007); Maxine Berg, 'New commodities, luxuries and their consumers in eighteenth-century England', in Maxine Berg and Helen Clifford (eds), *Consumers and Luxury: Consumer Culture in Europe, 1650–1850* (Manchester: Manchester University Press, 1999); Beverly Lemire, *Dress, Culture and Commerce: The English Clothing Trade Before the Factory, 1660–1800* (Basingstoke: Macmillan, 1997); Beverly Lemire and Giorgio Riello, '*East & west: Textiles and fashion in early modern Europe*', *Journal of Social History*, 41, 4 (2008), 887–916; Giorgio Riello and Prasannan Parthasarathi, *The Spinning World: A Global History of Cotton Textiles, 1200–1850* (Oxford: Oxford University Press/Pasold, 2009); Styles, *The Dress of the People*; Cissie Fairchilds, 'The production and marketing of populuxe goods in eighteenth-century Paris', in John Brewer and Roy Porter (eds), *Consumption and the World of Goods* (London: Routledge, 1993), 228–48.

33 Social emulation has been an influential theme in studies of consumption since the publication of Neil McKendrick, John Brewer and J. Plumb, *The Birth of a Consumer Society: The Commercialization of Eighteenth-Century England* (Bloomington: Indiana University Press, 1982). For critique and discussion of the concept of social emulation, see Hohti, '"Conspicuous" consumption and popular consumers'.

34 Joanne M. Ferraro, 'The manufacture and movement of goods', in John Jeffries Martin (ed.), *The Renaissance World* (New York: Routledge, 2007), 87–100, 98.

35 Fairchilds, 'The production and marketing of populuxe goods'.

PART I

Innovation and imitation

1

Transformations in textiles, 1400–1760

John Styles

Introduction

Previous generations of historians often assumed that, before the Industrial Revolution, the families of small farmers, craftspeople and labourers across rural England were clothed in a narrow range of coarse textiles, often homespun and largely unchanging.[1] It is a view that has increasingly been challenged by their successors, or, at the very least, been heavily qualified.[2] Between the fifteenth and the eighteenth centuries, western Europe witnessed a tide of novelty in textiles. A wave of innovations, embracing both fabrics and equipment, swept the continent. The impact of these innovations and how they extended to non-elite consumers are the subject of this chapter. It falls into three parts. First, it examines the character of the innovations in terms of materials and techniques. Second, it assesses the ways these innovations have been understood, using English evidence to question how effectively historians have integrated changes in production and changes in consumption. Third and finally, it considers the impact of these changes on the production of textiles by households for their own use: in other words, their impact on what has been termed 'auto-consumption', 'household self-provisioning' or, in an older historical literature, 'homespun'.[3]

Innovation

Early modern textile innovation in western Europe was underpinned by two linked developments during the later Middle Ages. The first was the shift from a two-fibre textile culture, with production and consumption monopolised by wool and flax, to a four-fibre textile culture, with woollens and linens supplemented by fabrics made in Europe from silk and from cotton. Introduced initially from the eastern Mediterranean, the production of silks and cottons had become established in parts of western Europe by the end of the Middle Ages. In the course of the next three centuries, their manufacture and consumption expanded, both geographically and socially, driven by a change in elite taste described by Patrick Chorley as a Europe-wide 'shift away from woollen broad cloth to silk that characterized this whole period up to the seventeenth century'.[4]

Following the dissemination across western and northern Europe of the horizontal treadle loom during the twelfth century, heavily napped woollen broadcloths and silk velvets dominated the European high-prestige textile market. From the fifteenth century, however, they were challenged by lighter, smoother, colour-patterned fabrics, often with a distinct sheen. Silks led the trend. Indeed, lighter silk fabrics were to lead European high fashion for the next three centuries, but they were quickly followed by textiles made to mimic patterned silks in wool, in mixed materials and eventually in cotton and linen. Interaction between the fibres stimulated dramatic expansion in the range of textiles available to consumers of modest means, with substitutions of one fibre for another, technology transfers between fibres and a proliferation of new, mixed-fibre fabrics. The cheapness of these lighter fabrics, including new lighter silks, as well as their fashionability, secured them far wider markets than their heavy medieval predecessors, extending deep into the middle and eventually lower ranks of western European society.

These wider markets were shaped by the second development which underpinned textile innovation in early modern Europe – the reinvigoration of European trade at the end of the Middle Ages. As John Munro argues, the changing pattern of trade in Europe saw the re-emergence of trans-European trade in medium- and low-quality coarse textiles, eclipsed during the contraction of European trade in the fourteenth and fifteenth centuries, as well as a reinvigorated trade in textile raw materials.[5] Munro, when discussing the fourteenth century contraction of trade in cheaper says and serges, suggests they went on being made, but for household use or for purely local consumption. He says little more about them. The implication is that there was a later medieval falling back to reliance on local textile resources, with a corresponding contraction in the diversity of textiles in everyday use. Chris Wickham goes further, arguing that, even at its zenith, around 1300, long-distance trade in woollen textiles from the urban cloth-making centres of Italy and Flanders served predominantly 'lords and their entourages'. Peasants in the surrounding areas could always 'make most of their necessary goods themselves'.[6]

With the expansion of European trade in lighter, cheaper textiles from the later fifteenth century, a relatively small number of major manufacturing centres emerged. They were to dominate production of these textiles for national and international markets during the sixteenth and seventeenth centuries. These centres were located in western Europe's most economically advanced regions: initially, in the sixteenth century, in Flanders, Italy and Picardy, and subsequently in Holland, England and other parts of France. They supplied markets across Europe, drawing on what Stuart Jenks has identified as a distribution revolution in the fifteenth century, and subsequently beyond Europe with the opening of intercontinental maritime trade in the sixteenth century.[7] For textiles, the consequences of European maritime expansion were to be profound, including new or previously unfamiliar dyestuffs (indigo, cochineal, logwood); new techniques, especially for decorating fabric with colour (colourfast painting and printing); new sources of textile raw materials (American long-staple varieties of cotton, Chinese and Indian raw silks); and new overseas markets in West Africa, Asia and the Americas, with distinctive tastes and unfamiliar competitors.

Two main trends in product innovation characterised Europe's new, four-fibre textile culture from the end of the fifteenth century. First, the shift already identified towards lighter, more colourful and more highly patterned fabrics, used both for clothing and for furnishings. Second, the dissemination of textiles employing new or relatively unfamiliar techniques, such as knitting, lace-making and colourfast printing. The impact of these innovations can be observed across the whole range of textile fibres, including wool, linen, silk and cotton. Their effects were felt at every level of the market, from the finest patterned silks worn by monarchs and their courtiers to the cheap ribbons worn by housemaids on their caps. These forms of product innovation were intimately linked to innovation in technology, fashion and marketing. They were associated with the invention, dissemination and refinement of new machines. They went hand-in-hand with an intensification and systematisation of fashion, culminating in the emergence of an annual fashion cycle for silks, at least, during the later seventeenth century.[8]

Pre-existing types of woven fabrics became lighter as well. Loom-patterned silks, produced principally in Italy, but widely exported, were the most costly and high-status textiles in sixteenth-century Europe. Between the mid-fifteenth century and the early seventeenth century, their weave density fell by a third, reflecting a shift to lighter, thinner cloths.[9] The new, light silks – grosgrains, sarcenets, satins and damasks – cost only half to three-quarters of the price of the traditional heavy brocaded velvets they superseded.[10] An equivalent change can be observed in fine woollen broadcloths, which could be almost as expensive as silks. Between the 1630s and the 1680s, the weight of a typical coloured broadcloth made in Wiltshire, in the west of England, also fell by a third.[11]

The reduction in the weight of established silk and woollen fabrics was accompanied by the dramatic commercial success of a variety of light woven fabrics. Most prominent were those made with combed, long-staple wool. By the eighteenth century, it had become common in English to refer to these fabrics collectively as worsteds, which remains the modern usage, but previously they went by a variety of names.[12] Lighter fabrics incorporating coarse, long-staple wools, such as says and serges, had long been produced in Europe. Nevertheless, the expansion of European commerce during the late fifteenth and sixteenth centuries saw huge increases in the production of new, attractive varieties in key centres, initially in Flanders, but extending over the course of the next two centuries to Holland, England, France, Italy and beyond.[13] By the later seventeenth century they too were facing competition in key markets from another category of lightweight woven textiles that was new to Europe – all-cotton fabrics, such as calico, imported from India, initially by the Portuguese in the sixteenth century and, after 1600, on an ever-larger scale by the English and Dutch East India Companies.

Many of the new, lightweight woven fabrics were made from mixed materials, especially combinations of long-staple wool with silk or other fibres. They included union fabrics, in which the fibre of the warp yarn differed from the fibre of the weft yarn, as well as blended or union yarns, in which the yarns themselves combined different fibres. Mixed-fibre fabrics were not, of course, new. Yet despite medieval precursors, the proliferation of new kinds of mixed-fibre textiles which accompanied the European shift to lighter textiles from the sixteenth century was unparalleled. Like medieval half-silks, the new mixed fabrics often mimicked more expensive textiles made from a single type of fibre, but at a lower price. So as the new, cheaper, lightweight Italian silks swept western Europe in the later sixteenth century, their patterning, colours and sheen were evoked for less affluent consumers by cheaper textiles combining expensive silk yarns with cheaper yarns made from combed wool, mohair, cotton or linen. They mimicked a wide range of costly silk piece goods – satins, damasks, velvets and taffetas – but at a much lower price.[14] Similarly, the expensive new lighter broadcloths made with Spanish wool were imitated by fabrics like serges and says, which combined warp yarns made from combed wool with weft yarns made from carded wool, as well as by heavily napped fustians combining linen warps with cotton wefts.

Lighter-weight fabrics tended to be less durable. The Venetian ambassador to the French court complained in 1546 that the satins and damasks made by the Tuscans and the Genoese were 'cloths that cost little and last even less'.[15] Norwich was the principal English manufacturing centre for the new, light fabrics made with combed, long-staple wool, known as Norwich stuffs. In 1606 it was claimed, perhaps with some exaggeration, that Norwich's traditional 'ancient worsteds' of the mid-sixteenth century would have lasted six times longer than the new Norwich stuffs, yet both were fabrics made with combed, long-staple wool.[16] More than a century later, Daniel Defoe famously dismissed Indian cotton calicoes as 'ordinary, mean, low-priz'd, and soon in rags'.[17] Indeed, cotton was not to replace linen for everyday shirting and sheeting until after the mechanisation of its spinning and weaving in the nineteenth century, when it became cheap enough to compensate for its inferior durability.[18] Qualms about the durability of the new lightweight fabrics made with long-staple wool account for some of the names given them by English manufacturers, such as 'perpetuana', 'durance' and 'everlasting'.[19]

Cheaper, less durable fabrics facilitated more frequent purchases of a wider array of items, which aligned with a heightened sensitivity among consumers to variety, novelty and fashion. An emphasis on design innovation was a corollary of this acceleration in turnover. Almost all the new fabrics were distinguished by the speed with which their patterns and colours were changed.[20] Unlike woollen fabrics made entirely or partly from carded, short-staple wool, many of the new fabrics were not fulled or napped. Consequently, their yarns were visible, enabling a huge array of woven patterns (Figure 1.1). In 1611, Norwich stuffs were already being described as being 'of infinite variety of sorts, figures, colours and prices'. The need for new patterns was constantly stressed. 'Our trade is most benefitted by our new inventions and the varying of our stuffes which is contynually profitable.'[21]

The shift to lightweight fabrics for outer garments was complemented, from the fifteenth century onwards, by a mass diffusion of linen undergarments. It was associated with expanding output of fine linens such as cambric and lawn, as well as coarser flax and hemp fabrics, produced in the countryside for both international trade and local consumption.[22] The proliferation of linen undergarments reflected the spread of new conceptions of cleanliness, semiotically privileging white bleached linen over cheaper unbleached brown linen.[23] At the same time, it contributed to the multi-layering of dress associated with wearing outer garments which were individually thinner and lighter, and consequently provided less thermal insulation.

These innovations in textile piece goods were accompanied by an equivalent transformation in the variety and quantity of textile trimmings and clothing accessories. Ribbons and tapes had ancient origins; knitted goods were familiar in medieval Europe; and both needle and bobbin-lace had medieval precursors – nevertheless, all of these saw a remarkable elaboration and proliferation after 1500. Ribbons, often made from inferior or even waste silks, became key decorative elements in European dress, even among the poor (see Chapter 2). The same was true of lace trimmings, especially those made from the cheaper bobbin-lace (Figure 1.2, see also Experiment in focus IX). In England, stockings, knitted with multi-ply yarns made from silk, worsted or wool – and also subsequently from linen or cotton – almost entirely replaced medieval hose made from woven woollen cloth (see Experiment in focus II).[24] By the seventeenth

Figure 1.1 Samples of calamancoes and striped and plain worsted stuffs, Norwich or Spitalfields, London, 1720. The National Archives, MFQ 1/134: samples of textiles submitted to the Commissioners for Trade and Plantations by the London Weavers Company, 1720, fol. 150.

century, moreover, decorative kerchiefs were widely worn by both men and women (Figure 1.3). Small in size compared to the textile lengths used for main garments, they were relatively affordable. The most expensive were made from silk, cambric or Indian muslin, but even the cheaper versions employed premium materials such as bleached linen and (by the end of the century) multi-coloured loom-patterned or printed fabrics.[25]

The shift towards lighter fabrics and mixed materials saw corresponding changes in the supply of key inputs, particularly fibres and yarns. Traded textile fibres were subject to ever more precise sorting and differentiation by quality and price, some according to specific attributes of the fibre, some according to their place of origin. Each grade of fibre had a particular use and price. Precise grading of materials facilitated an expansion of product ranges, each range differentiated by its quality and targeted at different price points in different markets.[26]

Grading also extended to the waste products generated as materials were processed. The distinction between legitimate materials and waste was especially stark for silk, the most expensive textile raw material. Most silk yarns were not spun. They were made by winding the long silk filaments off the cocoon and combining and twisting them into various grades of thread. However, at each stage in the process, short-fibre silk waste was

Figure 1.2 Charles Beale, *Susan Gill*, c. 1680. A servant wearing a cap with a lace trim. British Museum, London, 1981,0516.15.1-94: Charles Beale Sketchbook, fol. 34.

created, which could be transformed into an inferior-quality yarn only by spinning it. In mid-sixteenth-century Venice, it was estimated that about a quarter of the material derived from silk cocoons ended up as waste silk that could only be spun.[27] Spun silks were used as weft in cheaper silk fabrics, or in mixed fabrics like the *burates* made with spun silk warps and woollen wefts at Nîmes in France in the eighteenth century.[28] Different grades of spun silk were also employed extensively in haberdashery and trimmings. Similarly, different grades of waste wool arising from wool-combing and cloth-shearing were used to make coarse yarns for hand-knitted stockings, as well as low-priced cloths, such as plains, duffels and blankets.[29]

The new, lighter-weight fabrics required finer yarns than their heavier predecessors. In spinning yarn for superfine Wiltshire broadcloth warps in the 1720s, only half the length of yarn was spun from a pound weight of wool as compared with yarn spun for worsted stuffs in East Anglia during the following decade.[30] Yarns also became more uniform, because in the new fabrics the yarn was often visible, which was not the case in the silk velvets and heavily napped broadcloths that preceded them. Most yarns continued to be spun by hand, but finer, more uniform yarns took longer to spin. Consequently, the shift towards lighter fabrics required ever-increasing numbers of hand spinners. It also encouraged innovations in the equipment employed for spinning, although the pace at which they were adopted was uneven.

Figure 1.3 John Riley, *Bridget Holmes*, 1686 (detail). An aged servant in the royal household wearing a white linen kerchief, cap and apron. The Royal Collection, London.

The spindle spinning wheel had been introduced into Europe from Asia during the Middle Ages, providing a typical gain in productivity over the hand spindle of at least double for short-staple fibres. Nevertheless, it replaced the hand spindle only slowly and unevenly, especially outside the orbit of the core textile-manufacturing areas which dominated inter-regional trade. The hand spindle was cheap, it could be used while walking, and in the spinning of coarser linen yarns its productivity appears not to have been markedly inferior to spinning with a spindle wheel.[31] In many parts of continental Europe, as well as in the Highlands of Scotland, it remained in use for spinning flax and coarse woollen yarns long into the eighteenth century.[32] In England, too, it continued to be employed for similar purposes throughout the sixteenth and seventeenth centuries. In 1675, the Hampshire agricultural writer John Worlidge defined the hand spindle or rock as 'an instrument generally used in some parts for the spinning of flax or hemp'.[33] In 1687, a chapman at Forton in Lancashire, who dealt in turned wooden objects, stocked hundreds of cheap wooden hand spindles, with accompanying wooden distaffs and wooden spindle whorls. Stoneware whorls for hand spindles were imported into sixteenth- and seventeenth-century England from Germany, alongside Rhenish stoneware jugs and tankards.[34] In the eighteenth century, however, the hand spindle largely disappeared from England. It went on being used commercially only in parts of Norfolk for producing some of the finest worsted yarns for Norwich stuffs, because it excelled at spinning certain very fine, high-twist yarns, albeit slowly. Similarly, in the eighteenth-century Swiss Alps, ultra-fine cotton yarn for muslins – known as *Löthli* yarn – was spun on a hand spindle for a higher rate of pay than wheel-spun yarn.[35]

Meanwhile, spinning wheels themselves underwent significant refinement and elaboration. The major innovation was the substitution of a flyer mechanism for the simple spindle, first undertaken sometime in the late fifteenth century. It increased the spinner's productivity because it removed the need for winding on the yarn, making spinning a continuous and potentially faster process. It could also be adapted so that the wheel was turned by means of a foot pedal, thereby allowing the spinner to use two hands to draft the fibre continuously. However, the flyer wheel was two or three times the price of a simple spindle wheel, it was not well suited to short fibres, such as short-staple wool and cotton, and it was used predominantly for spinning flax. At the same time, the spindle wheel itself underwent a process of adaptation for different fibres, involving different sizes of wheel, and different spindles and drives. The Dutch wool wheel, used to spin yarn from short-staple Spanish Merino wool for the new, light *lakens* woven at Leiden in the seventeenth century, was distinguished by its sloping platform and wooden spindle, which produced a softer and less twisted yarn.[36] It was copied all over Europe. In England by the eighteenth century it was possible to distinguish between wool wheels, worsted wheels, jersey wheels, cotton wheels, fustian wheels, flax wheels, linen wheels and tow wheels, as well as great wheels, long wheels and small wheels, Dutch wheels and Saxony wheels, double wheels and single wheels.

The trend towards textiles made from thinner, more visible yarns also encouraged the use of devices for measuring the fineness and uniformity of yarn. After yarn was spun, it was customarily reeled into skeins or hanks. Simple reel staffs – wooden rods equipped with crossbars at each end to hold the yarn – had been widely used for this purpose in the Middle Ages. In the early modern period, circular reels, mounted on an axle, became increasingly common, speeding up the process. Where yarn was measured and sold simply by weight, reeling was a way of arranging it into convenient bundles. However, when establishing the precise fineness of yarn was a priority, reels could also be used as instruments for measuring quality. If the circular reels used in a particular branch of textile manufacture were of a standard circumference, and skeins or hanks were a standard number of revolutions of that reel, then the number of skeins or hanks per unit of weight provided a measure of the yarn's fineness or count. Reels were adapted to enhance this process. Snap reels made a sound after a certain number of revolutions, while clock reels had a clock-like face with a pointer indicating the number of revolutions.

The use of reels to measure quality in this way was widely adopted in the manufacture of the new light fabrics made from long-staple combed wool. Differentiation between yarns was important here, because much of the fabrics' decorative effect derived from the interplay of different yarns. In the production of yarn for short-staple woollens and coarser linens and linen-cotton mixes, use of the reel as a measuring device was less common, although it became more so. For instance, in the mid-eighteenth century it was a key element in the drive to apply English techniques to the manufacture of French cotton-linens.[37] But it was in the production and distribution of silk yarns that the use of the reel for measuring fineness was at its most sophisticated, extending in eighteenth-century Italy to the use of a specialised yarn-testing reel by silk dealers as well as silk manufacturers.[38]

Producers and consumers

The first section of this chapter approached transformations in early modern European textiles in terms of the materiality of fabrics and the tools employed to make them. In economic history, however, innovations in textiles have more often been approached through study of the organisation of production and the incentives which shaped it. For an older, 'stages' theory of European industrial development, first formalised by German historians in the later nineteenth century, the early modern period was the era of rural domestic handicraft production, a distinct stage in the development of manufacturing between the urban guild system of medieval Europe and the urban factory system of the nineteenth century.[39] Control of the means of production was a key defining factor here, as well as location. The materiality of products and their consumption did not figure prominently. Many of the assumptions that underpinned this interpretation of early modern European manufacturing have since been discarded.[40] Yet the idea of a sequence of developmental stages culminating in the Industrial Revolution has continued to shape histories of early modern European manufacturing, notably with the concept of 'proto-industrialisation', defined by its author Franklin Mendels as 'the first phase of the industrialization process'.[41]

The notion of proto-industrialisation grew out of Mendels's work on early modern linen production in Flanders. His aim was to understand the internal economic and social dynamics of the rural domestic system, its emergence having often been treated simply as the response of guild-constrained merchant-manufacturers to high urban wages. Mendels insisted that the technological innovations of the Industrial Revolution were preceded not by stagnation but by a process of economic growth through regional specialisation. He presented a picture of early modern western Europe increasingly divided between rural regions specialising in commercial agriculture and rural regions specialising in commercial manufacturing. Some regions, which previously combined agricultural production with part-time industry, now gave up their industries and began to purchase industrial products from other regions. At the same time, other regions began to specialise in rural handicrafts, not simply to supply local markets but for 'regional, national, or international trade'.[42] Building on this model of regional specialisation, Mendels offered an explanation of, first, why rural farmers and labourers were drawn into industrial production, couched essentially in terms of population pressure on resources, and, second, how that development led to factory industrialisation, via, in particular, capital accumulation and further demographic expansion.

The proto-industrialisation thesis has been much criticised. For many if not most of the early modern rural regions that specialised in commercial handicrafts, proto-industrialisation was not the first phase of the industrialisation process. The destiny of their manufacturing industries was attenuation and disappearance. Other regions experienced mechanised industrialisation without ever passing through a proto-industrial stage.[43] Mendels, moreover, largely ignored the relationship between changes in production and changes in consumption. He paid little attention to the products of the industries he studied, denied the importance of technical innovation, ignored product innovation and offered no account of the shape of demand for proto-industrial goods beyond general references to local, regional, national and international markets.

In more recent economic history, the relationship between changes in production and changes in consumption during the early modern period has been analysed most effectively using the concept of an 'industrious revolution', proposed by Jan de Vries. According to de Vries, an early modern 'industrious revolution', driven by new kinds of consumer goods, was a precursor to the Industrial Revolution, providing the context in which it could unfold.[44] Confronted with evidence for real-wage stagnation in north-west Europe between the sixteenth and the eighteenth centuries, de Vries asks how this can be reconciled with equally compelling evidence for ownership of growing numbers of material things, including, conspicuously, textiles and clothing. De Vries insists that consumer demand was transformed in parts of north-west Europe between 1650 and 1800. New kinds of commodities emerged, with the capacity to entice ordinary people out of self-sufficiency and idleness into working harder and buying more. They included new manufactured goods which were cheaper, less durable and more fashion-sensitive, such as lightweight, printed cotton clothing, and containers for food and drink made from decorated earthenwares. They also included new kinds of stimulants, both exotic (such as tea, tobacco and sugar) and domestic (such as gin and brandy).

De Vries resolves the paradox in two ways. First, he points out that if macroeconomic growth before 1830 was slower than previously believed, as recent studies of the English Industrial Revolution have suggested, then 'it leaves "pre-industrial" England as a rather richer economy than had earlier been assumed, for the simple reason that less growth in the 1760–1830 period means the pre-1760 economy must have possessed a per capita income closer to that found in the post-1830 period'.[45] Like Mendels's proto-industrialisation thesis, this represents a powerful challenge to the old belief that the European economy between the fifteenth and the eighteenth centuries was premodern and therefore growthless and poor.

Second, de Vries argues that the key dynamic during this era of economic expansion before the Industrial Revolution was the behaviour of the household as an economic unit, in particular the way time was allocated among members of the household to different activities. In order to buy more, ordinary households devoted increasing time and effort to paid work, especially the wives and daughters of those households. Wage rates might have stagnated, or even fallen, but more household members participated in paid work, and they worked for more hours and more days. Consumption was key. 'Consumer demand developed through an interaction of market and household productive systems.'[46] People were drawn into paid work by the lure of new commodities superior to those they could produce for themselves. They reallocated time and consumption from the household to the market. Like Mendels, de Vries assumes that the paid work available would vary according to developing regional patterns of comparative advantage but work in new forms of textile manufacturing was especially prominent, particularly those labour-intensive processes that employed predominantly women and children, such as spinning and knitting.[47]

De Vries's industrious revolution thesis has been enormously influential. Yet in the light of the transformations in textiles outlined in the first section of this chapter, it looks chronologically timid. De Vries traces the start of his industrious revolution only as far back as the mid-seventeenth century, when the new commodities he identifies as the spur goading women and children into paid work first began to be available. His central focus is what he terms the 'long eighteenth century', from 1650 to 1850.[48] His principal sources are taxation records, especially import duties, and post-mortem probate inventories of possessions. Whether probate inventories provide reliable evidence of the consumption choices of the families of small farmers, craftspeople and labourers has been much disputed, as de Vries acknowledges.[49] His list of the new commodities that played a key transformative role in consumer culture is dominated by the imported groceries (tobacco, sugar, tea, coffee), Asian-influenced ceramics and Indian and Indian-influenced cotton textiles familiar from historians' debates about an eighteenth-century consumer revolution in England. Yet before the mid-eighteenth century, none of these goods (apart from tobacco) achieved anything approaching a degree of market penetration among English families of small farmers, craftspeople and labourers which could realistically have encouraged them to greater industry.[50] Rather than preceding the Industrial Revolution of the second half of the eighteenth century, their dissemination appears to have accompanied it.

These exotic consumer goods hold a prominent place in historians' debates for the same reasons de Vries focuses on them: they occasioned copious hostile commentary in print, and, conveniently, their rapidly growing consumption can be traced through taxation records. Although de Vries acknowledges 'the centrality of clothing to the power of consumer demand', both probate and taxation records are notoriously poor guides to changes in English domestic consumption of textiles and dress.[51] In so far as de Vries discusses changes in their consumption, he concentrates on the proliferation during 'the long eighteenth century' of Indian calicoes and the European-made cotton textiles that copied them, which in Britain were subject to customs and excise duties.[52] The transformations in textiles of the later sixteenth and seventeenth centuries are touched on, but not explored, with passing references to 'the changing composition of wardrobes toward lighter woollens, linen, cotton, and mixed fibers', and to Joan Thirsk's pioneering studies of the burgeoning commercial production of bobbin-lace and knitted textiles, especially stockings, in poor English rural households.[53]

Perhaps de Vries is led astray by his insistence on a stark boundary between an old consumption regime at the beginning of the seventeenth century and a new consumption regime in the eighteenth, between old luxury and new luxury, between intrinsic value and fashion, between what he portrays as an inherited material world and a leap to a novel, transient world of goods.[54] This chronology of rupture, combined with a lack of equivalent sources for the earlier periods, diverts attention from the material, technological and commercial transformations in textiles that, by the mid-seventeenth century, were already reshaping the lives of plebeian consumers in England. After all, textiles comprised an especially large proportion of regular household consumption of manufactured goods, particularly among working people.

De Vries's industrious revolution thesis does offer a compelling framework for understanding the link between changes in early modern consumption and production. Yet as far as textiles and dress are concerned, the thesis would be chronologically more coherent if it placed less emphasis on the printed calicoes and cotton-linens of the late seventeenth and eighteenth centuries, and more on the earlier shift to lighter, brighter, cheaper, fashion-sensitive and less durable textiles, which were equally alluring to humble consumers. Two in particular stand out. First, the worsted and worsted-mix textiles which began to be produced in the Low Countries from the end of the fifteenth century, and in many other parts of Europe during the sixteenth century. Second, the new, fashionable clothing accessories highlighted by Joan Thirsk, which also rapidly captured wide markets during the sixteenth and seventeenth centuries, particularly ribbons, knitted hosiery and bobbin-lace.[55] These textiles possessed the key characteristics that de Vries associates with accessible innovations capable of luring purchasers into a new industriousness. They were semi-durable, visually attractive in ways that aligned both with an accelerating fashion cycle and with the sartorial rhythms of everyday life, relatively inexpensive and getting cheaper.[56] Yet they are poorly represented in the sources de Vries cites, especially probate inventories, where it is durable, inheritable household goods that feature most prominently. Moreover, as Thirsk demonstrated, the new clothing textiles and accessories provided novel opportunities for women's and children's paid work. They needed frequent replacement and, especially in the case of worsteds, required a higher ratio of female spinners to male weavers than older, heavier woollen fabrics, because their finer, lighter threads took longer to spin. Accurate statistics for early modern women's employment in England are almost entirely lacking, but there is general agreement that the numbers employed in commercial spinning grew dramatically between the sixteenth and the eighteenth centuries.[57]

De Vries lays particular emphasis on printed calicoes and cotton-linens because they held a special fashionable appeal for women, whose increased participation in commercial manufacturing is crucial to the industriousness thesis. Here he draws on the literature on the so-called 'calico craze' associated with the surge in imports of Indian cotton textiles that took place in late seventeenth- and early eighteenth-century western Europe. Yet in England before 1700 Indian cotton textiles were not extensively worn as women's main garments: gowns or petticoats. Indeed, gowns and petticoats made from cotton textiles, whether cotton-linens, calicoes or muslins, did not come to dominate women's dress until the second half of the eighteenth century.[58] Nevertheless, the textiles used for these most visible and fashion-sensitive of women's main garments did change radically during the sixteenth and seventeenth centuries. The evidence of criminal trials for theft in counties in south-east England and the English Midlands (Table 1.1) indicates a dramatic shift between the later sixteenth and the later seventeenth centuries away from the use of heavy, short-staple woollen cloth for gowns, petticoats and the earlier kirtle, towards silks and especially towards worsteds – stuffs, serges, tammies, mohair.[59] Indeed, russet, a cheap woollen cloth woven in the natural colour of the sheep's fleece, which had epitomised simple rural life in sixteenth century plays and ballads, disappeared entirely by 1660.[60]

Table 1.1 Numbers of indictments for thefts of gowns, kirtles and petticoats, by textile type, Assizes and Quarter Sessions for Essex and Kent, 1559–1688, and Oxfordshire and Worcestershire, 1700–79[1]

Type of textile	*Essex & Kent 1559–1603*	*Essex & Kent 1660–88*	*Oxfordshire & Worcestershire 1700–49*	*Oxfordshire & Worcestershire 1750–79*
Silks	7	37	5	12
Woollens	55	23	1	3
Worsteds / stuffs[2]	11	61	20	20
Linens	0	4	7	18
Cottons / calicoes	0	0	1	12
All	73	125	34	65

Sources: ESSEX (Assizes and Quarter Sessions): James Cockburn (ed.), *Calendar of Assize Records: Essex Indictments, Elizabeth I* (London: HMSO, 1978); Calendars of Essex Assize Files in the Public Record Office (typescript), vol. 4, 1660–1685 and vol. 5, 1684–1714, Essex Record Office; Calendars of Essex Sessions Rolls (typescript), vols 1–18, 1536–1610, and vols 22–5, 1654–1699, Essex Record Office. KENT (Assizes only): James Cockburn (ed.), *Calendar of Assize Records: Kent Indictments, Elizabeth I* (London: HMSO, 1979); James Cockburn(ed.), *Calendar of Assize Records: Kent Indictments, Charles II, 1660–75* (London: HMSO,1995); James Cockburn (ed.), *Calendar of Assize Records: Kent Indictments, Charles II,1676–88* (Woodbridge and London: Boydell, 1997). OXFORDSHIRE (Assizes and Quarter Sessions): Crown Minute Books, Assizes, Oxford Circuit, 1714–79, The National Archives (TNA), ASSI 2; Indictment Files, Assizes, Oxford Circuit, 1714–79, TNA, ASS I5; Oxfordshire Quarter Sessions rolls, 1700–79, Oxfordshire History Centre, QSR. WORCESTERSHIRE (Assizes and Quarter Sessions): Crown Minute Books, 1714–79, Assizes, Oxford Circuit, TNA, ASSI 2; Indictment Files, Assizes, Oxford Circuit, 1714–79, TNA, ASSI 5; Worcestershire Quarter Sessions Rolls, 1700–79, Worcestershire Archive and Archaeology Service, 110 BA1.

Notes

1 Gowns and petticoats could be men's garments in the first half of the sixteenth century, but increasingly rarely thereafter. The definition and form of petticoats changed appreciably in the course of the early modern period, but, from the mid-sixteenth century to the later eighteenth century, the word 'petticoat' generally described a woman's skirt, worn from the waist down, sometimes with and sometimes without a gown over it. A kirtle was a form of under-gown, which fell out of use early in the seventeenth century.

2 The category 'worsteds / stuffs' embraces the new, lighter fabrics, made all or in part from long-staple sheep's or goat's wool. Between 1559 and 1603, it includes five worsted kirtles, one 'chamlett' kirtle and one 'chamblet' gown. Some or all of these are likely to have been made from Norwich's traditional 'ancient worsteds' – not the newer, lighter types of what later became known as 'worsted stuffs', which were introduced to Norwich by immigrants from the Low Countries from the later 1570s. See Luc N. D. Martin, 'Textile manufactures in Norwich and Norfolk, 1550–1622' (PhD thesis, University of Cambridge, 1991), chapter 1; John Oldland, '"Fyne worsted whech is almost like silke": Norwich's double worsted', *Textile History*, 42 (2011), 181–99.

It is difficult to measure the overall performance of the English woollen textile industries across the seventeenth century, but a transformation in the mix of products is clear enough. Indeed, the seventeenth-century shift in women's outer garments from heavy woollen fabrics to lighter, colourful, and often patterned worsteds and worsted-silk mixes was no less dramatic than the shift to printed fabrics after 1740. By the start of the eighteenth century, according to a later pamphlet, 'our women among the *Gentry*, were then clothed with fine *English* Brocades [silks], and *Venetians* [a worsted-silk mix]; our common Traders' Wives with slight Silk Damasks [silk or a worsted-silk mix]; our Country *Farmers Wives*, and other good Country *Dames* with *woorsted Damasks*, *flower'd Russels* and *flower'd Calimancoes* [all worsteds], and the meanest of them with plain woorsted Stuffs, etc.'.[61]

Gowns and petticoats employed significant lengths of fabric. During the first half of the eighteenth century, an English working woman's gown might require 9 to 11.5 yd of fabric, her petticoat 4 yd. In the 1740s, a yard of worsted camlet bought by a Lancashire small farmer's family for a gown cost 14.5d, a yard of blue flowered worsted damask for another gown 20.5d, while 13.5d per yard was paid for tammy for a petticoat.[62] When made up, the petticoat cost 54d for the outer fabric alone, roughly a week's wages for a full-time, adult cotton spinner in mid-eighteenth-century Lancashire, the gowns from 130d to 236d.[63] Evidently, these lightweight, often decorative worsted fabrics were accessible to plebeian consumers for wear as main garments (Figure 1.4). However, they represented major items of

Figure 1.4 Paul Sandby, *Black Heart Cherries*, c. 1759. A London street-seller wearing a blue and white striped petticoat under a red worsted stuff gown. A pink silk ribbon is tied round her cap and around her neck she wears a blue and white check kerchief. Yale Centre for British Art, New Haven, B1975.3.206.

expenditure when acquired in the fabric lengths necessary for petticoats or gowns. Such purchases were intermittent and could be postponed.

When Jan de Vries turns to what he calls 'the luxuries of the poor' – the new consumer items most likely to encourage the poorest manual workers to greater industriousness – textiles for gowns and petticoats are not his principal focus. For manual workers, he insists, the key 'incentive goods' were colonial groceries and new alcoholic drinks, which were bought much more frequently, in small quantities and at a fraction of the price.[64] Yet many of the novel clothing accessories of the sixteenth and seventeenth centuries – ribbons, knitted stockings, kerchiefs – fell into a price range that resembled working people's purchases of extra-European groceries. By the mid-eighteenth century, when both categories of goods were being bought regularly by plebeian consumers, the pattern of purchases was not dissimilar in terms of expenditure and frequency.

During the 1750s, Stephen Hudson kept a shop at Thruscross, high in the Yorkshire Pennine hills between Skipton and Knaresborough. It served a humble clientele consisting overwhelmingly of small farmers and tradespeople. They made frequent, small purchases of colonial groceries – tobacco, sugar, tea – and bought clothing accessories – stockings, ribbons, kerchiefs, mitts – in just the same way (Table 1.2). Hudson's typical sale of colourful silk ribbon to decorate a hat or a cap was worth 8d, the cost of a yard or two of ribbon, depending on width and quality (Figure 1.5). This was cheaper than his average sale of small parcels of sugar or tea, although the typical tobacco sale was smaller still. Stockings and mitts, in woollen or worsted yarn, the latter frame-knitted, were sold by Hudson largely in single pairs, at prices similar to his regular sales of tea. Most of his kerchief sales were also of single items. His cheap linen kerchiefs were close in

Table 1.2 Purchases of clothing accessories and extra-European groceries at Stephen Hudson's shop, Thruscross, Yorkshire, August 1758 to July 1759

Item[1]	*Number of purchases*	*Average spend per purchase (d)*
Clothing accessories		
Stockings (pairs)	60	15.5
Mitts (pairs)	12	15.0
Ribbons	38	8.0
Kerchiefs	43	24.0
Extra-European groceries		
Tobacco / snuff	54	2.5
Sugar	120	9.5
Tea	16	14.0

Source: Shop book of Stephen Hudson of Thruscross, Yorkshire, 1751–59, West Yorkshire Archive Service Bradford, 33D80/7.

Note
[1] In Hudson's account book, kerchiefs are referred to as 'handkerchiefs' and stockings as 'hose'.

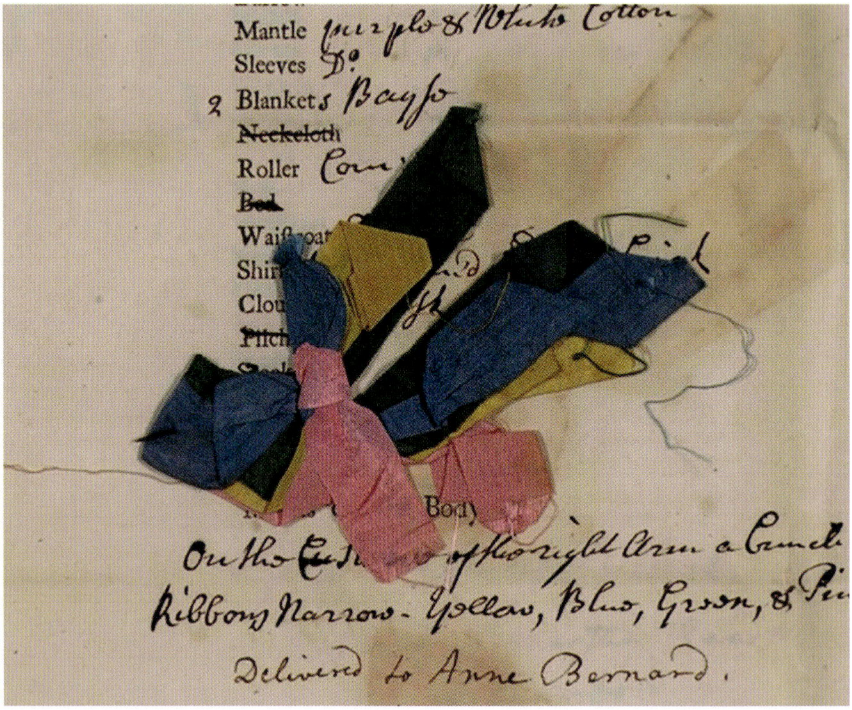

Figure 1.5 'A bunch of 4 ribbons narrow – Yellow, Blue, Green, & Pink', pinned to the billet for Foundling 170, a girl named Pamela Townley, 1743. These are plain, narrow, affordable silk ribbons, which would have sold for a few pence per yard. © Coram. London Metropolitan Archives, A/FH/A/9/1/3: Billet Book, 1742–46, London Foundling Hospital, which continues as the children's charity Coram (https://www.coram.org.uk/).

price to his stockings. His printed and loom-patterned kerchiefs, often with cotton or silk threads, could cost twice as much, yet they too were popular. Sales of kerchiefs, mitts and ribbons, like those of tea, were heavily skewed towards unmarried women, who made hardly any purchases of tobacco.[65]

Tobacco apart, it was only in the course of the eighteenth century that the extra-European groceries highlighted by de Vries became sufficiently affordable for regular purchase by the sort of people who patronised Stephen Hudson's remote Pennine shop.[66] Ribbons, stockings and kerchiefs became accessible much earlier. Knitted stockings had almost entirely replaced woven hose early in the seventeenth century.[67] Linen kerchiefs became everyday wear. As Margaret Spufford pointed out in her study of seventeenth-century pedlars, kerchiefs were 'the staple items of ready-made clothing amongst the chapmens' inventories … Even when a chapman carried no other finished goods, he was likely to carry these.'[68] Silk ribbons already appear in the probate inventories of English provincial shopkeepers in the sixteenth century, though not in large numbers. In the course of the seventeenth century, their numbers and variety increased markedly as manufacture in England grew and spread from London to the provinces, especially the area around Coventry.[69] In her extensive survey of haberdashery in retailers' probate inventories, Polly Hamilton notes that 'the quantities of ribbons … took an upturn in the 1670s, rising from roughly fourteen varieties noted per decade to thirty-two varieties per decade around the turn of the century'.[70] By 1701, in the small market town of Ambleside, deep in the mountains of the Lake District, a blacksmith could buy cheap silk ribbons for his servant maid at 2d or 4d a yard.[71] The wide appeal of the fashion for top-knot ribbons at this period was the subject of popular broadside ballads:

> Every Dragel'-tayl'd Country Girl,
> when once she comes up to the City,
> If she can get but a Ribbon-Fallal,
> O then she is wondrous pretty.[72]

The seventeenth century saw these cheap petty clothing luxuries emerge as essential components of everyday fashion. They became fixtures in the stock of the small shopkeepers and pedlars who served plebeian customers decades before most extra-European groceries. Coveted, inexpensive and universally available, they were equally potent incentives to industriousness.

The fate of homespun

Jan de Vries argues that his industrious revolution saw rural women and children drawn into various forms of home-based, income-generating manufacturing, especially spinning, by the lure of attractive new consumer goods, from textiles to tea. This move into market-orientated work was unprecedented, 'making effective market use of labor [previously] trapped in idleness and underemployment by the seasonal constraints of agriculture'.[73] Yet de Vries never explores this pre-existing world of female rural 'idleness and underemployment'. It serves merely as a counterpoint to a subsequent surge of industriousness.

So what were early modern women and children doing with their time before income-generating work in commercial manufacturing became widely available? English commentators towards the end of the eighteenth century were almost unanimous in insisting that one of the things they were doing was domestic self-provisioning with textiles, even in regions not typically associated with rural textile production. They repeatedly bemoaned its decline over the course of the century. In Staffordshire, for instance, it was reported in 1794 that 'there is no considerable public manufacture of linen, but a good deal of hurden [harden], hempen, and flaxen cloth, got up in private families', while 'a good deal of woollen cloth is got up in the country by private families, though in less quantity than formerly'.[74] Similarly, in the East Riding of Yorkshire:

> the domestic manufacture of coarse grey woollen-cloths, from a mixture of black and white wool, for the clothing of the farmer and his family, which was formerly not unusual, has now long ceased; but the careful housewife still spins and knits the stockings for her family of black or mixed wool (most farmers keeping one or two black sheep for the purpose,) and likewise during winter spins flax for a web of linen for sheets or shirts.[75]

Of course, household self-sufficiency in textiles could never have been complete in any era, because the different stages of production involved radically different time commitments. A single weaver working full-time required the output of several spinners, their precise number depending on the fibre, the preparatory processes, the fineness or count to which the yarn was being spun and the intensity at which they worked. So, when we speak of household self-provisioning in textiles, we are speaking principally about spinning by the women of the household. Weaving the yarn spun into cloth for the household's own use was generally undertaken by skilled specialists – men who identified themselves occupationally as weavers and owned looms and an appropriate range of loom gear. Such weavers worked on a jobbing basis for a variety of customers, weaving up the yarn spun by local women into cloth for a fee.[76]

In the mid-eighteenth century, when commentators suggested that textile self-provisioning was already declining in England, evidence for the presence of jobbing weavers in the population is provided by the lists of men balloted for the militia in three English counties, and in subdivisions of three others, between 1759 and 1777 (Table 1.3). These militia lists provide a reasonably comprehensive census of adult male occupations.[77] Together, they enable us to trace the presence of weavers in several widely separated parts of England, characterised by a variety of rural economies. Northamptonshire was a county with a significant but contracting worsted industry producing for distant markets. Hence it is not surprising that 9.5 per cent of the men listed there were weavers.[78] By contrast, none of the five other counties or subdivisions was noted for large-scale textile production. Nevertheless, the militia lists reveal the presence in all of them of weavers, most of whom must have worked on a jobbing basis. At one extreme was Hertfordshire, agriculturally prosperous with ready access to London and its many suppliers of textiles and clothing. Only 0.3 per cent of its men listed for the ballot were identified as weavers. At the other extreme was Northumberland in the far north of England, with remote upland areas characterised by poor pastoral agriculture. There the proportion of weavers was considerably higher at 4.3 per cent, but still less than half the percentage in worsted-manufacturing Northamptonshire.[79] The proportion of weavers in the three subdivisions, scattered across the

Table 1.3 Weavers in militia ballot lists, 1759–77

Locality and year[1]	*All men listed*	*Weavers (No.)*	*Weavers (%)*
Counties			
Hertfordshire 1759	12,360	38	0.3
Northumberland 1762	13,916	601	4.3
Northamptonshire 1777	11,206	1,065	9.5
Sub-divisions			
Dorchester, Dorset 1758	2,157	14	0.6
Wingham, Kent 1764	1,361	14	1.0
Soke of Peterborough 1762	876	14	1.6

Sources: HERTFORDSHIRE: 'Hertfordshire Militia Ballot Lists', PDF files on compact disk (Watford: Hertfordshire Family History Society, 2008); NORTHUMBERLAND: Joseph Barker, Peter M. K. Kitson, Leigh Shaw-Taylor and E. A. Wrigley, '1762 Northumberland Militia Ballot List Database', database (Cambridge: Cambridge Group for the History of Population and Social Structure, 2013); NORTHAMPTONSHIRE: Wendy Raybould, 'Open for business: Textile manufacture in Northamptonshire, c. 1685–1800' (PhD thesis, University of Leicester, 2005), 94–5; DORCHESTER: Mervyn Medlycott (ed.), *Index to Dorset Militia Ballot Lists, 1757 to 1799*, vol. 2 (Weymouth: Somerset and Dorset Family History Society, 1999); WINGHAM: Returns of men liable for service in the militia, made by each parish or 'borough', Wingham subdivision, 1764, Kent Archives, L/M/4/1.; SOKE OF PETERBOROUGH: Victor A. Hatley and Brian G. Statham, 'Nassaburgh militia lists, 1762', in Edmund King (ed.), *A Northamptonshire Miscellany* (Northampton: Northamptonshire Record Society, 1983), 109–46.

Note

[1] Figures for Northumberland exclude Newcastle-upon-Tyne, Berwick on Tweed and the enclaves of County Durham: Norhamshire, Islandshire and Bedlingtonshire. Figures for the Soke of Peterborough exclude the City of Peterborough.

south and east of England, varied from 0.6 per cent in south Dorset to 1.6 per cent in the Soke of Peterborough.

The proportion of weavers may appear small outside Northamptonshire with its worsted industry, but in Kent in the extreme south-east of England, in Northumberland in the far north and in the Soke of Peterborough in the east it did not differ dramatically from the proportion of other rural craft occupations in the listed population, particularly tailors and shoemakers.[80] In those areas, between a fifth and a half of parishes or townships hosted a resident weaver. The evidence of the militia lists suggests, therefore, that jobbing weavers were readily accessible to households self-provisioning with textiles across many parts of mid-eighteenth-century rural England, but not all. In Hertfordshire and south and west Dorset, access was possible, but more restricted. The Finnish botanist Pehr Kalm noted in 1748 that in Hertfordshire 'the women do not get sore fingers by much spinning, or arm-ache or back-ache from weaving. It is the part of the Manufacturers to make up for this, and the men's purses are punished in this matter.'[81]

What were these jobbing weavers actually weaving? In the case of Northumberland, we know, because in 1792 jobbing weavers around the town of Hexham circulated handbills listing the charges for their services weaving household-spun linen yarn of various finenesses for 'housewives'.[82]

All the textiles they wove were linens – yard-wide plain linen cloth, sheeting, ticking and huckaback – and mostly coarse. The pattern book of a jobbing weaver, Thomas Jackson, who lived in the vicinity of Kirkleatham in the north-east of Yorkshire, shows how these weavers adjusted standard designs to the requirements of local customers.[83] Alongside a draft for a simple linen damask diaper (Figure 1.6), Jackson notes that it was woven for one customer in 1756 with a yarn of 38 cuts or leas per pound (NeL 38), for another in 1761 with a coarser yarn of 28 cuts or leas per pound (NeL 28), and for another again in 1769 with a yarn of 40 cuts or leas per pound (NeL 40).[84] Jackson evidently adjusted to accommodate the fineness of yarn supplied by his customers, and yarns could vary considerably in fineness between different spinners. Nevertheless, these were all relatively coarse yarns. In the internationally competitive Irish linen industry, what was termed two-hank yarn (NeL 24) was not even exported, but used for local consumption, while three-hank yarn (NeL 36) was deemed just

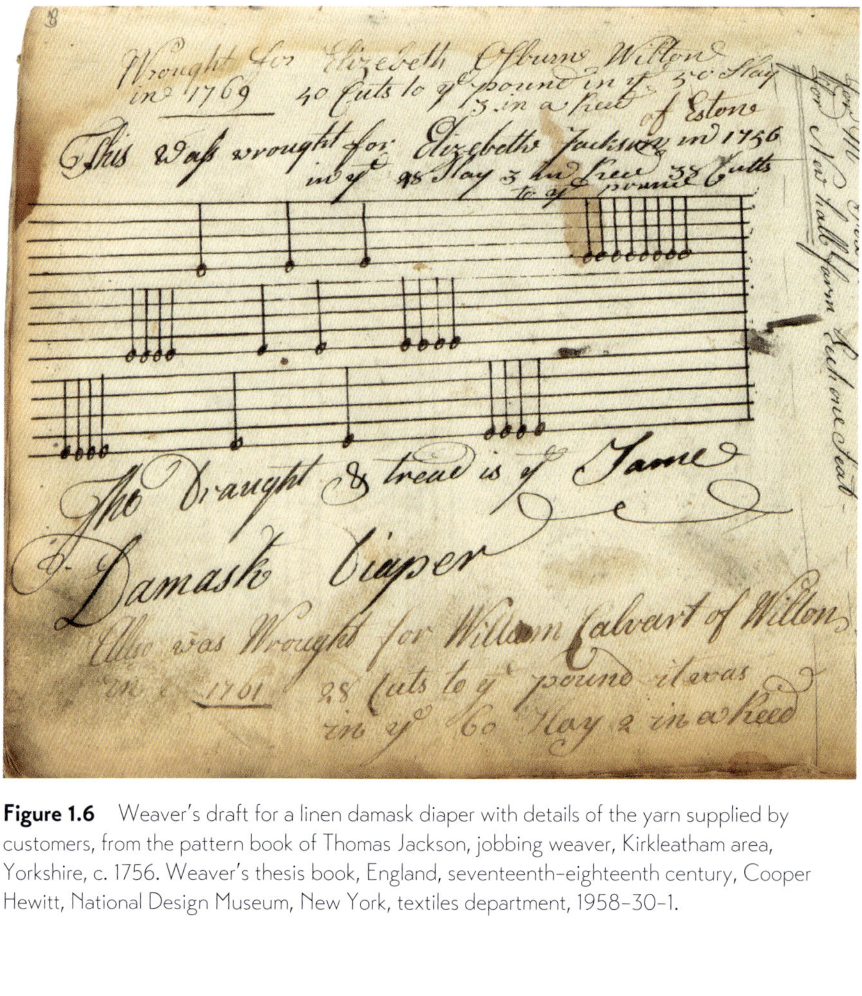

Figure 1.6 Weaver's draft for a linen damask diaper with details of the yarn supplied by customers, from the pattern book of Thomas Jackson, jobbing weaver, Kirkleatham area, Yorkshire, c. 1756. Weaver's thesis book, England, seventeenth-eighteenth century, Cooper Hewitt, National Design Museum, New York, textiles department, 1958-30-1.

about adequate for linen for labourers' shirts. It is notable that the 1792 Hexham weavers' price list included yarns even coarser than these, which in Ireland were used for sacking or coarse sheeting for the poor.[85]

We can observe this kind of auto-consumption from the household's perspective in the accounts kept by the Latham family in west Lancashire between 1724 and 1767. The Lathams farmed a smallholding of approximately nineteen statute acres in the rural township of Scarisbrick.[86] They are the most humble eighteenth-century English family for which a long run of household accounts survives. During the early years of married life, between 1726 and 1741, Richard and Ann (Nany) Latham produced eight children, seven of whom lived to adulthood; six of their surviving children were daughters.

From the very beginning of the accounts in 1724, there are payments for spinning wheels and their accoutrements, so it seems that spinning was performed both by Nany Latham and by her daughters as they grew beyond infancy. In 1724, shortly after her marriage, Nany Latham acquired a new flyer wheel for spinning flax, suggesting this was a priority for the couple. Payments appear in the accounts almost every year thereafter for raw flax. Raw hemp, raw cotton and sheep's wool were also bought from time to time, but far less frequently and in much smaller quantities. Between 1724 and 1767 the family bought more than 1,000 lb weight of flax, but only 34 lb of hemp, 36 lb of raw cotton and 113 lb of wool. In two years – 1745 and 1761 – the Lathams also bought seed to grow flax, which they then processed, but there is no mention of flax-growing in other years. The family also had a few sheep on their farm and owned sheep-shears.

The spinning the Latham family undertook on their own account was, therefore, overwhelmingly of flax. However, the women of the family undertook spinning not just to provision the household but also for wages as outwork. Scarisbrick was in the spinning zone for both the Lancashire cotton industry and the Lancashire-Yorkshire worsted industry. At the end of the 1730s, as the older daughters entered their teens, the family bought three wheels for spinning cotton, as well as cotton cards to prepare the fibre. Yet subsequently they bought only very small quantities of raw cotton. The presence of the specialised cotton wheels and a subsequent jump in the level of family expenditure suggests that cotton was being spun for wages under the putting-out system. Later, two more 'cotten wheels' were bought for the younger daughters as they grew up.

It is difficult to establish precisely how the textile fibre bought or grown by the Lathams was used, but much of it must have been spun into yarn and woven into linen cloth for the family's own use.[87] The first eighteen years of the accounts, up to 1741, cover the early, penurious period of the Lathams' marriage, when they were burdened with small children. They bought less than 29 yards of plain linen cloth ready woven, much of it described as 'fine'. Some ready-woven linen was also bought in the form of kerchiefs and occasionally aprons. It is inconceivable that a family which numbered nine by 1741 could have made do with such a small amount of plain linen, amounting to 1.5 yd per year, for clothing as well as for household and farm requirements (sheeting, sacking, etc.). Three yards of cloth was sufficient to make only one shirt for an adult man. Evidently the majority of their linen was self-provisioned – plain, relatively coarse, much of it probably unbleached, and used for shirts, shifts, sheets and sacking.

If the Lathams largely supplied themselves with linens during the early, challenging years of marriage when resources were limited, the same was not true of woollens. The family bought only relatively small quantities of wool to spin. Much of the yarn spun from that wool was probably knitted into stockings. They certainly owned knitting needles, at the modest cost of a halfpenny a pair. Some was also woven with their linen yarn to make linsey woolsey. However, the overwhelming majority of the woollen and worsted cloth they wore appears to have been purchased ready woven. During the years 1724 to 1741 the accounts record an average annual purchase of approximately 5.25 yd of woollen and worsted cloth per year, which was almost entirely used for making outer garments.

After 1742, when the family's income increased dramatically as the older daughters began to earn by spinning cotton yarn for the Lancashire cotton industry, the situation was transformed. Purchases of ready-woven woollen and worsted cloth for garments quadrupled. Annual spending on (often decorative) linen clothing accessories, such as kerchiefs, caps and aprons, jumped fivefold. The amount of plain linen bought ready woven also grew, albeit much more modestly, from 1.5 yd to 2.5 yd per year. Nevertheless, the Lathams continued to rely on self-provisioning for much of their plain linen, especially the coarser varieties. Their small annual purchases of coarse linens, valued at under 12d per yard, hardly changed and the same was probably true of the quantity of coarse linens they had woven from their own yarn. Between 1742 and 1754 the amount of flax and hemp fibre the family purchased declined by 20 per cent, but it was supplemented in the mid-1740s by some flax grown on the farm.

The transformation in the fortunes of the Latham family illustrates some of the principal elements in Jan de Vries's industriousness thesis. The booming Lancashire cotton industry of the 1740s and 1750s offered a family with six daughters ample opportunity to take on outwork spinning for wages, although they do not appear to have worked full-time at cotton spinning, combining it with dairying, work on the farm and domestic labour.[88] The family's spending choices suggest that fashion in clothing was a crucial incentive to industriousness, both for the daughters of the family and for their mother. A good part of their additional earnings was spent on what was, by the standards of Lancashire village life, expensive and fashionable female clothing, especially gowns and accessories. Nevertheless, it is important to emphasise that the increase in family spending required to make these additional clothing purchases between 1742 and 1754 was small, no more than an extra 20d a week, considerably less than the estimated weekly earnings of just one regularly employed outwork cotton spinner at the period.[89] Relatively small shifts in family income could produce dramatic transformations in material culture.

Although the Latham daughters worked at cotton spinning, their fashion choices were not dominated by cotton textiles. For gowns, they bought light, colourful worsteds and silk-worsted mixes – flowered damask and silk camlet – as well as the printed cotton-linen fabrics woven in Lancashire. Accessories included silk kerchiefs, aprons in fine white linens such as cambric and lawn, and lace borders for caps and aprons. In contrast to the regular purchases of sugar and treacle recorded in the family accounts, the entries for these clothing purchases were personal and individualistic, naming the recipient.[90] If the language of the accounts is any sort of guide, the Lathams invested their individual identities in clothes to the almost complete exclusion of the other goods the family bought, whether perishables or durables, cheap or expensive.[91]

Nevertheless, if the individualistic pleasures of fashionable consumption lured the Lathams into waged labour in the 1740s, in the manner de Vries suggests, it can hardly be said that the Latham family were previously trapped in rural idleness, when so much time and labour had gone into provisioning the young family with textiles. It was only during the early, impecunious years of married life, between 1726 and 1741, that the family paid others to spin for them, suggesting Nany Latham could barely fulfil her young family's own requirements for spun yarn, let alone undertake outwork or market spinning for money.

Five key attributes of textile self-provisioning emerge from the Latham accounts. First, fabrics made from yarn spun within the household made a major contribution to the quantity of textiles consumed by the Latham family, yet in no sense was the family self-sufficient in textiles. In other words, self-provisioning was important, but not paramount.

Second, spinning yarn for household use did not depend on access to land to grow flax or raise sheep. The Lathams purchased most of their flax and hemp, drawing on international supply chains that extended into the Baltic region. They grew flax on their small farm only briefly in the mid-1740s at a time when flax prices were rising, and subsequently repeated the experiment only once. In other words, we cannot assume that families – even landless labouring families – were excluded from spinning for household use simply because they did not have land to grow the raw material.

Third, the boundary between self-provisioning and market supply, between textiles made or processed within the household for use by its members and textiles sourced from outside the household, was drawn in terms of type of fibre and the quality of the cloth, as defined by yarn fineness and colour. The Lathams supplied themselves with textiles through a combination of self-provisioning and market transactions. The textiles they had woven from their homespun yarn were mainly coarser linens, some of them bleached, though probably not very intensely.[92] The family's production of poor-quality but quickly spun coarse yarn to make into brown or off-white cloth suggests a concern to save time, rather than a lack of skill. Fine spinning and thorough bleaching would have been much more time-consuming. As the family's income rose in the 1740s, when the daughters started to undertake outwork spinning for wages, these homespun coarse linens continued to be made and used, but the balance of the family's textile consumption shifted towards shop-bought, fine, light, visually attractive and often fashionable fabrics. In other words, fabric type, quality and fashionability were critical issues in the allocation of time between household production and production for the market.

Fourth, spinning for a wage and spinning for household use were not mutually exclusive within a family, although whether individual women of the Latham family specialised in one type of spinning or the other is not clear. Differing techniques depending on the fibre being spun may have been significant for divisions of labour within the household. In the accounts, Nany Latham's name tends to be associated with flax spinning wheels, her daughters' names with wheels for spinning cotton. Nevertheless, self-provisioning continued. We cannot assume that family self-provisioning with textiles necessarily declined because waged spinning increased.

Fifth, the experience of the Latham family confirms the view of eighteenth-century commentators that linen had come to predominate in textile self-provisioning, as production of homespun woollens declined, perhaps encouraged by a reallocation of homespun woollen yarn to self-provisioning with coarse hand-knitted stockings. A pattern of household self-provisioning dominated by purchases of flax or hemp to be spun into yarn for coarse linens, similar to the Lathams, can be observed in the domestic accounts of several better-off, middling-sort families at the same period, such as the curate Miles Tarn at Cheddington in Buckinghamshire, the Morley farming family at Breadsall, Derbyshire, and Hester Soame, the widow of a worsted manufacturer, who shared a house with her sister at Pytchley in Northamptonshire.[93] Even the household of the Lancashire farmer Clement Taylor, of Finsthwaite in Furness, who owned a flock of some two hundred sheep, generated as much homespun linen as homespun woollens.[94]

Earlier, in the sixteenth and seventeenth centuries, homespun woollens may have been more prominent. They certainly received as much attention as homespun linens in the many editions of Gervase

Markham's much-plagiarised *English Huswife*, first published in 1615, although Markham suggests that the woollen yarn spun by 'ordinary English House-wives' was undifferentiated and often coarse.[95] Homespun woollens appear regularly in the domestic accounts of gentry families at this period.[96] Yet in the 1640s, Sir Robert Filmer was recommending that the virtuous wife 'must seeke out such thinges as are profitable for her country, for in some places it is more available to buy cloath ready-made then to make it'.[97] Indeed, John Oldland has argued that homespun woollen textiles were already declining in importance in the fifteenth and early sixteenth centuries, with the rise in the English countryside of commercial manufacture of an increasingly diverse range of cheaper, colourful, woollen fabrics, at a time when the dress of the rural population was changing and demand for commercially produced woollen cloth was expanding.[98] Between the fifteenth and the eighteenth centuries, the market for commercially supplied linen also expanded, but England, as Europe's foremost producer of sheep's wool, was not noted for flax and hemp cultivation, or for commercial linen manufacture. Much of the increase in demand for linens, especially for finer, bleached linens, was supplied by imports from continental Europe. Indeed, for most of the early modern period, linen cloth was England's leading manufactured import. As Daniel Defoe commented early in the eighteenth century, 'England does as it were ravage the whole spinning world for linen and linen yarn and lace'.[99]

The period from the fifteenth to the eighteenth centuries witnessed diversification across the whole range of commercial textiles available to consumers of modest means. A concern with fashion and appearance, combined with falling prices and widening availability of waged work for women and children, encouraged purchase of the new, lighter, colourful varieties of English-manufactured fabrics and clothing accessories that swept the market in successive waves. The same imperatives drove purchases of the many varieties of French, Dutch and German fine bleached linens for the shirts and shifts, kerchiefs and caps that became typical contents of pedlars' packs.[100] Yet coarser linens did not simply disappear as imported finer varieties of linen became more widely available. Many English households continued to self-provision with coarser linens. They were assisted by access to raw material supply chains and technical innovations fostered by commercial textile manufacturing, in particular cheap Baltic flax and hemp, and the new, foot-pedalled flyer spinning wheel, which cut the time needed to spin flax and hemp.[101] Hardwearing, coarse, brown sheets, made from homespun harden or tow cloth, could satisfy the growing desire for domestic comfort and cleanliness which drove even the working poor to acquire increasing numbers of linen sheets.[102] After all, domestic linens such as bedsheets were not public sartorial signifiers of fashionability in the manner of the fine, white linens worn as caps and kerchiefs. The same was true of workaday shirts and shifts. In a household economy like that of the Latham family, the relationship between commercially sourced and domestically sourced textiles was more complementary than competitive, although there was always the potential for the kind of substitutions Jan de Vries describes, arising from shifts in employment opportunities, family incomes and textile prices.

Conclusion

The transformations in textiles outlined in this chapter extended across much of Europe, but their impact was uneven. Between the fifteenth and the eighteenth centuries, manufacturing the new, lighter textiles for national and especially international markets came to be associated with a limited number of major manufacturing cities and their rural hinterlands. Thus, international trade in ribbons was dominated by production centres such as Padua in Italy, Saint-Etienne in France, Coventry in England, Augsburg in Germany and Basel in Switzerland. Prominent centres for worsted stuffs were Amiens in France, Leiden in the Dutch Republic and Norwich in England.

Tracking consumption of the new, lighter textiles across Europe is more difficult. For early modern England, the pace and depth of innovations in dress are striking. England, however, was unusual in several respects. It was a large country with a highly integrated textile market, centred on London.[103] It lacked internal customs barriers, sumptuary laws were abandoned after 1604, and guild controls on product and process innovation were relatively weak. There is little evidence for anything that could be described as regional, folk or peasant costume.[104] More broadly, it was one of early modern Europe's most successful economies, internationally highly competitive in industry and commerce, with expanding opportunities for waged manufacturing work, especially for women and children, at a time of falling textile prices.

Elsewhere in Europe, as Sheilagh Ogilvie has pointed out, conditions could be much more hostile to innovation in textile consumption. Small territorial units could restrict the size of markets by enforcing controls or taxes on trade. Guilds or regulations could limit innovation, products, access to work and wages. Sumptuary laws could restrict what was worn by different kinds of people.[105] In many parts of continental Europe, moreover, the late sixteenth and seventeenth centuries were economically challenging, with falling living standards, contracting industries and reduced opportunities for waged work. Incentives to greater industriousness were more likely to derive from impoverishment than from the lure of attractive consumer goods. Where textile manufacturing for international markets became uncompetitive, as in parts of northern Italy, there could be a retreat into commercial production of low-quality cloth for local markets, with reliance on local raw materials.[106]

The notion of an 'industrious revolution', as proposed by Jan de Vries, offers a compelling framework for understanding the link between production and consumption of early modern textiles. Yet as de Vries acknowledges, it was mainly England and the Dutch Republic that initially, at least, experienced its full impact.[107] Both states were, of course, leaders in the trade in extra-European commodities – tea, coffee, tobacco, sugar, calicoes – so it is perhaps not surprising that de Vries places so much emphasis on their consumption as an incentive to industriousness. Yet England and the Dutch Republic were also two of the most successful and innovative textile-manufacturing territories in Europe, at a time when textiles were traded in larger quantities than any other manufactured goods. Well before most extra-European commodities achieved anything close to a mass consumer market, new, light, fashionable textiles and textile accessories were transforming what ordinary women in England wore, while at the same time providing them with new opportunities for waged work. De Vries insists on the dynamism of the consumer economy of early modern Europe. Yet the evidence he provides suggests it was propelled chiefly by engagement with goods imported from the world beyond Europe. The history of textiles and fashion between the fifteenth and the eighteenth centuries indicates that, on the contrary, the dynamism of the continent's early modern consumer economy was driven – initially at least – by textiles made in Europe.

Notes

1 James E. Thorold Rogers, *A History of Agriculture and Prices in England: 1583–1702* (Oxford: Clarendon, 1887), 551, 587; Michael A. Havinden, *Household and Farm Inventories in Oxfordshire, 1550–1590* (London: HMSO, 1966), 26; George D. Ramsay, *The English Woollen Industry, 1500–1750* (London: Macmillan, 1982), 32.

2 See, for example, John Styles, *The Dress of the People: Everyday Fashion in Eighteenth-Century England* (New Haven and London: Yale University Press, 2007), chapter 8.

3 For the ways historians and economists have conceptualised domestic labour and household self-provisioning, see Jane Whittle, 'A critique of approaches to "domestic work": women, work and the pre-industrial economy', *Past & Present*, 243 (2019), 35–70. For the distinctive associations of the term 'homespun' in histories of North America, see Laurel Thatcher Ulrich, *The Age of Homespun* (New York: Knopf, 2001). Also, Michael Merrill, 'Cash is good to eat: Self-sufficiency and exchange in the rural economy of the United States', *Radical History Review*, 3 (1977), 42–71; Christopher Clark, *The Roots of Rural Capitalism: Western Massachusetts, 1780–1860* (Ithaca: Cornell University Press, 1992); Allan Kulikoff, *Agrarian Origins of American Capitalism* (Charlotte: Virginia University Press, 1992).

4 Patrick Chorley, 'The "Draperies légères" of Lille, Arras, Tournai, Valenciennes: New materials for new markets?', in Marc Boone and Walter Prevenier (eds), *La Draperie Ancienne des Pays-Bas: débouchés et stratégies de survie (14e–16e siècles)*

(Leuven: Garant, 1993), 163. John Oldland argues that in England from 1400 to 1550 the main trend in clothing among the peasantry was a shift towards heavier, better-quality, short-staple woollen broadcloths and kerseys, as living standards improved. For the majority of the rural population, the turn to the newly fashionable lighter fabrics came only after 1550. John Oldland, *The English Woollen Industry, c. 1200–c. 1560* (Abingdon: Routledge, 2019), chapter 3. For the shift from two fibres to four, see Giorgio Riello, 'The world of textiles in three spheres: European woollens, Indian cottons and Chinese silks, 1300–1700', in Marie-Louise Nosch, Zhao Feng and Lotika Varadarajanpp (eds), *Global Textile Encounters* (Oxford: Oxbow, 2014), 93–106.

5 John Munro, 'Medieval woollens: Textiles, textile technology and industrial organisation, c. 800–1500', and 'Medieval woollens: The western European woollen industries and their struggles for international markets, c. 1000–1500', in David Jenkins (ed.), *The Cambridge History of Western Textiles* (Cambridge: Cambridge University Press, 2003), vol. 1, 181–227 and 228–324.

6 Chris Wickham, 'How did the feudal economy work? The economic logic of medieval societies', *Past & Present*, 251 (2021), 30.

7 Stuart Jenks, 'The missing link: The distribution revolution of the 15th century', in Carsten Jahnke and Angela L. Huang (eds), *Textiles and the Medieval Economy: Production, Trade and Consumption of Textiles, 8th to 16th Centuries* (Oxford: Oxbow, 2015), 230–52.

8 Carlo Poni, 'Fashion as flexible production: The strategies of the Lyons silk merchants in the eighteenth century', in Charles Sabel and Jonathan Zeitlin (eds), *World of Possibilities: Flexibility and Mass Production in Western Industrialization* (Cambridge: Cambridge University Press, 1997), 37–74; John Styles, 'Fashion and innovation in early-modern Europe', in Evelyn Welch (ed.), *Fashioning the Early Modern: Dress, Textiles, and Innovation in Europe, 1500–1800* (Oxford: Oxford University Press, 2016), 33–55.

9 Luca Molà, *The Silk Industry of Renaissance Venice* (Baltimore: Johns Hopkins University Press, 2000), 88, 146–52.

10 Elizabeth Currie, 'Diversity and design in the Florentine tailoring trade, 1560–1620', in Michelle O'Malley and Evelyn Welch (eds), *The Material Renaissance* (Manchester: Manchester University Press, 2007), 160.

11 Julia de Lacy Mann, *The Cloth Industry in the West of England from 1640 to 1880* (Oxford: Oxford University Press, 1971), 14, 312–15; Hugh Chevis, 'Why early modern English clothiers started using Spanish wool', *Textile History*, 52 (2021), 122–43.

12 These textiles are often referred to as the 'New Draperies' by historians of later sixteenth- and early- to mid-seventeenth-century England, but rarely later. This chapter does not use the term, because it was little used in early modern England outside of fiscal documents (as Eric Kerridge pointed out) and, confusingly, the meanings of equivalent terms in other European languages and settings were different. For an exhaustive discussion of the terminological issues, see, in particular, John Munro, 'The origins of the English "New Draperies": The resurrection of an old Flemish industry, 1270–1570', in Negley Harte (ed.), *The New Draperies in the Low Countries and England, 1300–1800* (Oxford: Oxford University Press, 1997), 35–127; Eric Kerridge, *Textile Manufactures in Early Modern England* (Manchester: Manchester University Press, 1985), 91.

13 Herman Van der Wee, 'The western European woollen industries, 1500–1750', in Jenkins, *Western Textiles*, vol. 1, 397–472; and Negley Harte, *New Draperies*; Robert S. DuPlessis, *Lille and the Dutch Revolt: Urban Stability in an Era of Revolution, 1500–1582* (Cambridge: Cambridge University Press, 1991), chapter 3; Charles Wilson, 'Cloth production and international competition in the seventeenth century', *Economic History Review*, 13 (1960), 209–21.

14 Chorley, '"Draperies légères"', 151–66.
15 Quoted in Molà, *Silk Industry*, 96.
16 Luc N. D. Martin, 'Textile manufactures in Norwich and Norfolk, 1550–1622' (PhD thesis, University of Cambridge, 1991), 4, 7.
17 Daniel Defoe, *A Brief Deduction of the Original, Progress, and Immense Greatness of the British Woollen Manufacture* (London, 1727), 50.
18 John Styles, 'What were cottons for in the Industrial Revolution?', in Giorgio Riello and Prasanan Parthasarathi (eds), *The Spinning World: A Global History of Cotton Textiles, 1200–1850* (Oxford: Oxford University Press, 2009), 315–21.
19 Donald C. Coleman, 'An innovation and its diffusion: "The new draperies"', *Economic History Review*, 22 (1969), 425.
20 Innovation in colour was, of course, encouraged by the availability of new dyestuffs from the Americas.
21 Quoted in Ursula Priestley, 'Norwich Stuffs, 1600–1700', in Harte, *New Draperies*, 278.
22 Stephen R. Epstein, 'The late medieval crisis as an "integration crisis"', in Maarten Prak (ed.), *Early Modern Capitalism: Economic and Social Change in Europe 1400–1800* (London: Routledge, 2001), 41; Angela Ling Huang, 'Hanseatic textile production in 15th century long distance trade', in Jahnke and Huang, *Textiles and the Medieval Economy*, 204–15.
23 Georges Vigarello, *Concepts of Cleanliness: Changing Attitudes in France since the Middle Ages* (Cambridge: Cambridge University Press, 1988), chapter 4. Chapter 9 below by Elizabeth Currie and Jordan Mitchell-King discusses the importance and proliferation of linen undergarments among rural women.
24 Andrea Caracausi, 'Textiles manufacturing, product innovations and transfers of technology in Padua and Venice between the sixteenth and eighteenth centuries', in Karel Davids and Bert de Munck (eds), *Innovation and Creativity in Late Medieval and Early Modern European Cities* (Aldershot: Ashgate, 2014), 131–60; Santina M. Levey, 'Lace in the early modern period, c. 1500–1780', in Jenkins, *Western Textiles*, vol. 1, 585–96; G. F. R. Spenceley, 'The origins of the English pillow lace industry', *Agricultural History Review*, 21 (1973), 81–93; Susan North, '"Galloon, inkle and points": Fashionable dress and accessories in rural England, 1552–1665', in Richard Jones and Christopher Dyer (eds), *Farmers, Consumers, Innovators* (Hatfield: University of Hertfordshire Press, 2016), 104–23; Carlo Marco Belfanti, 'Fashion and innovation: The origins of the Italian hosiery industry in the sixteenth and seventeenth centuries', *Textile History*, 27 (1996), 132–47; Joan Thirsk, 'The fantastical folly of fashion: The English stocking knitting industry 1500–1700', in Negley B. Harte and K. G. Ponting (eds), *Textile History and Economic History: Essays in Honour of Julia de Lacy Mann* (Manchester: Manchester University Press, 1973), 50–73.
25 Kerchiefs were worn at the neck and were variously referred to as kerchiefs, neck cloths, neckerchiefs, handkerchiefs and neck handkerchiefs, with neckerchief the more commonly used term in the seventeenth century and handkerchief in the eighteenth. I follow Sarah Bendall and Pat Poppy in using the term kerchief to refer to this group of accessories: Pat Poppy, 'The clothing accessory choices of Rachel, Countess of Bath, and other mid-seventeenth-century women', *Costume*, 54 (2020), 15–18; Sarah Bendall, *Shaping Femininity: Foundation Garments, the Body and Women in Early Modern England* (London: Bloomsbury, 2021), 257.
26 Michael L. Ryder, 'Fleece grading and wool sorting: the historical perspective', *Textile History*, 26 (1995), 3–22.
27 Molà, *Silk Industry*, 234.
28 Line Teisseyre-Sallmann, *L'industrie de la soie en Bas-Languedoc: XVIIe–XVIIIe siècles* (Paris: Ecole des Chartes, 1995), 90–1, 245.

29 John Brearley's Memorandum Book for 1772–73, University of Leeds, Brotherton Library Special Collections, Yorkshire Archaeological Society, MS 2022.
30 John Jeffries and John Usher, Cloth Making Book,1721–1726, Wiltshire and Swindon History Centre, 927/1; St Mary Ely, parish workhouse, spinning accounts, 1736–39, Cambridgeshire Archives, P68/12/32–7.
31 Walter Endrei, *L'évolution des techniques du filage et du tissage du moyen âge à la révolution industrielle* (Paris: Mouton, 1968), 85–7, 110.
32 For France, see Mémoire sur le commerce et les manufactures de la Provence, 1781, National Archives of Paris (ANP), F12 677A and Projet d'Etablissement de filature de Lins, Chanvres et Cottons dans les Villages qui avoisinent la Ville de Troyes, 1770, ANP, F12 1341; for Saxony, Dr Anton, 'Spinning in Lusatia', *Annals of Agriculture*, 10 (1788), 313; for Scotland, Thomas Pennant, *A Tour in Scotland; 1769* (London, 1776), 105.
33 John Worlidge, *Systema Agriculturae: The Mystery of Husbandry Discovered* (London, 1675), 321.
34 Inventory of Christopher Hewit/Hewet, Forton, chapman, 1687, Lancashire Archives, Archdeaconry of Richmond, Probate Records, WRW 1687. Stephen Moorhouse and John Hurst, 'An imported stoneware spindlewhorl, with some preliminary comments on stoneware spindlewhorls in England, their dating and origin', *Transactions of the London and Middlesex Archaeological Society*, 32 (1981), 124–8; Eleanor R. Standley, 'Spinning yarns: the archaeological evidence for hand spinning and its social implications, c. AD 1200–1500', *Medieval Archaeology*, 60 (2016), 266–99.
35 *Norfolk Chronicle*, 18 March 1786; Rudolf Braun, *Industrialisation and Everyday Life*, trans. S. Hanbury Tenison (Cambridge: Cambridge University Press, 1979), 140.
36 Lettre de l'inspecteur des manufactures à Rouen sur le mérite comparitif de la filature de la laine au rouet ou à la quenouille, 1762, National Archives of Paris, F12 677A. Confusingly, in the British Isles the term 'Dutch wheel' was also applied to the pedal-driven flyer wheel for spinning flax and hemp.
37 John Styles, 'Spinners and the law: Regulating yarn standards in the English worsted industries, 1550–1800', *Textile History*, 44, 2 (2013), 145–70; William Reddy, *The Rise of Market Culture: The Textile Trade and French Society, 1750–1900* (Cambridge: Cambridge University Press, 1984), 31–2.
38 Carlo Poni, 'Standards, trust and civil discourse: Measuring the thickness and quality of silk thread', *History of Technology*, 23 (2001), 1–16.
39 See William J. Ashley, *An Introduction to English Economic History and Theory, Part 2* (London: Longmans, 1893), 219–20; William Cunningham, *The Growth of English Industry and Commerce in Modern Times: The Mercantile System* (Cambridge: Cambridge University Press, 1903), 497, note 2; Ephraim Lipson, *The History of the Woollen and Worsted Industries* (London: A. and C. Black, 1921), 36–7, 72.
40 Manufacturing guilds and other corporate forms of organisation continued to flourish in many parts of Europe during the early modern period. Their reach frequently extended into the countryside. Yet for several of the new light textiles, the location of key production stages, especially weaving and dyeing, was predominantly urban, not rural.
41 Franklin Mendels, 'Proto-industrialization: The first phase of the industrialization process', *Journal of Economic History*, 32 (1972), 241–61.
42 Mendels, 'Proto-industrialization', 248.
43 Donald C. Coleman, 'Proto-industrialization: A concept too many', *Economic History Review*, 36 (1983), 435–48; Sheilagh C. Ogilvie and Markus Cerman, 'The theories of proto-industrialization', in Sheilagh C. Ogilvie and Markus Cerman (eds), *European Proto-Industrialization: An Introductory Handbook* (Cambridge: Cambridge University Press, 1996), 1–11. The terms 'proto-industry' and

'proto-industrial', now stripped of much of their theoretical content, are increasingly employed simply to describe any regionally specialised commercial handicraft industry of the early modern era, rural or urban, especially those that were export-orientated.

44 Jan de Vries, *The Industrious Revolution: Consumer Behaviour and the Household Economy, 1650 to the Present* (Cambridge: Cambridge University Press, 2008), ix. Also see Jan de Vries, 'The Industrial Revolution and the Industrious Revolution', *Journal of Economic History*, 54 (1994), 249–70; Jan de Vries, 'Between purchasing power and the world of goods: Understanding the household economy in early modern Europe', in John Brewer and Roy Porter (eds), *Consumption and the World of Goods in the Seventeenth and Eighteenth Centuries* (London: Routledge, 1993), 85–132. For de Vries's own assessment of the relationship between proto-industrialisation and his industrious revolution, see Jan de Vries, 'Rethinking protoindustry: Human capital and the rise of modern industry', in Kristine Bruland, Anne Gerritsen, Pat Hudson and Giorgio Riello (eds), *Reinventing the Economic History of Industrialisation* (Montreal and Kingston: McGill-Queen's University Press, 2020), 107–26.

45 De Vries, *Industrious Revolution*, 85–6, 7.

46 De Vries, *Industrious Revolution*, 122.

47 For women and children's employment in knitting, see Lesley O'Connell Edwards, 'The stocking knitting industry of later sixteenth-century Norwich', *Textile History*, 52 (2021), 144–64.

48 'My historical claim is that northwestern Europe and British North America experienced an "industrious revolution" during a long eighteenth century, roughly 1650–1850', De Vries, *Industrious Revolution*, 10.

49 Craig Muldrew, *Food, Energy and the Creation of Industriousness: Work and Material Culture in Agrarian England, 1550–1780* (Cambridge: Cambridge University Press, 2011); Gregory Clark, 'The consumer revolution: turning point in human history, or statistical artifact?', MPRA, Paper No. 25467 (September 2010); Joanne Sear and Ken Sneath, *The Origins of the Consumer Revolution in England: From Brass Pots to Clocks* (Abingdon: Routledge, 2020), chapter 9; De Vries, *Industrious Revolution*, 154–5, 177–8. De Vries has also been criticised for overestimating the importance of incentives to industriousness among the working poor offered by attractive new commodities, and underestimating the stimulus of economic hardship. See Robert C. Allen and Jacob Weisdorf, 'Was there an "industrious revolution" before the industrial revolution? An empirical exercise for England, c. 1300–1830', *Economic History Review*, 64 (2011), 715–29; and, for Sweden, Kathryn E. Gary and Mats Olsson, 'Men at work: Wages and industriousness in southern Sweden, 1500–1850', *Scandinavian Economic History Review*, 68 (2020), 112–28.

50 For tobacco, tea and sugar, see Carole Shammas, *The Pre-Industrial Consumer in England and America* (Oxford: Oxford University Press, 1990), chapter 4; for printed calicoes and cottons, see Styles, *Dress of the People*, chapter 7.

51 De Vries, *Industrious Revolution*, 138; Styles, Dress of the People, appendix 1.

52 De Vries, *Industrious Revolution*, 133–43, 154.

53 De Vries, *Industrious Revolution*, 133; 97–8. De Vries also mentions the earlier expansion in the consumption of linen fabrics.

54 De Vries, *Industrious Revolution*, 44–70, 144, 177.

55 Joan Thirsk, *Economic Policy and Projects; The Development of a Consumer Society in Early Modern England* (Oxford: Oxford University Press, 1978).

56 Carole Shammas, 'The decline of textile prices in England and British America prior to industrialization', *Economic History Review*, 47 (1994), 483–507.

57 For a review of the evidence and some estimates, see Craig Muldrew, '"Th'ancient Distaff" and "Whirling Spindle": Measuring the contribution of spinning to

household earnings and the national economy of England, 1550–1770', *Economic History Review*, 65 (2012), 498–526.

58 Styles, *Dress of the People*, chapter 7. Even during the third quarter of the eighteenth century, the market for printed gown fabrics was shared between prints on cotton-linens and prints on linens.

59 The indictments are not a reliable source for the social status of the owners of the stolen goods they list, so these figures represent all owners, rich or poor.

60 Russet also disappeared by 1660 from the expenditure on orphaned children recorded in probate accounts: see Margaret Spufford, 'Fabric for seventeenth-century children and adolescents' clothes', *Textile History*, 34 (2003), 50. The word russet could mean both a cloth and a colour. I follow Maria Hayward in assuming that references to russet in the criminal indictments were to the cloth, unless there is evidence to the contrary; Maria Hayward, *Rich Apparel: Clothing and the Law in Henry VIII's England* (Abingdon: Routledge, 2016), xxii. For russet as a symbol of rural simplicity, see Thomas Deloney, *A most Excellent and Vertuous Ballad of the Patient Grissell* (London, ?1624), Magdalene College, Pepys Ballads 1.34–35, EBBA 20160.

61 (Anon.), *The Weavers' True Case* (London, 1719), 38–9.

62 Account book of Richard Latham of Scarisbrick, 1723–1767, Lancashire Archives, DP 385.

63 Arthur Young, *A Six Months Tour through the North of England* (London, 1771), vol. 3, 192.

64 De Vries, *Industrious Revolution*, 150–4, 178–9.

65 Shop book of Stephen Hudson of Thruscross township, Yorkshire, 1751–59, West Yorkshire Archive Service Bradford, 33D80/7. For 1758–59, the ratio of male to female clients in Hudson's daybook is approximately three to one, suggesting that married women's purchases were entered under the names of their husbands.

66 See note 50.

67 Thirsk, 'Fantastical folly of fashion', 52–5.

68 Margaret Spufford, *The Great Reclothing of Rural England: Petty Chapmen and Their Wares in the Seventeenth Century* (London: Hambledon, 1984), 103.

69 Steve Hindle, The Social Topography of a Rural Community: Scenes of Labouring Life in Seventeenth Century England (Oxford: Oxford University Press, 2023), chapter 5.

70 Polly Hamilton, 'Haberdashery for use in dress, 1550–1800' (PhD thesis, University of Wolverhampton, 2007), 273.

71 Account book of an Ambleside shopkeeper, 1701–9, Cumbria Archive Service (Kendal), Browne of Troutbeck Mss., WD/TE/11/16, fol. 25.

72 (Anon.), *Advice to the Maidens of LONDON: To Forsake Their Fantastical TOP-KNOTS; Since they are become so Common with Billings-gate Women, and the Wenches that cryes Kitchin-Stuff: Together with the Wanton Misses of the Town* (London, 1685–88), Magdalene College, Pepys Ballads 4.365, EBBA 22029. Also see Angela McShane and Claire Backhouse, 'Top-knots and lower sorts: popular print and promiscuous consumption in late seventeenth-century England', in Michael Hunter (ed.), *Printed Images in Early Modern Britain: Essays in Interpretation* (Farnham: Ashgate, 2010), 337–57.

73 De Vries, *Industrious Revolution*, 97.

74 William Pitt, *General View of the Agriculture of the County of Stafford* (London, 1794), 162–3.

75 Henry E. Strickland, *A General View of the Agriculture of the East Riding of Yorkshire* (York, 1812), 283–84.

76 John H. Clapham termed them 'customer weavers', see John H. Clapham, *An Economic History of Modern Britain: The Early Railway Age, 1820–1850* (Cambridge: Cambridge University Press, 1930), 159.

77 For militia ballot lists as sources for occupations, see Paul Glennie, *'Distinguishing men's trades': Occupational Sources and Debates for Pre-Census England* (Cheltenham: Historical Geography Research Group, 1990).

78 Wendy Raybould, 'Open for business: Textile manufacture in Northamptonshire, c. 1685–1800' (PhD thesis, University of Leicester, 2005).

79 Peter Kitson, 'The male occupational structure of Northumberland, 1762–1871: A preliminary report', working paper, Cambridge Group for the History of Population and Social Structure, 2005, 18. Also see Clapham, *Economic History of Modern Britain*, 161.

80 Anne Kussmaul, drawing on militia lists and other sources, suggests that six craft occupations were common in eighteenth-century rural parishes – tailor, blacksmith, carpenter, shoemaker, weaver and wheelwright; Anne Kussmaul, *A General View of the Rural Economy of England, 1538–1840* (Cambridge: Cambridge University Press, 1990), 54.

81 Pehr Kalm, *Kalm's Account of His Visit to England on His Way to America in 1748*, trans. Joseph Lucas (London: Macmillan, 1892), 333. It may be significant that the two areas with the lowest proportion of weavers – Hertfordshire and south Dorset – were both in southern England. It was claimed in 1737 that self-provisioning with coarse linens was especially characteristic of England north of the River Trent; *Journals of the House of Commons*, 23 (1737–41), 77. Yet even in the deep south of England, near Chichester in Sussex, the Baptist minister and son of a local farmer, James Spershott, recorded in his memoirs that 'spinning of household linen was in use in most families' during his youth in the first half of the eighteenth century; Francis W. Steer (ed.), *The Memoirs of James Spershott* (Chichester: Chichester City Council, 1962), 15.

82 (Anon.), *At a meeting of the master weavers, living in Hexham, and parts adjacent, it was unanimously agreed, that the prices for weaving of plain cloth, sheeting, ticking, huckaback, etc. be as follows* (Hexham, 1792). For advertisements by jobbing weavers in other parts of England, see *Ipswich Journal*, 18 May 1771; *Cambridge Chronicle*, 24 January 1772; *Derby Mercury*, 6 March 1778; Handbill for Nicholas Forster, weaver, Alnwick, Northumberland, 1800, Castle Museum, York.

83 Weaver's thesis book, England, seventeenth–eighteenth century, Cooper Hewitt, National Design Museum, New York, 1958–30–1, fol. 63. For other examples of pattern books compiled by eighteenth-century jobbing weavers, see Pattern book of William Jones and William Jones junior, jobbing weavers, Holt, 1775–82, Denbighshire Record Office, PD/39/1/81; The weaver's guide, linen designs of Ralph Watson of Aiskew, late eighteenth century, North Yorkshire County Record Office, Z.371.

84 NeL is a standard indirect measure of yarn count (i.e., yarn fineness) for yarns made from flax or hemp, calculated according to units employed in linen manufacture in the British Isles during the eighteenth century. It is based on the number of leas or cuts of linen yarn, 300 yd in length, per pound weight. See Michael J. Denton and Paul N. Daniels, *Textile Terms and Definitions* (Manchester: Textile Institute, 2002), 403; David J. Jeremy, 'British and American yarn count systems: An historical analysis', *Business History Review*, 45 (1971), 336–68.

85 Arthur Young, *A Tour in Ireland: With General Observations on the State of that Kingdom*, second edition, vol. 1 (London: 1780), 175, 239, 277–8. According to Thomas Firmin, the linen yarn spun by paupers at his workhouse in London in the 1670s was also coarser than the yarn supplied for Jackson to weave; Thomas Firmin, *Some Proposals for the Imployment of the Poor* (London, 1681), 17–18.

86 Account book of Richard Latham of Scarisbrick, 1723–1767, Lancashire Archives, DP 385. Also see Lorna Weatherill, *The Account Book of Richard Latham, 1724–1767* (Oxford: Oxford University Press, 1990); Charles F. Foster, *Seven Households: Life in Cheshire and Lancashire, 1582–1774* (Northwich: Arley Hall

Press, 2002), chapter 5; Styles, *Dress of the People*, chapter 14; Andrew J. Gritt, 'The farming and domestic economy of a Lancashire smallholder: Richard Latham and the agricultural revolution, 1724–67', in Richard W. Hoyle (ed.), *The Farmer in England, 1650–1980* (Farnham: Ashgate, 2013), 101–34; Alice Dolan, 'The fabric of life: time and textiles in an eighteenth-century plebeian home', *Home Cultures*, 11 (2014), 353–74.

87 There are very few payments in the accounts for weaving, yet there is no evidence the family owned a loom. Charles Foster has plausibly suggested that weaving the family's linen yarn was paid for by barter; Foster, *Seven Households*, 159.

88 Gritt, 'The farming and domestic economy', 124–6.

89 Young, *Six Months Tour*, vol. 3, 192.

90 The Latham accounts record very few purchases of tobacco and none of tea or coffee, suggesting that their purchases of sugar and treacle were used for food preservation and flavouring, rather than sweetening hot drinks.

91 Styles, *Dress of the People*, chapter 14.

92 Dolan, 'Fabric of life', 366.

93 Account book of Reverend Miles Tarn, Rector of Dean, Cumberland, 1719–97, Lewis Walpole Library, Mss, vol. 212; William Smedley's executors, account book of farm receipts and disbursements, Derbyshire, 1741–52, Huntington Library, HM 31192; Personal and household account book of Hester Soame, 1753–62, Wiltshire and Swindon History Centre, 1720/744.

94 Janet B. Martin (ed.), *The Account Book of Clement Taylor of Finsthwaite, 1712–1753* (Liverpool: Record Society of Lancashire and Cheshire, 1997).

95 Gervase Markham, *Country Contentments, or The English Huswife* (London, 1623), 160.

96 See, for examples, Jane Whittle and Elizabeth Griffiths, *Consumption and Gender in the Early Seventeenth-Century Household: The World of Alice Le Strange* (Oxford: Oxford University Press, 2012), 65–6; Todd Gray, *Devon Household Accounts*, 1627–59, Part 2 (Exeter: Devon and Cornwall Record Society, 1996), esp. 169–70, 220.

97 Margaret J. M. Ezell, *The Patriarch's Wife: Literary Evidence and the History of the Family* (Chapel Hill: University of North Carolina Press, 1987), 183.

98 John Oldland, 'The economic impact of clothmaking on rural society, 1300–1550', in Martin Allen and Matthew Davies (eds), *Medieval Merchants and Money* (London: Institute of Historical Research, 2016), 247–9; Oldland, *English Woollen Industry*, especially chapter 10. Also see Mark Bailey, *Medieval Suffolk: An Economic and Social History, 1200–1500* (Woodbridge: Boydell, 2007), 264.

99 Daniel Defoe, *A Plan of the English Commerce* (London, 1728), 207. Also see Negley B. Harte, 'The rise of protection and the English linen trade, 1690–1790', in Harte and Ponting, *Textile History and Economic History*, 74–112; David Ormrod, *The Rise of Commercial Empires: England and the Netherlands in the Age of Mercantilism, 1650–1770* (Cambridge: Cambridge University Press, 2003), chapter 5.

100 Spufford, *Great Reclothing*, chapter 6.

101 The prices listed by Thomas Firmin suggest that in London in the 1670s Baltic Paternoster flax was considerably cheaper than English flax, and Baltic hemp cheaper still; see T. F., *Some Proposals*, 18.

102 For social differences in the ownership of different types of sheeting, see Paul Glennie, 'The social shape of the market for domestic linens in early modern England', working paper for the conference 'Clothing and Consumption in England and America, 1600–1800', Victoria and Albert Museum, 1992, table 8. Also, Darron Dean, Andrew Hann, Mark Overton and Jane Whittle, *Production and Consumption in English Households, 1600–1750* (London: Routledge, 2004), 108–11; Christopher Husbands, 'Standards of living in north Warwickshire in the seventeenth century', *Warwickshire History*, 4, 6 (1980–1), 203–15.

103 Oldland, *English Woollen Industry*, 335.
104 Anne Buck, 'Variations in English women's dress in the eighteenth century', *Folk Life*, 9 (1971), 5–28.
105 Sheilagh Ogilvie, 'Consumption, social capital, and the "Industrious Revolution" in early modern Germany', *Journal of Economic History*, 70 (2010), 287–325. For the uneven impact of sumptuary regulations on popular fashion in early modern Europe, see Giorgio Riello and Ulinka Rublack (eds), *The Right to Dress: Sumptuary Laws in a Global Perspective, c. 1200–1800* (Cambridge: Cambridge University Press, 2019).
106 Francesco Vianello, 'Rural manufactures and patterns of economic specialization: Cases from the Venetian mainland', in Paola Lanaro (ed.), *At the Centre of the Old World: Trade and Manufacturing in Venice and the Venetian Mainland, 1400–1800* (Toronto: Centre for Reformation and Renaissance Studies, 2006), 343–66.
107 De Vries, Industrious Revolution, x.

2

Ribbon culture in early modern Italy

Andrea Caracausi

Introduction

During the early modern period, the clothing and wardrobes of various social groups saw the appearance of a seemingly trivial object: the silk ribbon. Despite its apparent insignificance, this accessory became a significant object of fashion and technological innovation in the course of the sixteenth and seventeenth centuries, and it played an active role in the specialisation of production processes, reconfiguration of labour markets and evolution of brand strategies that are characteristic of early capitalism.[1] The importance placed on ribbons, as John Styles has shown in Chapter 1, also contributed to the emergence of new forms of fashion, as well as of cultural meanings and expressions of gender identity associated with ribbons. Yet the place of ribbons in early modern material culture is virtually forgotten.[2]

The purpose of this chapter is to address this oversight by portraying the emergence of a culture and fashion based on the consumption of ribbons in early modern Italy. It shows how, from the late Renaissance onwards, trimmings and ribbons became fashionable items of wear because of their physical attractiveness and their ability to represent personal identities, gender differences and sexual meanings. The chapter also demonstrates how, through mediation of mercers and merchants, ribbons encouraged fashionable dress in a wider cross-section of society, helping to stimulate the consumerism and the development of marketing strategies.[3]

The primary evidence for this chapter comes mainly from Italian sources, but comparisons with information from other areas are also included. The first section will examine the ribbons themselves. It will underline the variety of high-, medium- and low-quality wares intended for different segments of the market. The second section will focus on the uses of ribbons, showing how they became not just fashion accessories but distinctive material objects with specific symbolic meanings. The third section will concentrate on fashion and market evolution, demonstrating how the expanding market for ribbons stimulated demand and made ribbons more broadly available among all social groups. As lightweight items, ribbons were cheap and easy to transport in local and long-distance trade networks. Mercers and merchants disseminated these items through fairs, markets and shops, thereby stimulating demand across the various urban and rural social groups. The fourth section will analyse the marketing strategies developed by mercers and merchants in order to make ribbons more marketable and attractive. The great variety of ribbons available and rising consumption galvanised the diffusion of 'brands' and 'labels', and inspired processes of imitation (see Chapters 1 and 3).[4] Commercial strategies were linked to marketing strategies and market competition fostered rivalry between manufacturers in different regions of Europe. The final section touches on the developments in manufacturing technology and work organisation that made the increased production of ribbons possible.

The immense variety of ribbons

At the end of the seventeenth century, Jacques Savary des Brûlons, the French inspector-general of manufactures and author of the *Dictionnaire universel de commerce*, defined ribbon as

> very thin fabric, which serves many functions, depending on the material it is made from. There are ribbons of gold, silver, silk, waste silk [*fleuret* or *capiton*], wool, flax, etc. They are made in narrow, broad and intermediate widths, and produced in one piece, double-sided or with a reverse, stamped, embroidered, lacework, plain, extra-smooth; indeed, they are any colour and fashion that accords with the genius of the ribbon-maker, the taste of the shopkeeper, or the current vogue.[5]

Des Brûlons went on to highlight how the proliferation of different types of ribbons allowed for their widespread use in the clothing of vast numbers of customers – from the finest and most flamboyant silks that ornamented the breeches of royalty to the half-silks employed by less exalted folk for a plethora of more utilitarian purposes, such as binding hems, tying stays, preserving the style of wigs and trimming hats and bonnets (see Figure 1.5). Ribbons were integral to changes in fashion, to the creation of new styles and even new garments.[6] Their sheer variety merits consideration in more detail.

Although ribbons had been used to a certain degree in Europe in the fourteenth and fifteenth centuries, their dissemination across a broad spectrum of society in the Italian peninsula dates to the second half of the sixteenth century.[7] As John Styles shows in Chapter 1 above, this was a period with many innovations, such as knitted hosiery, second-hand silks and novel, light draperies spread across Europe, with these new trends often available at the most affordable prices.[8] Garment forms changed, too. Already in the first half of the sixteenth century, clothing had become more structured and shapelier, similar in a way to the contemporary evolution in architecture from the organic forms and vertical extension typical of the Gothic style to the simpler, more elegant, geometric proportions of Renaissance classicism. Unlike the traditional long, loose garments, some early modern fashion manufactures were commonly available ready-to-wear, such as in the case of breeches or stockings.[9] This, along with the use of cheaper and lighter plain fabrics, gave rise to the more general use of accessories in the form of a multitude of different ribbons and trimmings.[10]

The variety of ribbons divides into two main categories: patterned trimmings and plain ribbons. In the first group (Figure 2.1, see also 3.5), more expensive trimmings were normally woven from silk or, more rarely, silk mixed with wool, cotton or flax, and they featured the most extravagant patterns and greatest range of colours. The best-quality trimmings were wide decorative bands called 'galloons' (*guarnitioni* or *galons*). Made in widths ranging from 2 to 5 cm, they were commonly decorated with

Figure 2.1 Alessandro Allori, *Annunciation* (detail), 1603. Pillow with fine gold lace and large bobbins. Galleria dell'Accademia, Florence.

floral, diamond or chequered patterns and encompassed a kaleidoscope of colours, from red to white and blue to gold. Such galloons were normally a feature of luxury clothing, such as women's petticoats or velvet gowns of ecclesiastics, nobles and gentlemen. Lesser-quality trimmings included lace trims (*dentelles*, *dentelli*) and fleurets (*fiocchetti*), simple small bows around 1 cm wide that were worn in hair or attached to clothes (see Figures 1.5 and ix.1).[11]

The second group, plain ribbons, was generally simpler and less expensive, but it also varied widely, including a wide variety of strings, cords, fringes and bows. Merchants' account books and post-mortem inventories from early modern Padua – one of the most important production centres of plain ribbons in Italy – have allowed us to identify the differing compositions of high-, medium- and low-quality ribbons. In general, the highest-quality plain ribbons were made of pure silk, such as *cordelle ormesinade*.[12] A slightly lower quality of ribbon was made with pure silk wefts and waste silk warps (*filosello*). The lowest quality ribbons, known as *mezzeposte*, were made entirely from waste silk or silk mixed with wool or flax. The 'Refashioning the Renaissance' database shows that most of these types of ribbons were used by Italian artisans to decorate their garments, including the more expensive types, such as coloured ribbons with gold or lace.[13]

Appreciating the differences in cost and quality is crucial for understanding the significance of ribbons. It highlights the proliferation of an item that might otherwise, mistakenly, be considered marginal. From expensive luxury braids to cheap and cheerful plain ribbons, the price differentiation was up to 80 per cent, allowing a broad spectrum of market segments to be catered for.[14]

As some scholars have already noted for other products – especially socks and knitwear – the way that the purchase of silk ribbons was within the means of so many people prompted change not only in fashions but also in the cultural meanings associated with ribbons.[15]

The cultural meanings of ribbons

In apparel, trimmings and ribbons had several major uses. First, they were used to fasten clothes together or to attach them to parts of the body, such as sleeves to elbows and breeches to knees. This was common in both women's and men's fashions. Second, ribbons were used to finish luxury silk garments. Tailors hemmed sleeves, dresses and breeches with decorative bands and used braid to hide the seams of doublets. Third, plain ribbons decorated common garments and accessories: aprons, collars, skirts and shirts were all ornamented with ribbons as well as embroidery, as were handkerchiefs. Ribbons, too, were applied to items such as shirts, and to simple accessories such as collars. Finally, tailors, mercers and shoemakers employed plain ribbons to add a dash of flair to hats or caps, and to embellish hair accessories or shoes with colourful bows or rosettes.[16]

Ribbons were not simply fashionable accessories designed to beautify clothes. They were also objects in their own right, imbued with personal symbolism and cultural significance. As prominent elements of heraldic, military and religious insignia, they operated as formal signifiers of distinction and power.[17] Especially for men, they were conspicuous symbols of honour, valour and social distinction.

Ribbons, moreover, were particularly closely associated with women's involvement in conspicuous and leisured consumption, with women's fashions and with perceptions of female vanity. The Bolognese artist Giuseppe Maria Mitelli (1634–1718) seems to have shared this view: in one of the light-hearted scenes he engraved as a component for a gambling game – *The Game of Women and Their Business* (*Gioco delle donne, e sue facende*) – a woman is busy buying 'ribbons from the mercer' (Figure 2.2).[18] In early modern Venice, for instance, tapes and ribbons were referred to as 'female merchandise' (*mercanzia femminina*).[19] Similarly, when the shop boy Truffaldino, in Carlo Goldoni's 1741 comedy *The Bankruptcy, or, The Failed Merchant* (*La banca rotta o sia mercante fallito*), asks his master, 'What do you think, we should give as gifts to these women, Signor Pantalone?', the elderly merchant responds, 'We should give them some dress offcuts, some furbelows, ribbons, knick-knacks.' Truffaldino entirely approves: 'Women would like those exceedingly, sir.'[20]

Figure 2.2 Giuseppe Maria Mitelli, *Gioco delle donne, e sue facende* (*Game of Wives, and Their Chores*), 1654–1718. Metropolitan Museum of Art, New York.

Fashion and the markets for ribbons

As fashion accessories or independent objects, ribbons and tapes were essential parts of men's and women's clothing from the late Renaissance onwards.[21] A general overview of continuity and change in this fashion – and ribbons were especially subject to changes in fashion – can be traced for Italy. Already in the mid-sixteenth century, there is widespread evidence of the use of ribbons in various cities. In Naples, for instance, an observer emphasised that 'the ostentation and gallantries in both clothes and liturgical vestments even extended to coaches in the form of fringes and ornaments'.[22] In Venice, a sumptuary law issued in 1600 claimed that ribbons and braids were being used with 'immoderate profusion' in all types of garments, by men and women alike.[23] In the seventeenth century, Baroque fashion inspired even greater use of ribbons. In Venice, in the 1660s, the ribbon-makers' guild (*arte dei passamaneri*) testified to a reduction in clothing braids during the seventeenth century but an increase in ribbons which were made in a range of colours and also with gold and silver.[24]

Analysis of the 'Refashioning the Renaissance' database helps us see these qualitative cues, confirming the above-mentioned trend. The presence of ribbons increased during the 1550–1650 period. Over the decades there is a shift from thirty-one occurrences of ribbons as accessories on average per decade in 1550–1600 to ninety-three occurrences in the period 1600–1650.[25] In artisan homes, they are primarily owned by textile and clothing makers, as well as by artisans active in food, metalworking and retail sectors. A number of ribbons are found also in shops of merchants and tailors.

The growing presence and popularity of ribbons can be explained by many factors. First, ribbons were small, light and easy to display.[26] Second, they were luxury objects whose affordability and attractiveness made them popular with poorer social groups. Written descriptions suggest that ribbons were widely adopted for wear on feast days and other special occasions. For instance, a reference from the eighteenth-century *Storia documentata di Venezia* shows that women from lower social strata had their own holiday, called the *garanghelo*, when they left the city for the mainland. On this day, they usually wore 'a scarlet jacket over a cotton skirt, white linen, aprons from Persia and ... decorated them with gold, silver, ribbons and tapes'.[27] In 1775, in Turin 'ribbons in current fashion, or for women of the countryside' (*bindelli alla moda, o da paesane*) were much used, and the people from the countryside 'loved their shiny appearance and low price'.[28]

In addition, the apogee of Baroque court culture, combined with a rise in real wages during the second half of the seventeenth century, may have had a 'trickle-down effect'.[29] The market for silk ribbons, especially the cheapest varieties, seems to have become something resembling a mass market.[30] One notable aspect of this was the diffusion of ribbon-selling out from urban settings and into rural fairs and markets. In this process, merchants and mercers played an important role, travelling widely and reaching people deep in the countryside, to whom they sold many types of 'coloured tapes, ribbons and cheap tobacco', galloons, lace and combs.[31] Ribbon manufacturing could easily be adapted to the budgets and tastes of rural customers. In eighteenth-century Turin, the ribbon-makers' guild asked permission to use fake gold and waste silk in ribbons to sell to country people, who 'love their shiny appearance and low price' although 'their desires are not matched by their means'.[32]

Selling ribbons, convincing consumers

During the late sixteenth and early seventeenth centuries, the key figures selling and marketing ribbons were the mercers. From the mid-seventeenth century onwards, however, mercers no longer relied entirely on ribbons designed and made by other craftspeople, but they also began to manufacture ribbons themselves to their own unique patterns. This tendency explains why mercers' guilds sought to organise themselves into distinct groups (*colonelli*) – or requested permission to establish separate new guilds – with the right to make ribbons and braids. In Venice, in 1659, the mercers' guild asked for the freedom to sell imported ribbons (Figure 2.3) The Senate deemed that the requested privilege would damage the prerogatives of the ribbon-makers' guild and denied it.[33] In eighteenth-century Turin, however, master ribbon-makers were totally subject to the whim of the fashion merchants (*mercanti di moda*) who determined ribbon type and quality.[34] It was the same in Milan, where local merchants decided the quality of yarn to be used for ribbons.[35]

Among their sales strategies, aimed particularly at women, was the prominent use of fashion dolls to model their wares. The Venetian poet Franceso Algarotti (1712–64) described how women in his city gathered in droves to see a French doll, *piavola de Franza*, which had newly arrived from Paris, curiously noting every detail of its dress from its head to its toes. 'You will see them flocking before you', writes the poet, 'scrutinising piece by piece the *andrienne*, the cap, the ribbons, the vast hoop, and casting their eager looks even to the inside and underneath of each petticoat'.[36]

Merchants and merchant-manufacturers alike experimented with a range of approaches in order to compete with rivals in attracting consumers and conquering new markets for 'fashionable' ribbons. These included the development and promotion of something resembling brand names for their products. Of course, many ribbons were named after the cities where they were made, such as 'ribbons from Milan' or 'from Naples'. But other factors, only partly related to places of origin,

Figure 2.3 Ribbons could be sold door to door by hawkers. Gaetano Gherardo Zompini, *Le arti che vanno per via nella città di Venezia* (Venice, 1785), c. 28. Bibliothèque nationale de France (Paris).

contributed to the choice of name and the nature of the brand's identity. These were manufacturing processes, trading networks and marketing strategies. From the manufacturing perspective, names often depended on the choice of raw materials, on the tools and equipment used (particularly types of loom) and on the technological processes employed, especially in the finishing stages. For instance, 'Padua ribbon' (or *Padoue* in French) normally referred to a ribbon made with waste silk (*fleuret*, *filosello* and *capitone*) – and which was commonplace from the late seventeenth century onwards across Europe, but in France, Switzerland and Sweden in particular.[37] In eighteenth-century Venice, ribbons were called 'Lyons fashion' (*all'uso di Lione*) only if specific finishing processes were used.[38] Although these techniques originated in Lyon, they had spread to other manufacturing locations, and so 'Lyons fashion' ribbons were named after the processes used to create them, rather than necessarily in reference to their place of production.

Such transmission of production techniques and technology was enabled by Europe's trading networks. The mechanical or engine looms used for weaving silk ribbons in mid-eighteenth-century France were called 'Zurich looms' because the artisan who introduced them came from Switzerland.[39] The influence of trading networks was visible in the names given to the ribbons themselves. In Paris, 'Padoue de Lyon' ribbons were referred to as 'from Lyons' not because they were produced in that city but because it was from there that Parisian mercers imported the ribbons woven in the nearby towns of Saint-Etienne and Saint-Chamond.[40] It was the names of major silk centres and commercial hubs – Lyons or Zurich – that were bestowed on ribbons and the technology that produced them, rather than the towns where silks were actually made – Saint-Etienne and Saint-Chamond, or, in the case of Switzerland, Basel.[41]

Another factor influencing ribbon naming and branding was marketing strategy. Merchants tried to appeal to customers by promoting products which conjured up 'foreign' fashions. For instance, ribbons 'from London' were very popular in Paris, while in London it was ribbons 'from Paris' which were the best on the market.[42] At the same time, merchants monitored the chain of production and organised trade in order to maximise their profits. Eighteenth-century Milanese merchants, for example, exported their finer raw silk to France – importing high-quality patterned ribbons from Lyons in return – and distributed the poorer raw silk to local makers, so that they were able to produce only low-quality ribbons. The consequence was that the merchants made larger profits from the local market, selling foreign ribbons at higher prices while the craftspeople were unable to compete because of the poor silk that was all they could obtain.[43]

Technology and work organisation

At the same time as merchants and retailers were becoming one of the most significant factors determining the nature of the ribbon industry in Italy, technological developments were beginning to affect the structure and composition of its work organisation across the peninsula, too. Workforce composition and the institutional framework of the guilds were influenced by loom technology in particular. This was the case especially in cities which developed an export-oriented and specialist ribbon-focused manufacture, as distinct from those centres where ribbon production developed within the broader framework of a major silk-weaving industry.

Traditionally, the technology employed in the initial spinning process in ribbon manufacturing was the same as that used for the production of silk yarn more generally. Once twisted, the yarn was ready to be dyed, before being warped on to the loom and then woven into ribbons. Plain ribbons and tapes were normally made with 'small looms' (*telaretti*), which were very simple, easy to use and cheap to buy. Finer ribbons, such as gallons, were made on smaller versions of the looms used to weave silks for clothes. There were two basic types of these: the horizontal loom; and, from the late sixteenth century onwards, the vertical loom, which was used in particular for the very finest ribbons.[44]

Ribbon manufacturing experienced one major technological innovation during the sixteenth and early seventeenth century: the silk-ribbon engine loom, patented in 1605 by inventor Willem Dirckz Sonnevelt and utilized widely in Dutch cities like Leiden, Haarlem and Amsterdam.[45] A labour-saving innovation, the frame allowed a single worker to weave between six and twenty-four ribbons at the same time. But it was also capital-intensive, and technical limitations meant it was normally restricted to the production of low-quality plain ribbons. Although the engine loom was soon widespread across northern and central Europe, in both urban and rural areas, it did not arrive in Italy until the mid-eighteenth century, where its impact was limited. Its diffusion was restricted by a variety of factors, from the limited availability of raw materials and the size of the market for finished products to the structure of production and the nature of the workforce.[46]

The whole process of ribbon manufacturing was based on the well-known putting-out system, where work was subcontracted to individuals working in their own homes. This operated in both town and country, but there were major differences when it came to the weaving, which depended on the type of ribbons being produced and the looms being used. The 'small looms' used for plain ribbons were usually worked by female weavers, with the support of a young (usually) female assistant, and were common in houses, hospitals, orphanages and convents.[47] The horizontal and vertical looms used for high-quality ribbons were normally the preserve of male weavers, and were commonly installed in their own small room within the weaver's home.[48] There he was assisted in weaving by a woman, who generally warped the silk, and a male assistant or journeyman who was responsible for preparing warps of the right length.

In cities such as Milan, Turin, Naples and Venice, the institutional structure of ribbon production was normally guild-based; while in smaller centres, such as Padua and Vigevano – where ribbon-making was highly developed – weavers operated outside the guild system. Where guilds were absent, male artisans produced the more costly patterned braids, while cheap plain ribbons were woven by female master weavers and young female assistants. However, in cities where weavers' guilds were established, women were normally barred from the workforce outright, with the exception of a few wives, widows and daughters of master weavers. In both contexts, though, the labour market was normally divided between a fixed, permanent workforce (such as owners of silk mills, master weavers and dyers) and a flexible, often itinerant one (such as spinners, workshop assistants and journeyman weavers). And individual weavers were normally reliant on groups of merchants and merchant-manufacturers for the supply of raw materials and the selling of their output (as we have seen in the case of the Turin ribbon-masters); even when weaving in rural areas increased, as occurred around Padua and Vigevano in the second half of the seventeenth century, it remained firmly under the control of urban merchants.[49]

Conclusion

Ribbon consumption in early modern Italy enables us to draw some general conclusions on the introduction, use, popularity and cultural meaning of ribbons as fashion items. The dissemination of ribbons as innovative products provides an intriguing illustration of the way that styles in fashion and patterns of consumption undergo transformation. Still, today, understanding ribbon-wearing requires their complexity to be acknowledged and understood.[50]

This chapter has explored how in early modern Italy ribbons existed in concert with other fashions (especially those for silk products). They were accessories used to decorate clothes, but they also existed independently, as objects with their own cultural significations, drawing meaning especially upon gendered distinctions. While they were conspicuous symbols of honour, valour and social distinction for men, for women they were frivolous consumer products, emblems of vanity, pleasure and play.

The proliferation of ribbons in a plethora of fashionable styles stimulated developments in both manufacturing and distribution during the late seventeenth and eighteenth centuries. The use of 'brand names' and 'appellations of origin' to denote the distinctive styles and qualities of ribbons evolved under the influence of increasingly sophisticated commercial strategies into a major means of marketing products, by catering to demands for the foreign-sounding and the exotic. These same strategies encouraged craftspeople to innovate, copy and disseminate new techniques and methods of production to feed the hunger for new fashions. But they also resulted in attempts to restrict the supply of raw materials and the sale of finished products to inflate profits.

The demand fostered by this culture had a significant bearing on almost every aspect of ribbon production. First, it gave rise to a new industry in the mid-sixteenth century. Ribbon manufacturing had already begun in Italy in the fourteenth and fifteenth centuries, in Rome, Racconigi (near Turin), Florence, Naples and Genoa. But it was in the sixteenth century, as the silk industry grew, that centres specialising in the production of silk ribbons sprang up in the peninsula – notably in Milan, Venice, Padua, Verona and Bologna – and ribbon-makers' guilds began to appear.[51] These centres maintained their leadership in skills and product quality until the late seventeenth century, when the geography of ribbon production started to alter in accordance with changing patterns of demand and the growth of export markets. Increasingly fierce competition drove capital investment and technological innovation, exemplified by the 'Dutch' engine loom. Its dissemination, first into Germany and Switzerland between 1660 and 1710, and only later in the eighteenth century in France and Italy, was accompanied by a shift in the nucleus of ribbon production from southern to central and north-western Europe, where it concentrated in locations including Basel in Switzerland, Krefeld in Germany, Saint-Etienne and Saint-Chamond in France, and Coventry and the Spitalfields district of London in England.[52]

As ribbon-making burgeoned, it became an increasingly large-scale employer. In addition to the increased numbers of workers, the industry also stimulated the growing concentration of labour, and even the use of forced labour, too. In Padua, Florence and Turin, merchants and merchant-manufacturers financed or brought together production in hospitals, orphanages and charitable institutions, concentrating their labour forces in large numbers in order to make ribbons more efficiently, whether with or without mechanical looms.[53] Together with other textile-manufacturing sectors such as lace-making, embroidery, hosiery and silk-throwing, ribbon-making was one of the first 'proto-manufactures', with high concentrations of female labour often working in exploitative conditions. In the late eighteenth century, Marsilio Landriani proposed introducing these working practices to Milan so that industries could take advantage of the city's untapped child-labour resources. Ribbon-making, he suggested, would be an ideal candidate for this because it required neither strength nor skill.[54]

During the seventeenth and eighteenth centuries, the dissemination of the ribbon-making industry created a labour market strongly segmented by gender. The weaving of top-quality patterned braids remained under the control of male masters who were organised into traditional guild-based structures. On the other hand, the weaving of plain ribbons – especially in export-oriented centres of production – was traditionally the preserve of female masters, who were assisted by young workers either temporarily over the winter in rural environments, or year-round in urban areas. Where mechanical looms were introduced into plain-ribbon manufacturing, such as in Turin, groups of male workers tried to monopolise production, relegating women to hand-loom weaving.[55]

Innovation in popular fashions underpinned socio-economic changes that were linked to the evolution of early capitalism in Europe. The way in which a small object like the ribbon became established in clothing is a good example of this.

Notes

1 John Styles, 'Product innovation in early modern London', *Past & Present*, 168 (2000), 124–69; Maxine Berg, 'From imitation to invention: Creating commodities in eighteenth-century Britain', *Economic History Review*, 55, 1 (2002), 1–30; Maxine

Berg, *Luxury and Pleasure in Eighteenth-Century Britain* (Oxford: Oxford University Press, 2007); Maxine Berg and Elizabeth Eger (eds), *Luxury in the Eighteenth Century: Debates, Desires and Delectable Goods* (Basingstoke: Palgrave Macmillan, 2002).

2 See, however, David Mitchell, '"What d'ye lack ladies? Hoods, ribbands, very fine silk stockings": The silk trade in restoration London', in Luca Molà, Giorgio Riello and Dagmar Schäfer (eds), *Threads of Global Desire: Silk in the Pre-Modern World* (Woodbridge: Boydell Press, 2018), 187–224.

3 A great deal has been written on the subject of early modern consumerism. See especially John Brewer and Roy Porter (eds), *Consumption and the World of Goods* (London: Routledge, 1993); Jan de Vries, *The Industrious Revolution: Consumer Behavior and the Household Economy, 1650 to the Present* (Cambridge, Cambridge University Press, 2008); Carlo Marco Belfanti, *Civiltà della moda* (Bologna, il Mulino, 2008); Giorgio Riello and Peter McNeil (eds), *The Fashion History Reader: Global Perspectives* (London: Routledge, 2010); Glenn Adamson, Giorgio Riello and Sarah Teasley (eds), *Global Design History* (London: Routledge, 2011).

4 On the links between imitation, marketing and counterfeiting prior to the Industrial Revolution, see Carlo Marco Belfanti, 'Branding before the brand: Marks, imitations and counterfeits in pre-modern Europe', *Business History*, 60, 8 (2018), 1127–46; Corine Maitte, 'Imitation, copie, contrafaçon, faux: définitions et pratiques sous l'Ancien Régime', *Entreprise et Histoire*, 78 (2015), 13–26.

5 'Tissu très mince qui sert à plusieurs usages, suivant les matiéres don il est fabriqué. On fait des Rubans d'or, d'argent, de soie, de fleuret, ou capiton, de laine, de fil, & c. On est fait d'estroits, de larges, de demi larges, e façonnés, d'unis, à deux endroits & avec un envers, de gauffrés, de brochés, à raiseau, de simples, de doubles en lisse; enfine de toutes couleurs & de tous desseins suivant le génie du Rubanier, le gout du Marchand qui le commande, ou la mode qui court', Jacques Savary des Brûlons, *Dictionnaire universel du commerce* (Copenhagen, 1759–65), vol. 4, 593.

6 Natalie Rothstein, 'Silk in the early modern period, *c.* 1500–1780', in David Jenkins (ed.), *The Cambridge History of Western Textiles* (Cambridge: Cambridge University Press, 2003), vol. 2, 560.

7 Catherine Donzel and Sabine Marchal, *L'art de la passementerie et sa contribution à l'histoire de la mode et de la decoration* (Paris: Chêne, 1992), 36; Rosalba Ragosta, *Napoli, città della seta: Produzione e mercato in età moderna* (Rome: Donzelli, 2009), 15.

8 Carlo Marco Belfanti, 'Fashion and innovation: The origins of the Italian hosiery industry in the sixteenth and seventeenth centuries', *Textile History*, 27, 2 (1996), 132–47; Carlo Marco Belfanti, 'The civilization of fashion: At the origins of a western social institution', *Journal of Social History*, 43, 2 (2009), 261–83, 263–7; Hermann Van der Wee, 'The western European woollen industries, 1500–1750', in Jenkins (ed.), *Cambridge History of Western Textiles*, vol. 2, 435–6, 455; Joan Thirsk, 'Knitting and knitwear, c. 1500–1780', in Jenkins (ed.), *Cambridge History of Western Textiles*, vol. 2, 566; Frederick A. Wells, *The British Hosiery and Knitwear Industry: Its History and Organisation* (New York: Barnes & Noble, 1972), 15–16; Irene Turnau, 'The diffusion of knitting in medieval Europe', in Negley B. Harte and Kenneth G. Pointing (eds), *Cloth and Clothing in Medieval Europe: Essays in Memory of Professor E. M. Carus-Wilson* (London: Heinemann Educational Books, 1983), 368–73; Luca Molà, *The Silk Industry of Renaissance Venice* (Baltimore: Johns Hopkins University Press, 2000), 177–85.

9 Irene Turnau, 'La bonneterie en Europe du XVIe au XVIIIe siècle', *Annales. Economies, Sociétés, Civilisations*, 26, 5 (1971), 1120; Doretta Davanzo Poli, *Abiti antichi e moderni dei Veneziani* (Vicenza: Neri Pozza, 2001), 69, 74.

10 Andrea Caracausi, *Nastri, nastrini, cordelle: l'industria serica nel Padovano secc. XVII–XIX* (Padua 2004), 40–52.

11 Caracausi, *Nastri*, 55–6; Alfred Plummer, *The London Weavers' Company, 1600–1970* (London: Routledge, 2006), 457.

12 *Ormesino* was a high-quality silk normally made at the famous Bolognese silk mills: see Carlo Poni, 'Archeologie de la fabrique: La diffusion des moulins à soie "alla bolognese" dans les Etats vénitiens de xvii au xviii siècle', *Annales. Economies, Sociétés, Civilisations.*, 27, 6 (1972), 1475–96.
13 See, for example, 'Un gamurrino verde buio di rovescio con nastrino verde e di oro', inventory of the waterseller Francesco Ristori 12 September 1631, State archives of Florence (ASF), Magistrato dei pupilli, 2718, 2, fol. 193r; '1 traversa nuova diversa con cordela e merli', inventory of the weigher Domenico Giustinani, 24 March 1631, State archives of Venice (ASV), Giudice di Petizion, *Inventari*, 352, 88, fol. 7r; and other examples at the 'Refashioning the Renaissance' database, www.refashioningrenaissance.eu/database.
14 On prices, see Andrea Caracausi, 'Una compagnia di cordelle di seta nella seconda metà del Seicento: la "Eredi Giupponi & C." di Padova', *OS. Opificio Della Storia*, 2, 2 (2021), 8–21.
15 For knitwear, see Belfanti, 'Fashion and innovation'.
16 Donzel and Marchal, *L'art de la passementerie*, 66–70.
17 They could be worn as a symbol of adherence to a particular group or belief, or as totems invested with personal symbolism and significance. For example, the wearing of green ribbons by prisoners of the French Revolution was associated with loyalty to the Crown. Natalie Rothstein, 'Fashion, silk and the worshipful company of weavers', in Simonetta Cavaciocchi (ed.), *La seta in Europa secc. XIII–XX* (Florence : Le Monnier, 1993), 171; Donzel, *L'art de la passementerie*, 56. For blue and red ribbons as political symbols in seventeenth-century England and Scotland, see Edward Legon, 'Bound up with meaning: The politics and memory of ribbon wearing in Restoration England and Scotland', *Journal of British Studies*, 56, 1 (2017), 27–50, especially 31.
18 See panel 17: 'Compra nastri dal marzaro', at www.metmuseum.org/art/collection/search/630969, accessed 22 June 2024.
19 Carlo Goldoni, *La banca rotta o sia mercante fallito*, in *Opere Complete di Carlo Goldoni* (Venice: Municipio di Venezia, 1907), vol. 1, 345.
20 Goldoni, *La Banca Rotta*, 395. 'Truffaldino. Lassemo andar sti discorsi, che no serve gente: cossa pensela, sior Pantalon, de voler donar a ste donne? – Pantalone. (Eh, ti ghe cascherà, furbazzo !) (da sé) Se poderia donarghe qualche taggio de roba, qualche cavezzo de drappo, della cordela, delle galanterie de bottega. Truffaldino. Sior sì, ste cosse le donne le gradisse infinitamente.'
21 Davanzo Poli, *Abiti antichi*, 67.
22 Quoted in Rosalba Ragosta, 'Specializzazione produttiva a Napoli nei secoli XVI e XVII', in Cavaciocchi (ed.), *La seta in Europa*, 339–49, 341.
23 Molà, *The Silk Industry*, 307.
24 Museo Civico di Venezia, *Donà delle Rose*, No. 332, VI, 'per la povera arte dei passamaneri'.
25 Based on the 'Refashioning the Renaissance' data, excluding shop inventories, at www.refashioningrenaissance.eu/database. I thank Stefania Montemezzo for data analysis.
26 Martin, *Buying*, 171.
27 Samuele Romanin, *Storia documentata di Venezia*, 3rd ed., vol. 9, 19 (Venice: Filippi, 1972).
28 State Archives of Turin (AST), Materie economiche, 9, 22, November 1775.
29 Ulrich Pfister, 'Craft guilds and technological change: The engine loom in the European silk ribbon industry in the seventeenth and eighteenth centuries', in Stephen R. Epstein and Maarten Prak (eds), *Guilds, Innovation, and the European Economy, 1400–1800* (Cambridge: Cambridge University Press, 2008), 172–98, at 174.
30 Pfister, 'Craft guilds', 174.

31 Margaret Spufford, *The Great Reclothing of Rural England: Petty Chapmen and Their Wares in the Seventeenth Century* (London: Hambledon, 1984), 6, 145.
32 'Ama il contadino la comparsa, ma le facoltà non s'adattano al di lui desiderio', in AST, Commercio, mazzo 9, fasc. 22, 19 January 1756.
33 ASV, Inquisitorato alle Arti, 68, 30 December, 1659.
34 AST, Materie economiche, *Arti e Manifatture*, 9, 22, 23 December, 1755.
35 Paola Curatolo, *Struttura, crisi e trasformazione di un sistema produttivo urbano: le corporazioni auroseriche milanesi (1570–1720)* (Milan: Cooperativa universitaria editrice, 1996), 239–46.
36 Ribeiro, *Dress*, 62. The *andrienne* was a typical eighteenth-century dress, a fine example of *robe à la française*.
37 'Ceux de capiton qu'on appel *Padouë*, s'employent par les Tailleurs, Couturiéres, &c. les Rubands de laines & de fil par les Tapissiers, Fripiers, Selliers & autre femblable Ouvriers', Savary des Brûlons, *Dictionnaire*, vol. IV, 593.
38 See ASV, Inquisitorato alle arti, 68, 'Finocchi di Padova contro Passamaneri', IV–VI.
39 Pfister, 'Craft guilds', 193.
40 Savary des Brûlons, *Dictionnaire*, vol. IV, 594.
41 Pfister, 'Craft guilds', 193.
42 Savary des Brûlons, *Dictionnaire*, vol. IV, 593.
43 Curatolo, *Struttura, crisi e trasformazione*, 172–3.
44 Lilian Mottu-Weber, *Économie et refuge à Genève au siècle de la Réforme: la draperie et la soierie (1540–1630)* (Geneva: Droz, 1987), 332–3.
45 The story is well known. See Alfred P. Wadsworth and Julia de Lacy Mann, *The Cotton Trade and Industrial Lancashire, 1600–1780* (Manchester: Manchester University Press, 1931); Liliane Mottu-Weber, 'Production et innovation en Suisse et dans les états allemands', in Cavaciocchi (ed.), *La seta in Europa*; Pfister, 'Crafts guilds'.
46 Andrea Caracausi, 'Between Innovation and Resistance: The Role of Guilds in Early Modern Italian Ribbon Manufacturing', *Technology and Culture*, 65, 3 (2024), 791–817.
47 Andrea Caracausi, *Dentro la bottega: culture del lavoro in una città d'età moderna* (Venice: Marsilio, 2008), 38–9. For weaving by young girls (and, sometimes, male foreigners) in orphanages and hospitals, see Andrea Caracausi and Corine Maitte (eds), 'Le istituzioni caritative come luoghi di lavoro (secc. XVI–XX)', *Mediterranea: Ricerche storiche*, 48 (2020), 83–246.
48 Mottu-Weber, *Économie*, 332.
49 Andrea Caracausi, 'Fashion, capitalism and ribbon-making in early modern Europe', in Thomas Max Safley (ed.), *Labor before the Industrial Revolution: Work, Technology and Their Ecologies in an Age of Early Capitalism* (London: Routledge, 2019), 48–69.
50 Sarah E. H. Moore, *Ribbon Culture: Charity, Compassion, and Public Awareness* (Basingstoke: Palgrave Macmillan, 2008), 2.
51 Caracausi, 'Fashion', 55–7.
52 Pfister, 'Craft guilds', 179–94.
53 For Padua, see Caracausi, *Dentro la bottega*, 38–9; Andrea Caracausi, 'Fra sistema a domicilio e manifattura accentrata: l'istituto degli orfani nazzareni di Padova nella prima metà del Seicento', *Mediterranea: Ricerche Storiche*, 48 (2020), 123–42; for Turin, see AST, Consolato di Commercio, *Atti del Consolato in materia di arti e mestieri*, 37, 'Regia opera delle povere figlie raminghe', 24 March 1783, fol. 87r; for Florence, see Corine Maitte, 'Donner du travail aux pauvres: les logiques laborieuses dans les institutions charitables florentines aux XVIIe et XVIIIe siècle', *Mediterranea Ricerche Storiche*, 48 (2020), 97–122, 113.
54 Landriani, *Relazioni*, 88. See also, for seventeenth-century London, David Mitchell, 'What d'ye lack ladies?', 193.
55 For seasonal production in Padua, see Caracausi, *Nastri*, 95.

3

Imitation in early modern artisan fashion

Sophie Pitman

Introduction

In 1594, a precious cargo was transported from Florence to Transylvania. In a letter, dated 6 April, Grand Duke of Tuscany Ferdinando I de' Medici described the contents to Prince Sigismund Báthory as 'some sables and lynxes, not those real ones that come from other parts, but those that art imitates from my lands'. Ferdinando emphasised that it was not the economic value of these furs that was notable but rather the 'novelty of their artifice'.[1] This brief missive reveals much about the availability, interest in and production of imitation materials used in early modern fashionable dress. We discover that furs imitating sable (a small marten with fine dark-brown or black fur, trapped in Siberia) and lynx (a wild cat with a silky white belly and black-patterned coat found in Scandinavia and eastern Europe) were being made in Italy. Unlike imported 'real' furs, esteemed for their rarity and cost, these fakes were prized for the skilled craftsmanship they manifested.

This chapter explores these ingenious so-called 'artificial novelties' in order to discover how and what we can learn about imitation materials in early modern fashion, particularly when so few examples survive.[2] It uses hands-on experimental reconstructions, made and tested following early modern sources, to suggest how fakes might have operated aesthetically and culturally, and explores how mimetic materials enabled individuals who were economically or legally unable to wear the finest furs, fabrics, metals and jewels to dress fashionably. Using the 'Refashioning the Renaissance' database of 448 inventories that record the belongings of members of the artisan classes in Siena, Florence and Venice between 1550 and 1650, as well as sources from other European cities, it locates some of these materials among the possessions of artisans. It also challenges the assumption that imitations were simply inferior copies for the non-elite. In doing so, it reveals that the often overlooked category of imitations disrupted a culture of clothing that had relied on a hierarchy of materials to display social status and led to creative ways of making and dressing.

Imitation furs and the rarity of materials

We do not know how Ferdinando's fakes were made, but fine furs could be imitated by treating lesser skins. Common mustelids, such as pine martens and stone martens (from northern Europe), calabers (a kind of squirrel), other squirrels and even domestic cats were dyed black to appear like fine sable. In cities such as Rouen and London, skinners were forbidden from selling dyed furs, but, in 1533, records show that 'cats painted' were sold in England.[3] Jewish dealers were accused of selling 'fake' furs on the streets of Istanbul.[4] The 1563 Italian edition of Girolamo Ruscelli's *De secreti* – perhaps the most popular, copied and translated book of secrets – includes a recipe that promises to transform a cheap white fur into 'the colour of a leopard or panther'.[5] It explains how to make a mixture of lead and quicklime that will permanently tint readily available furs (such as rabbit) with black spots that simulate the patterning on wild cats' pelts. As the technique chemically weakens the skins, and survival bias privileges the finest materials, it is unsurprising that no existing examples of this fake leopard fur are known. But by reconstructing the process we can learn how it worked. An experiment on imitation fur suggested that visually a fake fur could be rather convincing, but the texture is much coarser than authentic leopard fur (Experiment in focus I and Figure i.3).[6] Those who bought fake furs might have been willing to compromise on quality or durability in return for affordability or availability. Ferdinando de' Medici himself experimented with counterfeit jewels in his alchemical laboratory at the ducal palace, and so the idea that exotic furs could be created in workshops as well as hunted in distant lands may well have appealed to his sense of wonder. The art historian Marlise Rijks suggests that educated consumers might have valued counterfeits more than the real thing due to what she terms 'process appreciation' – that is, valuing the skill of the artisan in making one material appear like another.[7] Such 'material mimesis' could be appreciated as a manifestation of the scientific and artistic fascination of early modern experimenters in pushing the limits of material properties.[8] Moreover, high-quality pelts took significant time and effort to source, as we see in letters from the merchant Hans Fugger detailing his attempts to acquire enough lynx for a Hungarian-style coat.[9] Fake furs provided an alternative option in the streets and princely courts of early modern Europe.

The fashion for furs, hitherto associated with elite luxury, widened in the sixteenth century as urban middling and labouring people aspired to fur clothing for aesthetics and utility.[10] We find more than six hundred fur items in the inventories of Italian artisans, mostly from Siena and Venice. A variety of pelts including otter, fox, squirrel, marten, rabbit, cat, wolf are described for trimmings, linings, muffs and occasionally outer garments. For example, a Venetian baker owned two old muffs made of rabbit, and a hemp-spinner possessed an old neck ornament (*cravatta*) made of purple-dyed fur lined with fox.[11] The majority of recorded furs – especially the finest sable and lynx – comprised the stock of furriers and clothing dealers, giving us a broader glimpse of what was available on the market, rather than representing the typical artisan wardrobe. Only two pieces of leopard fur are found among the inventories, both belonging to the Venetian furrier Baldissara Da Pozzo.[12] Given such rarity, it is perhaps unsurprising that people might have wished to imitate leopard fur using Ruscelli's *Secreti* or purchase Ferdinando de' Medici's artificial sable and lynx novelties.

Beaver fur was another relative rarity. With the creatures driven to extinction in most of Europe, pelts had to be imported from Russia, Scandinavia and through French-controlled Canada.[13] Its comparative strength and weatherproof qualities reduced the need for the adhesives and varnishes required to produce sturdy headgear from other kinds of animal hair. High felted beaver hats were among the most expensive accessories on the market, but we find thirteen of them among the belongings of Sienese, Florentine and Venetian artisans. Nine of these were owned by the Sienese shopkeeper Adriano Guagni and were described as 'half beaver' (*mezzo castor*).[14] The appearance of pure beaver could be imitated by blending it with rabbit fur or sheep's wool to produce 'demicastor hats', or even by felted wool alone.[15] The process of reconstructing a felted beaver-style hat using pure Merino wool glazed with rabbit-skin glue and napped with a brush reveals how effective this substitute material could be (Figures 3.1a, 3.1b).[16] Many surviving early modern felted hats are assumed to be beaver, but without fibre analysis it is hard to tell whether they are made from wool, a blend of furs or pure beaver.[17] Mixed materials and skilful finishing can conceal their status as imitation fashions.

Most furs belonging to the artisans studied by the 'Refashioning' project were used in linings, borders and trim, suggesting that fur was both a practical material for keeping warm and a luxurious decorative flourish. The magnificent effect of fine furs could also be achieved by the careful placement of small pieces. Garments could be 'purfled' – lined at the neck and sleeve openings – with costly furs, with the remainder made of a cheaper fur.[18] A Venetian rag-dealer's inventory from 1555 records a heavy *duliman* gown lined with 'red skins' and edged with stone marten.[19] Here, the imitation operated through the implication that an object included more than the bare minimum of fine materials.

Figure 3.1a Hat, 1590–1670. Victoria and Albert Museum, T.22-1938 T.

Figure 3.1b Brushing a reconstructed felted hat to create a smooth shiny surface, Sophie Pitman at the School of Historical Dress, London, 2014.

Imitation silks and the price of finery

Silk fabrics were coveted for their smoothness, lustre, drape and softness. The highest-quality and most expensive silk velvets and damasks were instant markers of luxury.[20] In the hierarchies of materials created by sumptuary laws, such silks were often the preserve of the elite, and yet silk was also enjoyed by those lower down the social spectrum through a lively second-hand market, and the careful use of small pieces in clothes and affordable accessories like ribbons and masks.[21] Spun silks, produced from the waste after the finest silk threads had been reeled from the cocoons, were another lower-cost alternative. A Florentine clog-maker, for example, owned a black *filaticcio* (spun silk waste) cloak trimmed with sarcenet, and a mask of black velvet.[22]

Silks were also imitated by lesser fibres, or blends which combined silk with wool, linen, hemp or cotton. One of the key transformations in the era, as John Styles in Chapter 1 above shows, was a shift towards lighter mixed-fibre fabrics, which were less expensive and less durable than pure wools and silks.[23] In the 'Refashioning the Renaissance' database there are more than 1,700 items made of these mixed materials, such as *brocatello*, *buratto*, *ciambellotto*, *mezzalana*, *mocaiardo*, *mussolo* and *panno misto*. While previous narratives about Renaissance textiles typically separate wools, silks and linens, the widespread use of these mixed blended fabrics, particularly among the middling classes, suggests that a more nuanced approach to textiles in the era is needed.[24] By examining one of these materials, *mocaiardo* – a 'mock' velvet known as 'mockado' in English – we can observe the widespread purchase, use and economic and legal impact of one of these novel blends.[25]

Mockado was a mixed-fibre textile with the structure of velvet, known for its hairy or fluffy pile created from looped supplemental warps that could be cut or left uncut. It was woven from some combination of wool, hemp, linen, camel or goat hair or spun silk, making it a far more affordable option than pure silk velvet.[26] The word is possibly derived from the Arabic *mukhayyar*, meaning 'select, choice', and possibly was etymologically confused with *mocajar*, a mid-quality camlet, and mohair. Mockado, and its many variant spellings and translations, were often elided with Naples fustian, *tripp* and later *moquette*.[27] Produced in the Levant, in Lille and across the Low Countries, and by immigrant weavers in Norwich, it was one of the most popular fabrics among artisans.[28] It appears 294 times in the 'Refashioning the Renaissance' database, from across the period – 238 of these were in Venetian inventories, only one from Siena and 55 from Florence. Mockado was used in large items of apparel such as cloaks, gowns, petticoats, hose and doublets and smaller pieces including sleeves and other accessories.[29] Mockado was worn by men, women and children alike, and dyed into a kaleidoscope of colours: most commonly black, but also brown, blue, green, red, white, pink, purple and yellow. Mockado items were often intended to be decorative and showy: among the most fashionable garments owned by the Florentine wine-seller Evandio Honesti, for example, were a tawny mockado doublet, a red mockado gown with yellow cords, a black mockado petticoat and another petticoat the colour of 'dried rose', decorated with yellow and black cords.[30]

Sumptuary laws suggest that mockado was regarded as a substitute for silk velvet, suitable for those lower down the social scale. A 1551 law in Mantua forbade artisans and their wives – as well as Jews and soldiers – from wearing clothes, caps and shoes with silk or velvet decorations 'of any sort, except ones made of *moccaiato* or *bavella* [cloth made from spun silk]'.[31] In Pistoia in 1562, regulations offered *mocaiardo* as an alternative option for those forbidden from wearing 'silk of any sort'. Women from artisan families were permitted *mocaiardo*, camlet, *catarzo* (a coarse wool), samite and other fabrics not included in the category of *drappo*.[32] A 1563 Pisan law stipulated that artisan wives were permitted a single gown of silk mockado or say. Their male relatives, and male peasants and porters, were forbidden from using mockado, silk, russet or grain-dyed wool or camlet in their dress altogether.[33]

Sumptuary laws drove a market in substitutes. When silk passementerie was banned in Spain in 1551, for example, artisans adopted wool trimmings.[34] Despite the cultural historian Daniel Roche's claim that sumptuary laws represented a 'sartorial *ancien régime*' marked by 'inertia and immobility', they in fact prompted imaginative creativity among craftspeople and nimble purchasing by consumers.[35] Imitation goods could provide individuals with products that achieved the effects they desired without breaking the law.

The weaving of mixed-fibre fabrics was also a response to vigilant guild control. In 1576, a London ordinance explained that because blended novel fabrics such as 'tufted mockado' were 'not called by any of the same names that the cloths of woollen or linen of the former ancient time were (although they consist of the same substance)', the makers and merchants were selling them without paying duties or following guild and city customs, a practice deemed 'naughty and deceitful'.[36] The ordinance stated that that mixed wool and linen fabrics had to be inspected, checked and charged duty. Violations were subject to hefty fines; that for tufted mockado was 20d (about one day of wages for a skilled tradesperson).[37] The association between mockado and deceit was reflected linguistically in the way that the term was often used to refer to fakery.[38] In 1577, for example, the English Puritan William Fulke denounced Catholic texts as being full of false claims, including 'mockadoe miracles, narrations, and relations'.[39] While the novel and often changing names of new fabrics mean that it can be hard for historians to decipher what each one was (weave structure, weight, size, composition, finish etc.), the 1576 ordinance makes clear that the proliferation of names was not just a marketing tactic but also could help immigrants and others excluded from markets by the guilds to skirt controls.[40]

Mockado was not only consumed by those forbidden from wearing silk. On 6 July 1574, the servant John Knyveton wrote from London to the Earl of Shrewsbury to tell him that 'all kyndes of tufted mockadoes be so deare because of the [15 June] proclamation for apparel, that now paye xd in every yarde more than before and therefore I staye to bye any till the price be better'.[41] While the proclamation permitted earls to wear silk velvet, Shrewsbury's request suggests that mockado was desirable in

Figure 3.2a Stamped wool velvet, sixteenth–seventeenth century, Italian? Metropolitan Museum of Art, New York, 07.62.5.

its own right. Surviving velvets made with wool, hemp or linen are often remarkably effective simulants (Figure 3.2a). Their dense plush and sheen can be so similar to pure silk velvet that it is hard to identify their fibre composition without close analysis or scientific testing.

Imitative decoration and the development of techniques

Many woollen velvets are stamped in a manner that reproduces the decorative effect of a more complex weave structure: that is to say, they were woven without pattern, and after the fabric was taken off the loom, a pattern was impressed into the surface with a metal stamp or roller press (Figure 3.2b).[42] Six mockados in the database are described as stamped, including a sea-green mockado petticoat owned by the Florentine delicatessen worker Angelo Fantacci in 1635.[43] Other stamped fabrics recorded in artisan inventories include velvet, mixed fabrics and light silks such as camlet, grosgrain, sarcenet. Hands-on experiments, such as the reconstruction of the stamped mockado doublet belonging to a Florentine waterseller (Experiment in focus III) demonstrate that stamping is an easy and relatively quick process, whereas weaving patterned mockado is a highly skilled and time-consuming task.[44] Imitation fabrics might cut costs through the use of inferior materials or quicker processing, but the successful end result required ingenuity on the part of the maker.

Imitative decorative techniques could also mimic the shimmering designs of the finest woven silks. A velvet damask woven with crimson silk and gold thread in a pomegranate pattern represents perhaps the most iconic and costly fabric available in Renaissance Europe (Figure 3.3a). The high value of such a textile could be justified both on the grounds of its expensive raw materials (silk, gold, crimson dye) and the slow and highly skilled labour of the weaver. But the colours, patterns and visual effects could be replicated quickly by block printing on to linen, as a fragment

Figure 3.2b Stamping mockado using a heated four-petalled flower stamp by Jordan Colls and Jenny Tiramani.

in the Victoria and Albert Museum shows.[45] In this example, crushed mica (shiny silicate scales) was added to pigment to give it a glimmer that recalls fine silk velvet, especially in candlelight (Figure 3.3b). A recipe for an imitation damask, found in a French manuscript from the 1580s, suggests that the effect might otherwise be mimicked through a form of resist-dyeing.[46] Linen could be glazed using a smoother, starches and heat, which polished the surface into a silky sheen, as one fine surviving doublet demonstrates.[47] A gown of green satin *camuffada* striped with gold is mentioned in one artisan inventory.[48] The word *camuffada* could mean 'fake' (as in 'camouflage'), suggesting that the satin was woven with materials other than silk. But John Florio defined *camuffare* in his sixteenth-century English–Italian dictionary as 'to filch or steale craftily … Also to smooth between stones, to polish', so perhaps the gown was made of polished satin.[49] Such linguistic elision demonstrates how such finishes might be associated with fakery.

Buttons cast from gold or silver or made of precious metal and silk threads wrapped around a wooden bead became increasingly fashionable in the sixteenth and seventeenth centuries. A single outfit might require hundreds of buttons, so it makes sense that such a lucrative and highly

Figure 3.3a Damask of silk and metal thread, sixteenth century, Italy. Metropolitan Museum of Art, New York, 2002.494.127.

Figure 3.3b Block printed linen with crushed mica, c. 1550–99, French? Victoria and Albert Museum, London, 1546-1899.

visible element of dress would spur innovation.[50] While some artisans owned buttons of 'real' gold, silver and silk, many made use of imitations. A grosgrain gown and petticoat, owned by the linen merchant Filippo di Sforzo Guerrieri, had 244 buttons, made of both real and imitation materials. The finest buttons were carefully deployed in the most visible places: 1.5 oz of real gold were used for the buttons on the cloak, while those in the petticoat were made of false gold.[51]

We also find iron, copper and brass buttons in artisan inventories. Cast-metal buttons might have been polished to look like real gold or silver, but others were shaped to imitate the three-dimensional designs of silk-wrapped buttons (Figures 3.4a and 3.4.b). Patterns and techniques

Figure 3.4a Detail of silk and metal thread wound buttons on doublet, c. 1645. Metropolitan Museum of Art, New York, 1982.183.2a-c.

Figure 3.4b Pewter buttons, cast with pattern that imitates wrapped thread patterns, on jerkin, c. 1555–65. Museum of London, 36.237.

were translated across materials as artisans were inspired to imitate desirable products.[52] Imitative designs responded to consumer tastes which associated fashion not just with materials but with shape, colour, sheen and design.

Imitation metals and the anxiety of fraud

Cheaper metals treated to look like gold or silver were used in all kinds of dress, wrapped around threads, and woven into cloth, or hammered and cast into accessories. To take just three cases from over thirty examples in the 'Refashioning the Renaissance' database: the aforementioned mockado-wearing Evandio Honesti owned a doublet striped with red silk and fake silver; a Sienese linen weaver had a doublet of fake cloth of gold; while a Florentine carpenter possessed two copper chains, one of which was gilded.[53] Gold or silver-gilt lace was an incredibly fashionable and expensive trimming, made of thin wire wound around a silk core. It could be imitated by using copper instead of gold wire (Figure 3.5).

Copper substitute was so widely used in theatrical costume that it was associated with acting and pretence. In the satirical comedy *Poetaster*, first staged in 1601, actors are described as 'copper-lac't scoundrels'.[54] Another satire, the advice manual *The Gul's Horne-booke* (1609), even suggests that stylish men sit on the stage where they can 'examine the play-suits' lace, and perhaps win wagers upon laying 'tis copper'.[55] While audiences expect theatrical devices, fictive materials were more threatening offstage.

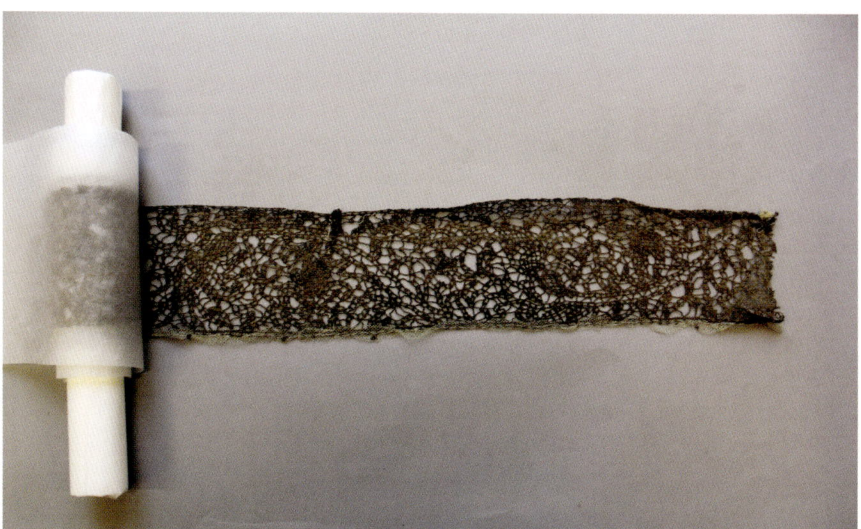

Figure 3.5 Copper lace border, c. 1700. Victoria and Albert Museum, London, T.24-1980.

Contemporaries worried about duplicitous vendors who might fool buyers into confusing fake with 'real'. The process of covering a silver or copper core with a thin layer of gold was described in detail in 1540 by the metallurgist Vannoccio Biringuccio, who deemed it a 'fraud'.[56] Imitations of all kinds were discussed with unease, but valuable gems and gold were particularly concerning. For the German writer Sebastian Franck, writing in 1541, false religion could be best explained through the simile of fake gold: 'a thing is all the more wicked and dangerous the more closely it approximates a real thing and yet is not that thing, as "conterfei" [an alloy] and brass approximate silver and gold'.[57] Consumers were warned to beware fakes: two Dutch travellers cautioned that on a visit to a Parisian second-hand market:

> one sees some very fine things but it is dangerous to buy unless one knows the trade well, for they have marvellous skill in restoring and patching up what is old so that it appears new ... you think you have bought a black coat but when you take it into the daylight, it is green or purple or spotted like leopard-skin.[58]

Already in the twelfth century many guilds forbade their members from working with inferior and fake materials; but, as global trade increased, the provenance of materials became even harder to judge.[59] In order to reassure buyers, guilds marked goods with signs of quality and authenticity such as distinctive selvedges and lead seals.[60] They also carried out workshop searches and punished dishonest workers. In October 1559, for example, the London dyers' guild reported to the City authorities that it had levied fines on 'false coloured cloths being a browne blewe, condemned by Judgement'.[61] In Milan, two men convicted of falsifying gems were shaved and branded in public in 1469, and three years later a Jew in Mantua was arrested and nearly executed for attempting to sell fake gems to local jewellers.[62]

Vendors were also held accountable. When the City of London's Court of Common Council passed an act in 1611 against dyeing black silk in a way that increased its weight and therefore price, the court stated that it would hold not just the dyers responsible but also the silkmen who wove and sold ribbons and lace made with 'silks so deceiptfullie dyed, [which] is so rotted corrupted and spoiled that it is not fit to be worne or used & yet is falslie made faire to the eye to deceave or abuse the buyer & wearer'.[63] But there must have been an economic incentive to make and sell fake goods. In Bruges, for example, buttons, buckles, clothing hooks and rings made of gilded pewter, brass, copper, iron and tin have all been excavated, despite a 1497 law forbidding the gilding of any metals other than silver.[64]

Imitation gems and pearls and the fascination with fakery

Historians of science have argued that early modern experimenters gained knowledge about nature when they artificially imitated metals, stones and other matter. Surviving objects and texts express how their findings shaped their understanding of nature, an 'artisanal epistemology', as Pamela Smith has called it, gained through the manipulation of matter.[65] Two recipes for making amber from the 1595 English translation of Ruscelli's *Secreti*, for example, suggest the process was about more than merely simulating appearance (Experiment in focus VI). One recipe instructs the reader to mix turpentine with cotton; by following the recipe, we see that the cotton gives the turpentine structural integrity and mimics the natural striations found in true amber. Another recommends making a paste of egg yolks, gum arabic and cherry-tree gum, which 'will become hard, and shine like glasse, and when you rub them, they will take up a straw unto them, as other amber stones doe'.[66] Ruscelli pointed out that the stones possessed all the properties of naturally occurring amber – even its triboelectric effect – implying that he regarded them as 'true' synthetic amber and not just visually convincing substitutes.[67] He also did not term these stones 'fakes' but entitled the recipes 'To make cleere stones of Amber'. This perspective may have been widely shared. False amber was presumably not considered negatively by Bernardino Ciampi, a cutlery-maker from Siena whose devotions were assisted with a fake amber rosary.[68]

Many men and women were keen to buy imitations advertised as such, and the fake diamonds, emeralds, rubies, turquoises, pearls and coral we find in the inventories of Italian artisans could be highly prized. For those legally prevented from wearing the real thing, a convincing imitation could enable them to participate in fashions. 'Because they are forbidden to wear pearls,' the artist and writer Cesare Vecellio explained in 1590, immigrant women in Venice 'wear what they call *tondini* [little round balls] of silver or gold, and other jewellery that imitates pearls'.[69] In Venice, fake pearls were so successful that they led to the founding of a new corporation of *supialume*-makers who controlled the manufacture of blown-glass imitation pearls from 1672. In Tallinn, an unusual 1706 regulation listed the sumptuary taxes that allowed people to wear certain materials if they were happy to pay for the privilege. Those who wished to wear real diamonds in a brooch or earrings had to pay 100 daler silvermynt annually, while fake ones were taxed at the lower rate of 50 d sm.[70]

Even those legally and financially able to wear the real thing might find benefits in choosing imitations. In one recipe, Isabella Cortese suggests that a fake pearl might surpass a real one in shape and sheen: 'when you compare it with a real pearl, it will always seem more beautiful to the eye for being more lustrous and more round'.[71] When King James I was crowned at Westminster in 1603, he wore a mix of real and fake gems, including 'Stones lyke topasses ... lyke saphyres ... lyke emaraldes ... and other made stones' along with their genuine counterparts.[72] Such material substitutions were probably a logistical necessity as much as an economic one when rare gems of high quality and appropriate size and shape were costly and challenging to source, especially in large quantities. As Timothy McCall has argued, courtly splendour required such an extravagant display of magnificence that members of the elite regularly relied upon 'material fictions' such as *oricalco* (a brass substitute for gold) or paste gems backed with coloured foils in order to create the required overwhelming effect of shimmering brilliance.[73] Imitation gems could cost almost as much as real ones because they required so much time, effort and expertise to manufacture.[74]

The social expectation of material magnificence was felt even among the middling or artisanal levels of society. Locating low-cost luxuries in non-elite Italian Renaissance homes, Paula Hohti discovered that most imitative products related to the social act of dining, such as jugs and glass goblets painted like gold, brass or bronze.[75] Clothing and accessories were an even more public expression of material abundance and discernment than tableware, and we find over four hundred examples of imitation and low-cost substitute materials in the wardrobes and jewellery boxes of Italian artisans in the 'Refashioning the Renaissance' database (excluding the aforementioned 1,700 mixed material fabrics).[76]

The motivation for imitation

Law-makers struggled with the economic, legal and social implications of imitation materials. The 1551 Spanish law specifically noted that hats could not be adorned with false gold or silver decorations.[77] In Bologna, laws in 1568 banned both real and imitation gold and silver on clothing or accessories ('neither good nor false'); and both high- and low-quality real pearls, along with counterfeit ones, were also expressly forbidden ('neither good nor bad nor false').[78] Such cases suggest that law-makers were often less concerned with material distinctions than with regulating appearances.

Some legislators saw benefits in allowing imitation materials. In October 1634, the Lord Keeper of the Great Seal refused a London Goldsmiths' Company's petition to prohibit counterfeit pearls, stones and gilt or enamelled metals. He believed not only that imitations did not devalue real gems but that fakes also benefited the economy, providing taxes and enabling individuals to fulfil their desire to appear bedizened in 'rich jewels' without bankrupting themselves or the country.[79] Nevertheless, Charles I issued a proclamation in 1636 prohibiting the 'wearing, buying, or selling of Counterfeit Iewels': fake fashions 'exhausted' his subjects' wealth and 'exported' money out of the country in return for things that were materially worthless, 'carrying onely a shew and semblance of Precious stones, Pearles and Jewels'.[80] In this legal controversy, we can see how imitative goods challenged economic policy, social distinction and the culture of appearances.

A culture of imitation

At once desirable and inferior, ingenious and deceptive, frugal and superfluous, imitative goods occupied an uneasy position in the cultural consciousness of early modern Europe. The economic historian Guido Guerzoni has noted the diverse Italian lexicon of imitation, finding the terms *imitante* ('imitating'), *falso* ('false, fake'), *finto* ('sham'), *apparente* ('illusory'), *contrafacto* ('counterfeit') and *camuffo* ('camouflage') in commercial, statutory and juridical sources.[81] The historian Corine Maitte turned to French dictionaries to establish the precise meanings of *imitation*, *copie*, *faux* and *contrefaçon*, but found slippery and indistinct terminology, with words being used synonymously and evoking both positive and negative traits.[82] We must resist the temptation to interpret this language of imitation through twenty-first-century associations with scams, acknowledging instead the central role of imitation in early modern culture generally, and in craft practice in particular.

Apprentices of all crafts gained skills through copying, and imitation was one of the primary means of developing expertise.[83] Giorgio Vasari encouraged artists to imitate nature in their paintings and sculpture, while Leon Battista Alberti stated that the painter who could imitate gold with his pigments merited greater praise than one who used gold leaf.[84] Crafts were improved, revitalised and diversified through the imitation of techniques. The success of the Florentine wool industry, for example, was probably based on the adoption of imported techniques.[85] Economic historians have written about the way that 'import substitution' and imitative processes led to product innovation.[86] The skills developed by Muranese glass-makers in their quest to imitate luxury materials resulted in new 'populuxe' products valued by an emerging urban middle class.[87]

Invention and imitation, therefore, were not diametrically opposed; they were part of the same craft practice. As global markets expanded and travel increased, craftspeople were exposed to novel fabrics, colours, cuts and silhouettes which inspired them to imitate fashions from other cities and countries. While some of the clothes listed in the inventories of the 'Refashioning the Renaissance' database are described as imported from elsewhere (such as 'from Holland', or *d'Olanda*), hundreds of items are referred to as being 'in the style of' foreign fashions. These are recorded by appraisers as generally 'foreign' as well as more specifically in the English, French, Genoese, German, Greek, Lucchese, Moorish, Moroccan, Neapolitan, Persian, Roman, Syrian or Turkish style (see Chapter 7 below).[88] Where the *buratto* gown cut 'in the French style' owned by the Venetian dyer Marin Marini in 1632 was actually made is unknown – but it was associated with French fashion.[89] Other entries explicitly suggest that artisans were imitating foreign products. Thirty-one pairs of leather shoes made 'in the fashionable French style' were found in Santi Biancardi's Sienese shop when he died in 1646.[90]

The imitation and adoption of foreign goods has been understood not only as an act of 'material appreciation' but also as a means of competition and even 'material conquest' of other nations.[91] Among contemporaries there was some anxiety about dressing in foreign styles. The clergyman William Harrison, for example, lamented that in late sixteenth-century England 'we do seem to imitate all nations round about us, wherein we be like to the *polypus* [octopus] or chameleon; and thereunto bestow most cost upon our arses'.[92] Imitations fuelled the fashion industry, and shaped narratives about identity, materiality and corporeality.

Conclusion

Acknowledging the widespread use of imitation materials has implications for the ways we understand surviving archival, visual and material sources as well as our understanding of the culture of appearances. This is important because many scholars rely on images to learn about what was in the wardrobes and on the backs of early modern people, believing that 'the literal truth in dress is what we do see'.[93] But weavers, dyers, cloth finishers, goldsmiths and other craftspeople were adept at making one material imitate the optical effects of another. We cannot be certain that a portrait sitter is really wearing a silk velvet rather than a wool imitation, or that their pearls were sourced from the sea rather than moulded in a workshop.

Archival sources revealing the scale of production and interest in fake materials and reconstructions demonstrating how simulant materials imitated fashionable effects should make us question what we are really seeing in Renaissance portraiture. We also need to interrogate surviving objects, for some early modern imitation materials are so successful that they hide in plain sight, misidentified in museum catalogues. While the use of fake materials at court, on stage and in the streets of Renaissance Europe was an open secret, we must also keep in mind that imitations and fakes might have fooled contemporaries too, further complicating the already challenging task of interpreting the descriptions of items in archival texts such as inventories and letters.[94] More than four hundred items in the 'Refashioning the Renaissance' database are explicitly described as being fake, but many more might have been missed by those who drew up the records. For instance, the appraiser of the Sienese bookseller Nicolò Ormandini's belongings in 1591 noted that 'a plain necklace was said to be fake', suggesting that he had not made the assessment himself but was perhaps informed by a witness.[95]

Early modern fakes made fashions available to a wider range of men and women than ever before – and the material traces and reconstructed examples of these imitations testify to the ingenuity of craftspeople in the sixteenth and seventeenth centuries. Imitation materials used in dress and accessories span a wide range of objects. At one end of the spectrum are poor-quality silk blends dyed with fugitive colours and deceptively sold to ordinary town- and countryfolk; at the other are fake diamonds and lynx furs which impressed early modern princes. By focusing on imitation as a category – or subculture – of early modern fashion, this chapter has brought together sources for the making, wearing and meaning of simulated materials with wider discussions of artifice, craft skill and innovation, and the power of appearance in early modern Europe. In doing so, it represents a first examination of early modern fake fashions as a discrete category, arguing that imitations should be recognised as a key feature of clothing in the period.

Notes

The research leading to these results was done as part of the 'Refashioning the Renaissance' project that has received funding from the European Research Council (ERC) under the European Union's Horizon 2020 research and innovation programme (grant agreement No. 726195).

1 'Et, per mia parte, le presenterà certi zibellini et lupi cervieri, non di quei veri che vengono da altre bande, ma di quei che l'arte immita ne luoghi miei. Et gliene invio non per il valore della cosa, ma per la novità dell'artifitio', Letter of 6 April 1594, ASF, Archivio Mediceo del Principato, 5080, 1202, cited in Gianluca Masi, 'I rapport tra il Granducato di Toscana e il Principato di Transilvania (1540–1699)' (PhD thesis, Ca' Foscari Venezia, 2013), 39.

2 I thank the 'Refashioning' team, in particular Stefania Montemezzo, for identifying, transcribing and making available the archival inventories of artisans used in this chapter, available open access at www.refashioningrenaissance.eu/database.

3 Elspeth M. Veale, *The English Fur Trade in the Later Middle Ages* (London: London Record Society, 2003), British History Online, www.british-history.ac.uk/london-record-soc/vol38/pp101-132, accessed 30 October 2023.

4 Beverly Lemire, *Global Trade and the Transformation of Consumer Cultures: The Material World Remade, c. 1500–1820* (Cambridge: Cambridge University Press, 2018), 46.

5 Girolamo Ruscelli, *De' secreti del reuerendo donno Alessio Piemontese* (Venice: Antonio de gli Antonii, 1563), book II, 36v. On the genre, see William Eamon, *Science and the Secrets of Nature: Books of Secrets in Medieval and Early Modern Culture* (Princeton: Princeton University Press, 1994).

6 See Sophie Pitman,'Experiment in focus I: Imitation of fur' in this volume for a detailed account.

7 Marlise Rijks, 'Gems and counterfeited gems in early modern Antwerp: From workshops to collections', in Michael Bycroft and Sven Dupré (eds), *Gems in the Early Modern World* (Cham: Springer International Publishing, 2019), 309–42.

8 Marta Ajmar, 'The Renaissance in material culture: Material mimesis as force and evidence of globalization', in Tamar Hodos (ed.), *The Routledge Handbook of Archaeology and Globalization* (London: Routledge, 2017), 669–86; Marjolijn Bol and E. C. Spary, *The Matter of Mimesis: Studies of Mimesis and Materials in Nature, Art and Science* (Leiden: Brill, 2023).

9 Ulinka Rublack, *Dressing Up: Cultural Identity in Renaissance Europe* (Oxford: Oxford University Press, 2010), 52–3.

10 Lemire, *Global Trade*, 461.
11 Inventory of Maffio Truscardi, 25 October 1645, ASV, Giudice di Petizion, Inventari, 359, 93, 3r; inventory of Cosmo di Battista, 28 November 1618, ASV, Giudice di Petizion, Inventari, 347, 19, fol. 1v.
12 Baldissara Da Pozzo, 12 November 1580, ASV, Giudice di Petizion, Inventari, 337, 21, fol. 2v.
13 Harold Adams Innis, *The Fur Trade in Canada: An Introduction to Canadian Economic History* (Toronto: University of Toronto Press, 1999), 74–5.
14 Adriano Guagni, 16 August 1638, ASS, Curio del Placito, 280, 68, fol. 32r.
15 David Corner, 'The tyranny of fashion: The case of the felt-hatting trade in the late seventeenth and eighteenth centuries', *Textile History*, 22, 2 (1991), 153–78.
16 For more, see Sophie Pitman, 'The making of clothing and the making of London, 1560–1660' (PhD thesis, University of Cambridge, 2017), 114–19.
17 For example, felt hat 1590–1680 at the Victoria and Albert Museum, No. 752–1893, https://collections.vam.ac.uk/item/O98558/hat-unknown, accessed 22 June 2024.
18 Jonathan Faiers, Fur: A Sensitive History (New Haven: Yale University Press, 2020), 153; Veale, English Fur Trade, www.british-history.ac.uk/london-record-soc/vol38, 22–35, accessed 22 June 2024.
19 Inventory of Antonio Rossati q. Bernardini, 3 September 1555, ASV, Cancelleria inferiore, Miscellanea, 39, 44, fol. 9r.
20 Lisa Monnas, 'A study in contrasts: Silk consumption in Italy and England during the fifteenth and sixteenth centuries', in Dagmar Schäfer, Giorgio Riello and Luca Molà (eds), *Threads of Global Desire: Silk in the Pre-Modern World* (Woodbridge: Boydell Press, 2018), 149–86.
21 See Chapter 2 by Andrea Caracausi above. See also Evelyn Welch and Juliet Claxton, 'Easy innovation in early modern Europe', in Evelyn Welch (ed.), *Fashioning the Early Modern: Dress, Textiles, and Innovation in Europe, 1500–1800* (Oxford: Oxford University Press, 2017), 87–110.
22 Inventory of Jacopo di Bastiano, 1609, ASF, Magistrato dei pupilli, 2716, fol. 271r.
23 See Chapter 1 by John Styles above. See also Alfons K. L. Thijs, 'Les textiles au marché anversois au XVIe siècle', in E. Aerts and J. Munro (eds), *Textiles of the Low Countries in European Economic History* (Proceedings of the Tenth International Economic History Congress, Leuven, August 1990), 76–86.
24 See, for example, Eleri Lynn, *Tudor Textiles* (New Haven: Yale University Press, 2021).
25 Also called moccaiaro, mochaiaro or mochagiaro. John Florio, *Queen Anna's New World of Words* (London: Edward Blount and William Barret, 1611), 317.
26 'Mockado', in Nancy Cox and Karin Dannehl, *Dictionary of Traded Goods and Commodities, 1550–1820* (Wolverhampton: University of Wolverhampton, 2007), at British History Online, www.british-history.ac.uk/no-series/traded-goods-dictionary/1550–1820/mocha-mohair-yarn.
27 Phyllis Ackerman, 'A note on suf or camlet', *Bulletin of the American Institute for Iranian Art and Archaeology*, 5, 3 (1938), 254–6. See also Florence Montgomery, *Textiles in America, 1650–1870* (New York: W. W. Norton, 1984), 295, and Nancy Cox and Karin Dannehl, 'Mocha – mohair yarn', in *Dictionary of Traded Goods and Commodities 1550–1820* (Wolverhampton: University of Wolverhamptton, 2007), British History Online, www.british-history.ac.uk/no-series/traded-goods-dictionary/1550–1820/mocha-mohair-yarn, accessed 23 January 2020.
28 Luc Martin, 'The rise of the New Draperies in Norwich, 1550–1622', in N. B. Harte (ed.), *The New Draperies in the Low Countries and England, 1300–1800* (Oxford: Oxford University Press, 1997), 245–74; Eric Kerridge, *Textile Manufactures in Early Modern England* (Manchester: Manchester University Press, 1985), 68–9; David French, 'A sixteenth century English merchant in Ankara?', *Anatolian Studies*, 22 (1972), 241–7.

29 Similar uses for mockado are found in England: see Martin, 'New Draperies', 252.
30 Evandio Honesti da Castiglione fiorentino, 1601, ASF, Magistrato dei pupilli, 2715, fol. 124r–v.
31 Cited in Eugenia Paulicelli, *Writing Fashion in Early Modern Italy: From Sprezzatura to Satire* (Farnham: Ashgate, 2014), 34.
32 Statuti et ordini della magnifica città di Pistoia sopra il vesire delle donne pubblicati il di diciotto settembre MDLVIII (n.p., 1563).
33 Lorenzo Cantini, *Legislazione toscana: raccolta e illustrata dal dottore Lorenzo Cantini, tomo quinto* (Florence: Albizziniana da S. Maria in Campo per Pietro Fantosini e figlio, 1803), 65–74.
34 Amanda Wunder, 'Spanish fashion and sumptuary legislation from the thirteenth to the eighteenth century', in Giorgio Riello and Ulinka Rublack (eds), *The Right to Dress: Sumptuary Laws in a Global Perspective c. 1200–1800* (Cambridge: Cambridge University Press, 2019), 259.
35 Daniel Roche, *The Culture of Clothing: Dress and Fashion in the 'Ancien Régime'*, trans. Jean Birrell (Cambridge: Cambridge University Press, 1996), 56.
36 *'By the Mayor'* (London: John Day, 1576), 1r–v. Short Title Catalogue 508176. Tufted mockado was probably voided to form designs in the pile.
37 Currency conversion from www.nationalarchives.gov.uk/currency-converter.
38 'Mockado, n.1 and adj.', *Oxford English Dictionary Online*, www.oed.com, accessed 27 February 2022.
39 William Fulke, *Two Treatises Written against the Papists* (London: Thomas Vautrollier, 1577), 287.
40 Sjoukje Colenbrander, 'Woollen velvets and their terminology: Panne, peluche, moquette, tripp and velours d'Utrecht', in Anna Jolly (ed.), *Furnishing Textiles: Studies on Seventeenth- and Eighteenth-Century Interior Decoration* (Riggisberg: Abegg-Stiftung, 2009), 211–22.
41 Letter from John Knyveton to the Earl of Shrewsbury, 6 July 1574, Lambeth Palace Archives, Shrewsbury Papers MS.697, fol. 47.
42 Denis Diderot and Jean le Rond d'Alembert, *L'Encyclopédie: Recueil de planches sur les sciences, les arts libéraux et les arts méchaniques, avec leur explication – Arts de l'habillement* (Paris, 1757), vol. 7, 'Découper et gaufrer', https://gallica.bnf.fr/ark:/12148/bpt6k9978d/f29.item.r=gaufrer, accessed 30 October 2023.
43 Angelo Fantacci, 31 January 1635, Magistrato dei pupilli, ASF, 2719, fol. 145r.
44 See Experiment in focus III; and Sophie Pitman and Paula Hohti (eds), *Remaking Dress History: Applying Reconstruction Methods to Early Modern Textiles and Clothing*, Special Issue, *Textile History*, forthcoming 2024.
45 Elizabeth Currie, 'Ink, print, repeat: A hands-on approach to early printed textiles', Victoria and Albert Museum blog (October 2019), www.vam.ac.uk/blog/museum-life/ink-print-repeat-a-hands-on-approach-to-early-printed-textiles.
46 See Experiment in focus I.
47 Doublet, c. 1635–40, Victorian and Albert Museum, 177–1900. https://collections.vam.ac.uk/item/O77444/doublet-unknown.
48 Inventory of Zuanne q. Battista Morana, 13 April 1619, ASV, Giudice di Petizion, Inventari, 347, 35, fol. 5r.
49 Florio, *New World of Words*, 78.
50 Barbara Bettoni, 'Usefulness, ornamental function and novelty: Debates on quality in button and buckle manufacturing in northern Italy (eighteenth to nineteenth centuries)', in Bert De Munck and Dries Lyna (eds), *Concepts of Value in European Material Culture, 1500–1900* (London: Routledge, 2015), 171–206.
51 Inventory of Filippo di Sforzo Guerrieri, 5 September 1627, ASF, Magistrato dei pupilli, 2718, 2, fol. 17v.

52 Jan Baart, *Opgravingen in Amsterdam: 20 Jaar Stadskernonderzoek* (Haarlem: Fibula-Van Dishoeck, 1977), 184–94; Ajmar, 'The Renaissance in material culture'; Helen Clifford, 'Concepts of invention, identity and imitation in the London and provincial metal-working trades, 1750–1800', *Journal of Design History*, 12, 3 (1999), 250–3; Reed Benhamou, 'Imitation in the decorative arts of the eighteenth century', *Journal of Design History*, 4, 1 (1991), 7–9.
53 Inventories of Evandio Honesti da Castiglione fiorentino, 1601, ASF, Magistrato dei pupilli, 2715, fol. 120v; of Antonio Biagi, 1598, ASS, Curia del Placito, 271, 1171, fol. 121v; Domenico di Dardiano, 7 July 1557, ASF, Magistrato dei pupilli, 2651, 19, fol. 147r.
54 Ben Jonson, *Poetaster; or The Arraignment* (London: M. Lownes, 1602) 3.4.161 in *The Cambridge Edition of the Works of Ben Jonson Online*.
55 Thomas Dekker, *The Guls Horne-booke* (London: R.S., 1609), 29.
56 Vannoccio Biringuccio, *The Pirotechnia of Vannoccio Biringuccio: The Classic Sixteenth-Century Treatise on Metals and Metallurgy*, ed. and trans. Cyril Stanley Smith and Martha Teach Gnudi (New York: Dover Publications, 1990), 379–80.
57 Cited in Sharon Gregory and Sally Anne Hickson (eds), *Inganno – the Art of Deception: Imitation, Reception, and Deceit in Early Modern Art* (Farnham: Ashgate, 2012), 6.
58 Cited in Evelyn Welch, 'The senses in the marketplace: Sensory knowledge in a material world', in Herman Roodenburg (ed.), *A Cultural History of the Senses in the Renaissance* (London: Bloomsbury Academic, 2014), 67.
59 Sven Dupré, 'The art of glassmaking and the nature of stones: The role of imitation in Anselm De Boodt's classification of stones', in Isabella Augart, Maurice Saß and Iris Wenderholm (eds), *Steinformen: Materialität, Qualität, Imitation* (Berlin: De Gruyter, 2019), 209.
60 Lisa Monnas, 'Loom widths and selvedges prescribed by Italian silk weaving statutes, 1265–1512: A preliminary investigation', *Bulletin de Liason du Centre International d'Etude des Textiles Anciens*, 66 (1988), 35–44; Luca Molà, *The Silk Industry of Renaissance Venice* (Baltimore: Johns Hopkins University Press, 2000), 129; Christof Jeggle, 'Copier et contrefaire des textiles au temps de la Renaissance', in Pascale Mounier and Colette Nativel (eds), *Copier et contrefaire à la Renaissance: faux et usage de faux* (Paris: Honoré Champion, 2014), 293–306.
61 Corporation of London book of fines, 1517–1628, London Metropolitan Archives, COL/CHD/CM/10/001, 138r.
62 Timothy McCall, 'Material fictions of luxury in Sforza Milan', in Catherine Kovesi (ed.), *Luxury and the Ethics of Greed in Early Modern Italy* (Turnhout: Brepols, 2018), 246.
63 Acts of the Court of Common Council, 1611, London Metropolitan Archives, COL/CC/01/01/29/01, fols 293v–294v.
64 Isis Sturtewagen, 'All together respectably dressed: Fashion and clothing in Bruges during the fifteenth and sixteenth centuries' (PhD thesis, University of Antwerp, 2016), 105.
65 Pamela Smith, *The Body of the Artisan* (Chicago: University of Chicago Press, 2004), 8.
66 Girolamo Ruscelli, *The Secrets of the Reuerend Maister Alexis of Piemont, Containing Excellent Remedies against Diuerse Diseases, Wounds, and Other Accidents*, trans. William Ward (London: Thomas Wright, 1595), part III, fols 251v–252r.
67 For more on the experimental reconstruction and a video of the triboelectric effect, see Sophie Pitman, 'Una Corona di Ambra Falsa: Imitating amber using early modern recipes', 'Refashioning the Renaissance' blog (April 2020), www.refashioningrenaissance.eu/imitating-amber/.
68 Bernardino Ciampi, 1646, ASS, Curia del Placito, 283, 273, fol. 134v.
69 Cesare Vecellio, *The Clothing of the Renaissance World: Europe, Asia, Africa, the Americas: Cesare Vecellio's Habiti Antichi et Moderni*, ed. and trans. Margaret

F. Rosenthal and Ann Rosalind Jones (London: Thames and Hudson, 2008), 137. On glass pearls, see Barbara Bettoni, *Perle di vetro e gioie false: produzioni e cultura del gioiello non prezioso nell'Italia moderna* (Venice: Marsilio, 2017).

70 Cited in Astrid Pajur, 'Costume and control: Sumptuary laws and social order in seventeenth-century Tallinn' (Master's thesis, Uppsala University, 2014), 44.

71 Isabella Cortese, *I secreti* (Venice: Giacomo Cornetti, 1584), 205. See also Experiment in focus VI.

72 Thanks to Michael Pearce. John Spilman and William Herrick Bill for repairing the Crown Jewels for the Coronation of King James the First of England, June 4 1604, Edinburgh University EU Laing II 525, as reproduced in Historical Manuscripts Commission, *Report on the Laing Manuscripts Preserved in the University of Edinburgh*, Vol. 1 (London: HMSO, 1914), 94.

73 McCall, 'Material fictions', 245.

74 Rijks, 'Gems and counterfeited gems', 309–42.

75 Paula Hohti, 'Cheap magnificence? Imitation and low-cost luxuries in Renaissance Italy', in Kovesi (ed.), *Luxury and the Ethics of Greed*, 283–4.

76 Seven are used in garments, 6 in accessories (hat, fan, sleeves, hair ornament, scarf), 2 in religious objects, 12 as ribbons or laces, 378 in jewellery.

77 Wunder, 'Spanish fashion', 258–9.

78 'ne buono, ne falso' ... 'ne buone ne finte, ò false', cited in Erin J. Campbell, *Old Women and Art in the Early Modern Italian Domestic Interior* (Abingdon: Routledge, 2016), 107, 130–1.

79 *Calendar of State Papers, Domestic Series, of the Reign of Charles I: Addenda – March 1625 to January 1649*, ed. William Douglas Hamilton and Sophia Crawford Lomas (London: HMSO, 1897), British History Online, www.british-history.ac.uk/cal-state-papers/domestic/chas1/addenda/1625-49/pp479-482, accessed 23 October 2023,

80 *By the King: A Proclamation Prohibiting the Wearing, Buying, or Selling of Counterfeit Iewels* (London: John Bill, 1636), STC 9062. Accessed via EEBO.

81 Guido Guerzoni, 'Social emulation, product imitation and technical innovation', paper presented at the Center for Science and Innovation Studies, University of California, Davis, 2 April 2014, 1, available at www.academia.edu/7833070, accessed 23 October 2023.

82 Corine Maitte, 'Imitation, copie, contrefaçon, faux: définitions et pratiques sous l'Ancien Régime', *Entreprises et Histoire*, 78, 1 (2015), 13–26.

83 Maarten Prak and Patrick Wallis (eds), *Apprenticeship in Early Modern Europe* (Cambridge: Cambridge University Press, 2019), 21–4.

84 Angela Cerasuolo, *Literature and Artistic Practice in Sixteenth-Century Italy* (Leiden: Brill, 2017), 136–7.

85 Francesco Ammannati, 'Craft guild legislation and woollen production: The Florentine Arte della Lana in the fifteenth and sixteenth centuries', in Karel Davids and Bert De Munck (eds), *Innovation and Creativity in Late Medieval and Early Modern European Cities* (Farnham: Ashgate, 2014), 67.

86 Maxine Berg, 'From imitation to invention: Creating commodities in eighteenth-century Brtiain', *Economic History Review*, 55, 1 (2002), 1–30; Carlo Marco Belfanti, 'Branding before the brand: Marks, imitations and counterfeits in pre-modern Europe', *Business History*, 60, 8 (2018), 1133.

87 Rachele Scuro, 'Shaping identity through glass in Renaissance Venice', in Susanna Burghartz et al. (eds), *Materialized Identities in Early Modern Culture, 1450–1750* (Amsterdam: Amsterdam University Press), 99–136.

88 *alla*; or *forestiero*.

89 Inventory of the dyer Marin Marini, 20 March 1632, ASV, Giudice di Petizion, Inventari, 353, 101, fol. 4r.

90 'alla francese ... e alla mode', inventory of Santi Biancardi, 4 March 1646, ASS, Curia del Placito, 283, 279, fol. 136v.

91 Anna Grasskamp, 'Unpacking foreign ingenuity: The German conquest of artful objects with "Indian" provenance', in Oosterhoff et al., *Ingenuity*, 205.
92 William Harrison, *The Description of England: The Classic Contemporary Account of Tudor Social Life*, ed. Georges Edelen (Washington, DC, and New York: Folger Shakespeare Library and Dover, 1994), 147.
93 Aileen Ribeiro, *The Gallery of Fashion* (London: National Portrait Gallery, 2000), 11.
94 Giorgio Riello, '"Things seen and unseen": The material culture of early modern inventories and their representation of domestic interiors', in Paula Findlen (ed.), *Early Modern Things: Objects and Their Histories, 1500–1800* (New York: Routledge, 2013), 125–50; Thomas Ertl and Barbara Karl (eds), *Inventories of Textiles – Textiles in Inventories: Studies on Late Medieval and Early Modern Material Culture* (Göttingen: V&R Unipress, 2017).
95 'una collana piana dissero esser falsa', inventory of Nicolò Ormandini, 1591, ASS, Curia del Placito, 266, 558, fol. 2r.

EXPERIMENT IN FOCUS I: Imitation of fur

Sophie Pitman

Introduction

Italians wore fur to keep warm and stylish. Because fine fur was expensive, fur appeared among the possessions of ordinary Italian most commonly in smaller accessories like hats and muffs, or linings and collars of outer garments. Fur was sourced from a wide range of animals, including cat, wolf, rabbit, marten, beaver, fox, ermine, foin (stone marten) and lynx (Figure i.1). As I have shown in Chapter 3 above, while the 'Refashioning the Renaissance' database shows that some artisans could afford smaller accessories made of more expensive marten fur, none of the artisans owned garments of fine ermine fur, let alone exotic leopard or panther fur, apart from the furrier Baldissara Da Pozzo, who had two pieces of leopard fur in his Venetian workshop in 1580.[1]

If any of our artisans owned fake furs, they were not noticed by the inventory appraisers. Yet recipes that promise to provide imitation leopard furs suggest that there was an interest in mimicking rare and pricy furs. Across the social spectrum, we see a desire for fake furs; in 1594, Grand Duke Ferdinando I de' Medici wrote that his agent Cosimo Bottegari was headed to Transylvania with 'some sables and lynxes, not those real ones that come from other parts, but those that Art imitates'.[2] While we do not know how Ferdinand's fakes were made, printed early modern instructions offer possibilities to discover more about these imitations that do not survive in material form.

At a workshop focusing on imitation materials and techniques in early modern clothing in March 2020, organised by the 'Refashioning the Renaissance' project, we explored some of the mimetic methods recommended in early modern recipes books. We wanted to see how imitative effects functioned. Could fake fur provide the effect of a real fur close or at

Figure I.1 Fine furs such as this lynx lining and sable *zibellino* signified fashion and great expense. Paolo Veronese, *Portrait of Countess Livia da Porto Thiene and Her Daughter Deidamia*, 1551. Walters Art Museum, Baltimore, 37.541.

a distance, and was it a convincing substitute? Did recipe books make rare furs, such as leopard, accessible within the reach of an ordinary artisan?

The experiment

These instructions, taken from the 1563 Italian edition of the popular recipe book *De secreti* by the pseudonymous Alessio Piemontese, promise to transform a humble white pelt by tinting it black to appear like leopard or panther fur:

> Take one ounce litharge of silver [white lead or the by-product of separating silver from lead], two ounces of quicklime, and in three ladles of water put on the fire in a new pot so that it gets warm, then take it from the fire and with a wooden stick mix it; then take a brush and tint the white hide as it seems to you, one spot here and the other there, and according to the material make them thick, then dry it in the sun and when it's well dried, hit it with a rod and you will see dark spots tawny in colour. And if it is not well coloured in this way, you could tint it another time, giving the strikes where you did the first time, and the colour will become stronger and in this way you will have your intent. And this colour is always maintained and gives a good odour; and also putting the said material on hair or a beard will make it become roan and beautiful.[3]

At first glance, the recipe seems unlikely: white lead ((PbCO3)2·Pb(OH)2), and quicklime (CaO) are not typical dyestuffs. But rather than dismiss these instructions out of hand, we decided to follow the recipe step-by-step. As lead is toxic, and when the caustic quicklime is added to water it forms slaked lime and creates a rapid exothermic reaction, a safety protocol was required. We used the Biofilia biological art laboratory at Aalto University, a wet bench research space equipped with fume hoods, and wore PPE including lab coats, masks, heatproof gloves and goggles. Parsing the recipe into modern parlance, with reference to Material Safety Data Sheets, and having a clear waste management plan, were crucial, to ensure the safety of all participants. The modernised interpretation of the recipe, which I reinterpreted with the guidance of biotechnician James Evans, looked like this:

Materials and tools: White Lead, Quicklime, Deionised Water, Metal Pan / Glass beaker, Heat Source, Stirring Stick, Hitting Stick, Brush, White Fur or Hide, Paper to lay down in fume hood

Process:
1. Weigh all ingredients and prepare all tools
2. Take 3 scoops of water and place in a metal pan or glass beaker, slowly add 2oz quicklime, and 1oz white lead
3. Heat on a hot plate until warm
4. Remove from heat and stir
5. Using a brush, tint the white hide with spots using the mixture
6. Leave to dry
7. Once dried, hit with a stick (work in fume hood)
8. Repeat if necessary

As the recipe does not specify a white fur, we used two kinds – rabbit and sheep's wool – to compare the effect on differently textured pelts. Having measured out the ingredients, we slowly added the quicklime to the water and used a magnetic stirrer to reduce bodily contact with ingredients (Figure i.2). We observed a temperature rise, then added lead, which transformed the mixture into a milky pink solution. We painted spots on to the two furs, which dried into pale hard lumps, and left overnight.

As we returned to the lab the following morning, we assumed the experiment had failed. The spots had not turned brown or black, and we wondered what we had done wrong. We then revisited the text and noticed the instruction to 'give strikes' to the spots. Upon hitting, to our surprise, the spots crumbled away to reveal brown hair beneath (Figure i.3).

Under a digital microscope, and with reference to research, it was clear that this is no ordinary dye recipe where colour is bound to the surface of a fibre, but rather a reaction that occurs within the core of hair fibres (Figure i.4).[4] The blackening occurs because quicklime creates an alkaline environment and weakens the hair fibre, allowing the lead to bind with sulphur in the amino acids of the shaft, creating galena (PbS) crystals, which are black. The hard surface blotches are likely excess solution and

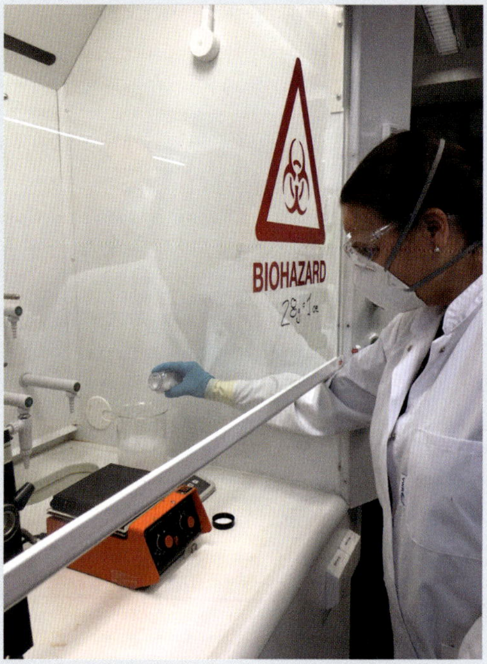

Figure I.2 Paula Hohti adds quicklime to the water at Aalto University biolab.

Figure I.3 Sheep and rabbit fur treated with the solution after beating (a few spots have been left with dried pink solution for comparison).

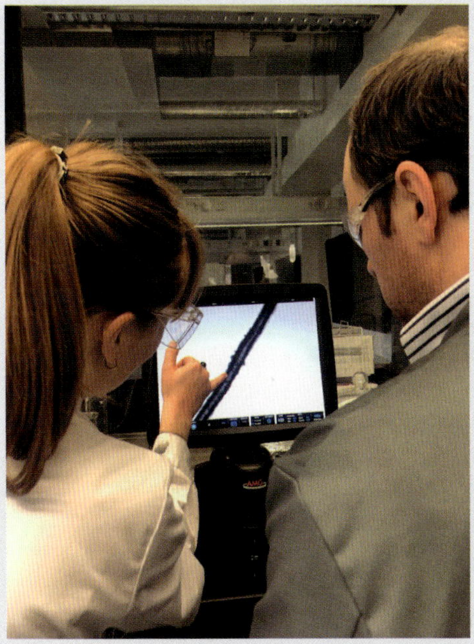

Figure I.4 Sophie Pitman and James Evans examine the formation of galena crystals in the hair shafts using a microscope.

waste salts from the reaction which need to be removed from the furs to reveal the transformation within the core of the hair.

The recipe was a success – the spots became dark and created a leopard-like pattern, but the shafts of hair become brittle and weak. Either this recipe requires a shorter application time and weaker solution, and therefore an experienced dyer, or the owner must sacrifice material quality for visual effects.

Piemontese's suggestion that this technique can be used to create imitation leopard or panther fur is a novel one, but similar recipes using quicklime and lead to create a 'very beautiful black' dye for wool and hair can be found in early modern manuscript and printed collections, including Gioanventura Rosetti's *Plichto* (Venice, 1548) and the Toulousian BnF Ms. Fr. 640 (c. 1580) examined by the 'Making and Knowing' project.[5] The Bolognese surgeon Leonardo Fioravanti even reasoned that the transformation worked because litharge is a species of lead, and lead 'by nature' (*di sua natura*) makes black. The lye, he claimed, dissolves it (*lo solve*).[6]

Conclusion

Piemontese's recipe might only represent one contemporary technique for fake fur, but through its reconstruction we gain a sense of a successful visual effect offered by a fake fur, albeit one that fails to imitate the soft fine furs of real leopard or panther. The experiment suggests how non-elite early modern men and women might have been able to participate in the fine fashions of exotic furs at less expense and reminds us that there was a widespread interest in manipulating materials to mimic rarer, finer or costlier ones.

The fake fur experiment was iterative, building upon the research and practical experimentation of the 'Making and Knowing' project, and nuancing findings based on the 'Refashioning the Renaissance' project archival research about early modern artisanal clothing. It demonstrates the dual reward in reconstructing recipes. On the one hand, material findings can enrich our understanding of visual and archival sources, by giving a material presence to objects that no longer survive, showing how they were made and operated in multi-sensory ways. In addition, the reconstruction of recipes also encourages us to interrogate the text itself. It is a means of close reading that demands we pursue in-depth research about the materials, processes, workshop conditions, artisanal skills and tacit knowledge as well as the social, economic and cultural context of making processes and finished objects.

Notes

The 'Refashioning the Renaissance' workshop 'Imitation in Early Modern Clothes and Accessories', organised by Sophie Pitman at Aalto University on 10–12 March 2020, was funded by the European Research Council (ERC) under the European Union's Horizon 2020 research and innovation programme (grant agreement No 726195). See www.refashioningrenaissance.eu/experiments/imitation. We thank our collaborator Tim McCall and the 'Refashioning the Renaissance' team members for participation, presentations and discussions.

1 Baldissara Da Pozzo, 12 November 1580, Archivio di Stato di Venezia, Giudice di Petizion, Inventari, 337, 21, fol. 2v.
2 'certi zibellini et lupi cervieri, non di quei veri che vengono da altre bande, ma di quei che l'arte immita ne luoghi miei', Letter of 6 April 1594, State archives of Florence (ASF), Archivio Mediceo del Principato, 5080, c. 1202. For more examples, see Pitman, Chapter 3 above.
3 'A far una pelle bianca con macchie negre in color di Leopardo ò di Pantera, e i capelli bianchi negri' in Girolamo Ruscelli, *De' Secreti del reverendo Donno Alessio Piemontese* (Venice: Antonio de gli Antonii, 1559), Book II, 36v.
4 Sophie Pitman, 'Black color for dyeing, and the place of textiles in ms. fr. 640', in Pamela H. Smith et al. (eds), *Secrets of Craft and Nature in Renaissance France: A Digital Critical Edition and English Translation of BnF Ms. Fr. 640* (New York: Making and Knowing Project, 2020), available online at edition640.makingandknowing.org/#/essays/ann_036_sp_16.
5 Pitman, 'Black color for dyeing'; Gioanventura Rosetti, *The Plictho: Instructions in the Art of the Dyers Which Teaches the Dyeing of Woolen Cloths, Linens, Cottons, and Silk by the Great Art as Well as by the Common*, ed. Sidney M Edelstein and Hector C Borghetty, translation of the first edition of 1548 (Cambridge, MA: MIT Press, 1969), 102.
6 Leonardo Fioravanti, *Del compendio de i secreti rationali: libri 5* (Venice: Vincenzo Valgrisi, 1564), fols 124v–125r.

EXPERIMENT IN FOCUS II: Knitted stockings

Piia Lempiäinen and Paula Hohti

Introduction

In the 1920s, during the restoration of Turku Cathedral, several seventeenth- and eighteenth-century coffins were opened, and well-preserved burial clothes belonging to local burgher and noble families were placed in the storage of the Turku Cathedral Museum. Among the items, the conservators found a pair of knitted silk stockings. These were discovered from a storage coffin that had once belonged to Elisabeth Bure (d. 1668), the wife of the Vice President of the Court of Appeals.[1] The stockings were catalogued as male stockings, probably because of their length, and dated to 1650. Their present-day colour is greyish brown, there is a narrow welt at the top of the stocking, a false zig-zag seam in the back, a decorative clock in the ankle and a zig-zag decoration at the edges of the heel gusset (Figure ii.1).

Knitted silk stockings, such as the surviving pair found in Turku, became a desired and widespread product and one of the key fashion innovations among European elites during the early modern period. As Andrea Caracausi and other historians have shown, by the end of the sixteenth century, extensive networks of makers and sellers produced and sold a wide range of knitted garments in a broad spectrum of qualities.[2]

Archival records confirm the growing presence of knitted stockings in the early modern period, not just among the privileged rich but also at the lower ranks of society. Family probate inventories from the sixteenth and seventeenth centuries recorded knitted socks and stockings in a range of qualities and textures and wide variety of designs and colours so that even common artisans, such as barbers, shoemakers and innkeepers, might eventually appear in stockings of silk A Milanese wool merchant complained that, by the late seventeenth century, 'it seems that even people of

Figure II.1 Pair of seventeenth-century hand-knitted silk stockings, with a false zig-zag seam in the back and a decorative clock, discovered in a storage coffin that had once belonged to Elisabeth Bure (d. 1668). Turku Cathedral Museum, Finland.

the lowest grade, carried by their ambition, take shame in using stockings of stamen or ordinary wool, using instead those made of *filusello* [spun waste silk thread] or silk made on frame'.[3]

The merchant's statement is confirmed by the 'Refashioning' data of artisan clothing. The database of Italian artisan fashions from 1550 to 1650 includes several pairs of stockings made of *filusello* and *bavella*, waste silks typically used for lower-grade knitted silk stockings.[4] One Venetian papermaker, Andrea, for example, who died in Venice in 1611, owned a dashing pair of yellow silk stockings of silk *bavella*, dyed in golden yellow.[5]

But how were the colourful knitted stockings that were described in the documents and depicted in contemporary visual images made, what did they look and feel like, and what level of skill and sophistication was required to knit different types of silk and wool stockings?

In order to understand the materials, techniques, visual look and sensory qualities of early modern knitting, the 'Refashioning the Renaissance' project designed a participatory research experiment in order to reconstruct knitted socks and stockings.[6] Carried out in 2019–22, this project based on citizen science had two aims: first, to generate new knowledge on the materiality and techniques of these popular fashion accessories, and second, to develop a collaborative framework for historical research by engaging the general public – in this case volunteer knitting experts – and museum professionals, in scientific research. To guarantee the success of a project involving volunteers, the project had to be planned well both scientifically and ethically.[7]

The experiment

The project began in 2019, when we placed a call on our website and social media inviting volunteers and collaborators to join our historical participatory research experiment. The call explained the scope of the project and the required skills, and we presented a detailed plan of the project for the volunteer participants during the first meeting. It was important to make sure that all volunteers understood the aims of the project and what they could expect from the collaboration, and that they had the agency to decide how much time and energy they wanted to contribute to the project. This included accepting the possibility that some volunteers might drop out during the project.

Since the early modern stocking industry produced a wide range of stockings of different qualities, we decided to reconstruct three different types of stockings: an artisanal wool sock based on archaeological evidence, a stocking based on the first known knitting instruction *The Order How to Knit a Hose* (1655), and a fine silk stocking replicating an extant seventeenth-century silk stocking at Turku Cathedral Museum.[8] The decision to focus on a range of stockings, each of which involved a different skill set and making process, served two purposes. First, it made it possible to bring to life the variety of stockings that were listed in our data of artisan fashions and compare their visual look, sensory qualities and how they were made. Second, the range of experiments allowed the voluntary participants to select a project that suited their skill level and felt meaningful to them.

This meant that, at every stage and action of the project, we had a double aim: to gain new scientific knowledge and to engage and empower the volunteers by inviting them to share their craft expertise. The volunteer knitters participated closely in planning, from designing the workplan to communicating the results, and we kept in regular touch with them both at a group and at an individual level to share knowledge, updates and results.[9]

The process of reconstruction

We started the project with thirty-five volunteer participants. At the beginning of the project, we provided all volunteers with appropriate knitting materials and tools and organised a series of skill-building sessions for the participants where we explored and discussed the cultural meanings of early modern stockings, features of extant historical stockings and technical aspects of early modern knitting. We also explored methods of notetaking and documentation in experimental research and gave the volunteers notetaking templates, advising them to record the knitting process step-by-step, why they made specific decisions, how much time it took them to knit, and what failed or succeeded. We also encouraged the volunteer knitters to take notes on other aspects that contributed to the knitting experience, such as the space they worked in, their experiences and feelings, how hard it was to concentrate, the similarities and differences between modern and historical knitting, the tactile experiences and the effect of the weather or time of day on their knitting. In addition to descriptive data, the notetaking template included smiley-face Likert scales which volunteers could use to record how difficult they found casting on, knitting round, the clock pattern and the heel.[10]

During the training sessions, the volunteers tried different yarns and needles and decided which project out of the three options they wanted to pursue. The first option, the recreation of a simple woollen artisanal stocking, was based on surviving material evidence from early modern Denmark and involved a level of the knitters' own creativity and problem-solving skills.[11] The aim of this experiment was to study the process of making a simple artisanal woollen stocking, and to consider role of creativity and personal preferences in the knitting process at a time when simple knitted artisanal socks both for domestic consumption and for the commercial market were typically made in rural homes without instructions to follow.

The second experiment, the reconstruction of a stocking based on the first known knitting pattern *The Order How to Knit a Hose* (1655), was designed to explore how early modern recipes can be used in reconstruction, how demanding was it was to follow the instructions and what kind of prior knowledge the knitter was expected to have to complete the task.

The final and the most challenging experiment, the replication of the fine seventeenth-century silk stocking from Turku Cathedral Museum, was the most complex of the experiments. We have chosen to describe this process here in more detail.

The reconstruction of the silk stocking

The replication experiment of the silk stocking began with a close technical examination of the original stocking at the Turku Cathedral Museum. The stocking was measured and photographed and, together with the conservator, we were allowed to take a tiny sample of the yarn from one of the unravelled edges of the stocking to carry out analysis of the fibres and dyes.[12] The scientific tests, carried out a few months later, suggested that this was a stocking of a fine calibre. The fibre analysis, conducted with SEM microscope and EDS analysis, demonstrated that the silk thread used for the stocking was made of the finest silk yarn available at this date, prepared from long and very fine slightly twisted Bombyx mori silk filaments (Figure ii.2). The analysis of the organic colourants showed that the stocking had been originally black, the colour of high fashion of the period, dyed using sticky alder bark and possibly mixed with gallnuts.[13]

In parallel with the scientific testing, a small group of experienced knitters began drafting the pattern. The group decided to split in three and each work on a different part of the stocking: the opening and the back seam, the clock and the foot. They relentlessly counted the stitches of the original stocking, knitted test swatches and tried out different techniques to figure out the pattern (Figure ii.3). Their close reading of the stocking revealed important preliminary information about its construction. The stocking was knitted in round from the top down, and all decorative stitches were made with purl stitches. The group also noticed that there is a change in the pattern on the back seam at the height of the clock, indicating that a more experienced knitter might have been responsible for the decorative clock, while the repetitive stitches and rounds were possibly made by their assistant or apprentice. The knitting also included mistakes, suggesting that, although this was a luxurious product, mistakes and slight asymmetry were considered acceptable or even inevitable.

While the pattern was being made by the knitting experts, we began to look for a yarn that would resemble the original Bombyx mori silk as

Figure II.2 SEM microscopic image of the Bombyx mori silk fibres. Aalto University. The sample is taken from a seventeenth-century knitted silk stocking, conserved at Turku Cathedral Museum, Finland.

Figure II.3 Volunteer knitters studied the high-resolution images of the stockings found from the Turku Cathedral and counted every stitch to create the knitting pattern for the reconstruction. Aalto University, Helsinki, 2020.

closely as possible. This was important because our aim was not only to replicate the size, shape and texture of the stocking but also to gain access to the visual and sensory properties of such a fine early modern silk stocking.

One of the immediate difficulties was that Bombyx mori silk is not easily available commercially, because the industry was almost completely destroyed in the course of the nineteenth century, due to the Pebrine epidemic that made infected silkworms unable to produce silk thread. In 2019, however, we were fortunate to discover a small co-operative in Calabria, Nido di Seta, which had recently taken an initiative to revive the traditional Bombyx mori silkworm breeding. After visiting the silk farm to study the processes of silk making, we decided to commission yarn samples of varying thickness from the co-operative (Figures ii.4a, ii4b).[14]

The volunteers knitted several small swatches, comparing hand- and machine-spun yarns, both to examine the differences and to find out whether it was possible to achieve a similar gauge and appearance as in the original using the Nido di Seta hand-spun yarn. The hand-spun and commercial yarns felt very different. The machine-spun yarns had more twist and the dried sericin made the yarns very stiff, while sericin in the freshly reeled hand-spun silk yarn was much softer, requiring a shorter degumming time. Yet, after the sericin had been removed from the silks during the degumming, we noted that the hand-spun swatches had shrunk much more than the machine-spun samples.

Figure II.4a Bombyx mori silkworms and cocoons at the co-operative Nido di seta, Calabria, Italy.

Figure II.4b Silk cocoons being processed by the 'Refashioning' team at Nido di seta, Calabria, Italy.

After several test batches, we selected the yarn that had been made of long silk filaments using 150 cocoons from the 2019 harvest and given a slight twist to reinforce the thread. Unfortunately, the rainy autumn in Calabria prevented the yarns from drying properly, and the sericin glued some of the strands of yarns together into clumpy skeins, which our volunteers lovingly called 'the tangle yarn'. This yarn was quite stiff and scrunchy, but, once the clumps were opened by reeling the skeins into cones and balls, it was surprisingly easy to knit. This experience made us fully aware of the extent to which weather conditions affected the quality of the silk yarn. Perhaps the ability to work with different qualities of silk thread had been a valuable skill for an early modern knitter.

The volunteers were also given the option to knit the Turku silk stocking using thin wool which matched the gauge of the silk. This allowed us to compare the differences in the process of knitting the same stocking with silk and wool, and to examine differences in the visual and sensory qualities of the finished products. We know from early modern documentary evidence that decorative and fine stockings, such as that from Turku, were made of wool as well as of silk, using the same pattern.[15] Fine worsted wool stockings might have provided a visually attractive alternative to silk stockings for artisan families and others who could not afford silk.

Some of our volunteers, who were uncertain of their skills, chose the more familiar wool yarn. But those who tried both silk and wool reported that they found silk easier to knit, because it was easier to distinguish individual tiny stitches of silk yarn and pick up the dropped stitched with the very thin 1mm needles compared to the fuzzy wool. Yet, whether using wool or sericin-coated silk, working with very fine needles was a different knitting experience from what the volunteers had got used to. Almost everyone found it challenging to cast the stitches, but after a while they found a flowing rhythm and knitting became easier. The clock pattern on the ankle as well as the heel offered variation and satisfaction for the challenge-seeking knitters, although many knitters noted that their purl stitches were looser than the knitted stitches, which made the stocking looser around the ankle. This, however, seems to be a feature in the original stocking too (Figure ii.5).

The experiment showed that making hand-knitted stockings was very time-consuming. The replications of the Turku stocking in silk, made exactly to the measure of the original stocking on 1mm needles, took between 215 and 260 hours to make per stocking, making it up to 520 hours per pair.[16] The same stockings could be made slightly faster using wool, but the volunteers reported that making fine stockings of wool was only slightly less laborious than working with silk. Two stockings reconstructed in fine wool instead of silk took between 42 and 135 hours respectively to knit. Even considering that the early modern professional knitters were probably able to knit faster than our knitters, all of whom were

Figure II.5 Reconstruction of seventeenth-century knitted stockings in progress, in linen, wool and silk. Made by volunteer participants at 'Refashioning' citizen science knitting project. Aalto University, Helsinki, 2020.

experienced and highly skilled knitters but not professionals or used to knitting fine silk or wool and historical patterns, it is likely that even a skilled hand knitter in the early modern period could not have been able to produce many pairs of fine decorative stockings in a month.

Having worked over two years with our volunteer knitters, it was very exciting to see and feel completed stockings and to try them on. While the wool stockings felt immediately comfortable on the skin, the silk stockings looked initially stiff and dull. However, after the stockings had been degummed, dyed using natural dyes and historical recipes, and stretched on a wooden sock block, the silk stockings transformed into lustrous items that had a beautiful sheen. The biggest surprise was the unprecedented and incomparable bodily sensation of comfort that we all felt upon trying on the degummed silk stocking. The smooth, cool and soft sensation of the stockings next to the skin left us speechless (Figures ii.6a and ii.6b).

While we know that fine hand-knitted silk stockings were considered a luxury product in the early modern period, our experiment left us wondering how early modern consumers might have experienced the sensation of luxury.[17] Already back then contemporaries pondered about the relationship between suffering and luxury, and we know that many early modern stockings were made by poor men, women and children in harsh conditions.[18] For us, the experience of wearing a stocking carefully made by a skilled maker who finds joy in the process was quite different.

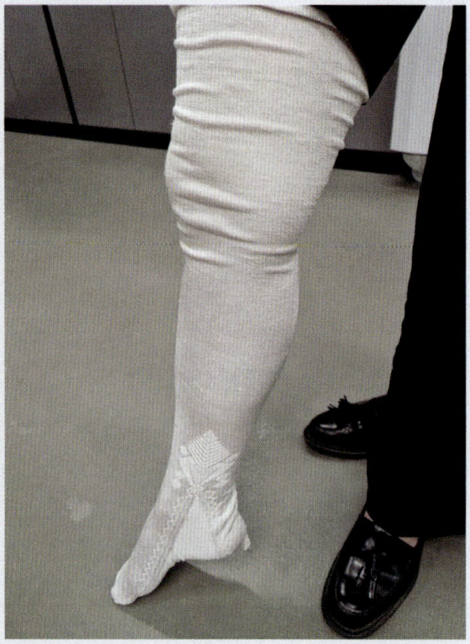

Figure II.6a Reconstruction of an extant hand-knitted silk stocking conserved at Turku Cathedral, Finland, being tried on. The stocking fitted a surprising range of feet sizes and the beautiful clock pattern came to life when the stocking was worn compared to lying flat.

Figure II.6b Reconstruction of an extant hand-knitted silk stocking conserved at Turku Cathedral, Finland. Made by Liisa Kylmänen and dyed with fustic by the 'Refashioning the Renaissance' project.

Conclusion

What did we learn from the experiments of making different types of early modern hand-knitted stockings? While working with craft experts allowed us to gain a profound understanding of the materials, tools, craft skills and the value of labour associated with making different stockings, the experiments also provided an explanation why the finest knitted stockings, such as the silk stocking from Turku, were still largely beyond the means of non-elite families. Their stunning visual look and the sensory qualities, together with the materials and the time invested in making them, made the stockings knitted from the finest silk an exceptional elite product worn mainly by men and women from the nobility or wealthy burgher class. Those further down the social scale had to be content with ordinary socks or stockings made of waste silk threads or wool, at least until machine-knitted stockings made stockings cheaper and more widely available.[19]

At the same time, as we generated new historical knowledge, this project allowed us to explore how to support new, open and socially responsible ways of doing and promoting research. Our experiment offered not only us but also the volunteers a chance to learn about history and to engage with academic researchers through conversations, trainings and workshop sessions. They expressed joy and pride over participating in historical research and having their skills recognised and valued. Many knitters enjoyed the social and collaborative aspect of the project and one even shared how participating in the project gave her solace after a tragedy in her family.

By involving craft experts in our research and creating a collaborative research framework that was designed to meet both the scientific aims and the interest of the volunteers, our project shows how historians can connect with new communities in order to find new ways to understand historical objects, materials and techniques associated with early modern fashion.

Notes

The ‘Knitting reconstruction project’, designed by the ‘Refashioning the Renaissance project’ and co-ordinated by Piia Lempiäinen, was funded by the European Research Council (ERC) under the European Union’s Horizon 2020 research and innovation programme (grant agreement No. 726195). See www.refashioningrenaissance.eu/historical-reconstructions/knitted-stockings. We thank most warmly our 35 voluntary knitters and project administrator Lena Kingelin, our postdoctoral researcher Sophie Pitman and the many museum and academic collaborators who gave advice and provided support on the research and implementation of the project, including Maj Ringgaard, Christel Brandenburgh, Elina Ovaska, Päivi Allinniemi, Miraim Pugliese and Krista Wright.

1 Riitta Pylkkänen, *1600-luvun kuolinpukuja Turun tuomiokirkkomuseossa* (Turku: Turun kaupungin Historiallinen Museo, 1955), 17–18.

2 Andrea Caracausi, 'Beaten children and women's work in early modern Italy', *Past & Present*, 222, 1 (2014), 95–128; Joan Thirsk, 'Knitting and knitwear, c. 1500–1780', in David Jenkins (ed.), *The Cambridge History of Western Textiles* (Cambridge: Cambridge University Press, 2003), 562–66; Carlo Marco Belfanti, 'Fashion and innovation: The origins of the Italian hosiery industry in the sixteenth and seventeenth centuries', *Textile History*, 27, 2 (1996), 132–47; Carlo Marco Belfanti, *Calze e maglie: moda e innovazione nell'industria della maglieria del Rinascimento a oggi* (Mantua: Tre Lune Edizioni, 2005); Riitta Pylkkänen, *Barokin pukumuoti Suomessa 1620–1720* (Helsinki: Suomen Muinaismuistoyhdistys, 1970), 373.

3 Belfanti, *Calze e maglie*, 45. Some scholars have argued that the new fashion captured rich and poor alike, see, for example, Thirsk, 'Knitting and knitwear', 566; and Joan Thirsk, 'The fantastical folly of fashion: The English stocking knitting industry, 1500–1700', in Negley B. Harte and Kenneth B. Ponting (eds), *Textile History and Economic History: Essays in Honour of Miss Julia de Lacy Mann* (Manchester: Manchester University Press, 1973), 51.

4 Of 1,477 pairs 270 were made of silk, including 45 made of filusello or bavella. All but nine pairs of all silk stockings were recorded after 1590s, see 'Refashioning the Renaissance' database at www.refashioningrenaissance.eu/database.

5 'Un paro di calce di bavelle dorete usade', inventory of the paper-maker Andrea, 26 May 1611, State archives of Venice (ASV), Giudice di Petizion, *Inventari*, 55, fol. 3v. The most common colour among ordinary Italians' stockings was black, then white and red, but the colour range becomes much wider in the late sixteenth century.

6 For a methodological and scientific analysis of the project, see Paula Hohti, 'Knitting history through reconstruction: The making and meaning of early modern stockings', in Sophie Pitman and Paula Hohti (eds), *Remaking Dress and Textile History: Applying Reconstruction Methods to Early Modern Textiles and Clothing*, Special Issue, *Textile History*, forthcoming, 2024.

7 Experimental archaeology projects on early modern knitting have provided an inspiring model and point of comparison for our work, in particular 'The Texel Stocking Project', led by Chrystel Brandenburgh in Textile Research Centre Leiden, and Jane Malcolm-Davis's citizen science initiatives within her KEME project. See Chrystel Brandenburgh, 'Een paar zijden kousen', in Arent D. Vos, Birgit van den Hoven and Iris Toussaint (eds), *Wereldvondsten uit een Hollands schip: Basisrapportage BZN17/Palmhoutwrak* (Leiden: Provincie Noord-Holland, 2019); Jane Malcolm-Davis, 'An early modern mystery: A pilot study of knitting, napping and capping', *Archaeological Textile Review*, 58 (2016), 57–74; and 'Knitting in early modern Europe', https://kemeresearch.com, accessed 8 May 2023. For ethical guidelines, see for example ECSA (European Citizen Science Association), *Ten Principles of Citizen Science* (Berlin, 2015), https://osf.io/xpr2n, accessed 8 May 2023; and Amy Twigger Holroyd and Emma Shercliff, *Stitching Together: Good Practice Guidelines: Advice for Facilitators of Participatory Textile Making Workshops and Projects* (Bournemouth: Stitching Together, 2020). Published in Philiatros, Natura Exenterata: Or Nature Unbowelled by the Most Exquisite Anatomies of Her (London: H. Twiford, 1655), 417–19.

8 Published in Philiatros, *Natura Exenterata*, 417–19.

9 See 'Renaissance Knitting Project', at www.ravelry.com, accessed 8 May 2023.

10 We thank Sophie Pitman for sharing her experience of taking field notes.

11 We would like to extend our gratitude to Maj Ringgaard, conservator at the National Museum of Denmark and a specialist in early modern knitting, who advised us and led a special training session on early modern knitted stockings for our knitters.

See also Maj Ringgaard, 'Hosekoner og sålede strixstrømper', *Dragtjournalen*, 8, 11 (2014); Lesley O'Connell Edwards, 'Knitted wool stockings in the Museum of London: A study of 16th century construction', *Archaeological Textiles Review*, 60 (2018), 42–50; and Sylvie Odstrčilová, 'Early modern stockings in museums in the Czech Republic', *Archaeological Textiles Review*, 60 (2018), 51–63.

12 We would like to thank Intendant Elina Ovaska and Conservator of Church Textiles Päivi Allinniemi for all their help.

13 The fibre analysis was conducted at the Aalto Nanomicroscopy Center by Krista Vajanto and the chromatographic analysis by Art Proaño Gaibor at Cultural Heritage Agency of the Netherlands. Proaño Gaibor's analysis shows that the chemical elements found in the dye analysis indicate that alder bark might have been used to produce iron acetate through fermentation in the so called 'lasagne method'.

14 See Nido di Seta, www.nidodiseta.com, accessed 8 May 2023. For more information about Bombyx mori silk and the use and production of silk yarn and floss silk in the early modern period, see Luca Molà, *The Silk industry of Renaissance Venice* (Baltimore: Johns Hopkins University Press, 2000), chapters 9–10.

15 There are surviving examples where the clock of the woollen under stocking is very similar to that of the silk stockings worn over it. Riitta Pylkkänen, *Säätyläispuku Suomessa Vanhalla Vaasa-ajalla 1550–1620* (Helsinki: Suomen Muinaismuistoyhdistys, 1956), 262–3.

16 This is in line with the experiences of the 'Texel Stocking Project'. See Brandenburgh, 'Een paar zijden kousen'.

17 Scholars working on reconstruction, such as Serana Dyer, Jenny Tiramani and Sarah Bendall among others, have emphasised that we cannot today 'accurately capture all the nuances of sensations or ideas of comfort in the same way as people five hundred years ago'. See for example Sarah Bendall, *Shaping Femininity: Foundation Garments, the Body and Women in Early Modern England* (London: Bloomsbury Publishing, 2021), 15.

18 One contemporary pondered whether pearls were so popular 'because they are brought from another world' or whether it was 'because they cost the lives of men'. See Molly A. Warsh, *American Baroque: Pearls and the Nature of Empire, 1492–1700* (Chapel Hill: University of North Carolina Press, 2018), 81. We thank Michele Robinson for this reference.

19 For further analysis, see Hohti, 'Knitting history through reconstruction'.

EXPERIMENT IN FOCUS III: Stamped mock-velvet doublet

Sophie Pitman

Introduction

Very few garments survive that represent and fully convey the fashionable achievements of Renaissance artisans. Worn in harsh working or living conditions, repurposed into new outfits or household rags, and even sold to make paper pulp, the vast majority of items of clothing from the past no longer survive in material form. Survival bias has privileged the extraordinary and elite above the ordinary and everyday. Yet, it is clear from the visual records and archival data of close to a hundred thousand records of artisan fashions from sixteenth- and seventeenth century post-mortem inventories from Florence, Siena and Venice, gathered by the 'Refashioning the Renaissance' project, that 'middling sorts' consumed clothing, textiles and other household goods in order to participate in fashionable Renaissance culture. Could we bring some of this data to life and reconstruct a garment, by hand, using historically appropriate raw materials and methods, that would represent and spur on further research about artisan fashions?

In order to fill in the gap in the material record, and understand some of the garments that were used, owned, and worn by the artisans uncovered during archival research, the 'Refashioning the Renaissance' project initiated a major reconstruction project: the reconstruction of a seventeenth-century male doublet from fibre to finished garment (Figure iii.1). Based on rigorous archival and visual research, scientific testing and close object analysis, the doublet was created in material form by skilled craftspeople and researchers using natural and historically appropriate materials, and in the digital realm using 3D animation (See Experiment in focus v).[1]

Our approach was inspired by prior successful reconstructions of early modern dress, but these tend to begin from a single rich textual, visual or material source that details dress of the elites.[2] Given the very different

Figure III.1 The finished doublet, mounted with sleeves, on display at the *Reconstructing Everyday Fashion* Exhibition at Aalto University, September 2021.

sources available about everyday artisan dress, we were challenged to imaginatively combine information spread thinly across a wide range of textual, visual and material sources, in order to be able to reconstruct a garment that does not exist.

The experiment

Our experiment began by selecting a source from which to begin. We searched the inventory database for an artisan who was representative of his peers, and whose wardrobe epitomised the fashionable creativity we could see quantitatively through our data.[3] After considerable discussion, we selected one entry, taken from the 1631 inventory of the Florentine waterseller Francesco Ristori. Among his family's wardrobe, totalling 122 items of clothing and accessories, one garment stood out: a doublet of black stamped mockado, described as nasty (*Un giubbone d' mucaiardo nero cattivo stampato*) (Figure iii.2).[4] Doublets, or in Italian *giubboni*, were upper body garments worn by some women and most men across the social spectrum. Worn as part of everyday dress, they often showcased extraordinary cutting, shaping and decorative techniques. The use of

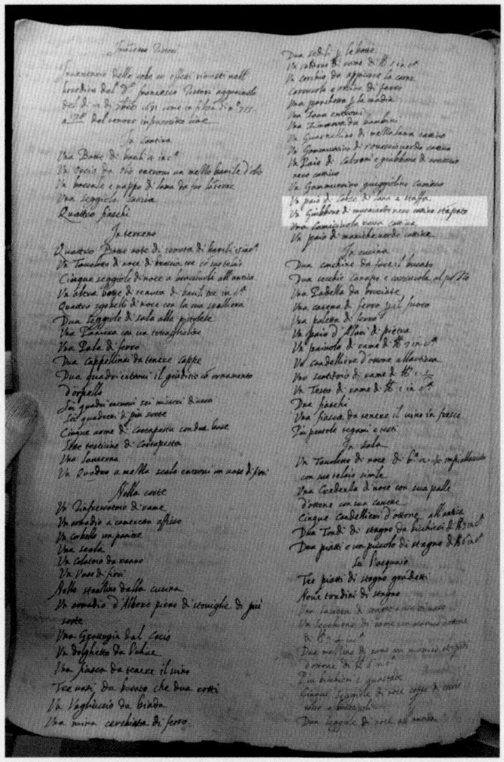

Figure III.2 The brief description of the black stamped mockado doublet, found in the Household Inventory of Francesco Ristori, Archivio di stato, Florence, Magistrato dei pupilli, 2718, 2, 1631.

mockado fabric, stamped decorations, and black dye in Ristori's doublet seemed to suggest a particularly artisanal and urban form of Renaissance fashionability that has been largely overlooked.

The process of making was inextricably linked to research. We worked with a team of makers at the School of Historical Dress in London (hereafter SHD). The team, led by Jenny Tiramani, comprises skilled hand-makers who have closely studied and reconstructed extant garments from the early modern period.[5] Their knowledge of historically appropriate materials and tools made them the ideal collaborators. Each step in the process raised new research questions – we constantly referred between archival documents, visual depictions of watersellers and other artisans, and surviving objects in order to select materials, choose decorations and make practical decisions. For example, the tailor and cutter Melanie Braun and I examined many depictions of Italian watersellers and noticed that many of them wore collars that fold down and fall open at the neck, and so this informed the cut and construction methods used in our doublet (Figure iii.3). Braun also took much inspiration from a rare example of a doublet associated

Figure III.3 Jacopo del Conte, *Il facchino*, c. 1580 originally located on Via del Corso, now on Via Lata in Rome (since 1872).

with a working artisan. Formerly owned by a mason, its survival is thanks to its later use by the Dutch humanist Hugo de Groot as a disguise when he fled the Netherlands in 1621 (see Figure 10.3).

The first step in the making process was to commission the woven fabrics from a skilled weaver. This launched a major research question: what was mockado, and how was it made? Guild records, sample books and lexicons suggest that mockado was an imitation fabric, woven with supplementary warp loops in the same structure as silk velvet, but using a blend of wool, linen, hemp and/or silk.[6] In order to get specific information about thread count and fibre and dyes, samples of the warp, weft and pile warp of a stamped wool velvet were taken from a surviving example in the SHD collection, which was examined under microscope and using Ultra-High Performance Liquid Chromatography, High Resolution Mass Spectrometry and Scanning Electron Microscopy by Cristina Carr at the Metropolitan Museum of Art, New York, Art Proaño Gabor at the Rijksmuseum, Amsterdam, and Krista Vajanto at Aalto University, Finland. From these results, we established that our mockado would be made of a linen ground with supplementary wool pile.

To weave the mockado, we first considered the structure. Velvet is more commonly woven with silk, and so first we attempted to work with a silk weaver. However, wool and linen have different properties from silk, requiring different hand skills and loom parts. Instead, the wool and linen hand-weaver Ruth Gilbert was willing to experiment with her technique and tools, improvising at her loom, first using plastic straws and knitting needles for the rods that create the supplementary pile loops, and later sourcing hollow brass rods. This prompted us to wonder about how mockado, one of the novel blended fabrics that reshaped the textile trade in early modern Europe, was invented: would it have been made by wool or linen weavers who learned velvet technique, or was it created by silk weavers who experimented with different fibres?[7] Without inventory information about the lining and structural materials for the doublet, we settled on fustian, linen and hemp interlinings, also woven by Gilbert, following examples found in early seventeenth-century sleeve linings in the SHD collections. Hand-woven linens and naturally dyed changeable pink and green silk taffeta were repurposed for interlinings and facings (Figure iii.4). Most fabrics required some level of post-loom treatment, which was all done by hand: the fustian was napped with a hand carder, and the hemp stabilised with a gum arabic size. Such processes remind us that cloth taken from the loom was rarely ready to be used straight away by

Figure III.4 Samples of mockado, fustian, hemp and linen linings woven by Ruth Gilbert, and loop manipulated braids made by Beth Trapnell. These were sent to Melanie Braun, who drafted the doublet at The School of Historical Dress, London in 2021.

a tailor or seamstress, and that many processes and makers were involved in transforming natural fibres into fashionable textiles.

We cannot know which dyestuffs were used for Ristori's black doublet, so we used the results of the UHPLC-PDA-HRMS dye analysis of the SHD black wool velvet as inspiration for our experiments. As the fabric was dyed in one piece (rather than as unwoven threads), we followed the same method, sending the woven length to the dyer Karl Robinson. As mockado consists of both vegetable and protein fibres, it needed to be dyed in multiple steps, to ensure colour adherence to different materials. The dye analysis suggested that woad or indigo was used in the first dyebath, and then the blue cloth was probably treated with an alum mordant before being dipped in a kermes bath. This would likely have dyed the wool, but not the hemp or linen ground. The dye analysis suggested that the fabric was dyed with black alder and oak gallnuts, a tannin and iron process that would turn the vegetable fibre dark black. A final dip in a logwood and potash dyebath probably gave the fabric a final deeper black hue. This complex recipe, combining more traditional expensive (blue and red dyes), cheap (iron and tannins) and state-of-the-art (logwood) methods, shows that early modern dyers used all the techniques available to achieve a good black. Due to limits in budget and material availability, we could not dye with kermes or alder, so Robinson used indigo and cochineal to approximate the first step in the process, rendering a beautiful deep blue-black tone, rather than a true deep black. Robinson also dyed loop-manipulated braid, made from six looped como silk threads by Beth Trapnell, which decorated the seams and edges of the doublet (Figure iii.5).

Drawing upon her experience as a trained tailor, and close observations of many early modern doublets, Melanie Braun cast pattern shapes using a compass and yardstick, and proportional measurements. This geometrical and mathematical approach meant that the shape of the doublet accounted for the physical body requirements of the model, the historian Valerio Zanetti, while giving him an elegant and idealised form (see Experiment in focus iv). After dyeing, Claire Thornton carefully laid out Braun's paper pattern pieces on the woven mockado, leaving barely any offcuts and even overlapping shapes which could be pieced with scraps (both an early modern practice of economy and a contemporary necessity when we had limited funds to commission mockado) (Figure iii.6). During fittings, we discovered an issue with the dyed fabric, which was so fluffy and saturated with excess dye that pile and dye were transferring onto the hand-made white linen shirt. This problem was overcome by the application of silk inkle lace to bind the armhole edges.

The fabric continued to present challenges throughout the construction process. Underneath the mockado, the doublet is a complex construction of linings and padding made of hemp, wool (woven and roving), linen, fustian and synthetic whalebone. Much structure also comes from

Figure III.5 The mockado and braids hanging after being dyed by Karl Robinson, 2021.

v-shaped pad stitching and hand manipulation as the tailor works. Jordan Colls, who assembled the doublet, explained that the 'trickiest and most time-consuming' element of work was responding to the hand-woven fabrics, particularly the mockado, which had a very tight weave in certain areas.[8] It required significant amounts of manipulation and finishing as the doublet was assembled; seam allowances were shaved down flat and pressed with a hot iron, and the fabric had to be combed and shaved to remove felting that had occurred during the dye process and to ensure an even appearance. Not only was this slow work, but it was also messy: the dye kept transferring on to hands, and a lot of wool dust was created, reminding us of the often dirty and dangerous working conditions endured by early modern textile workers.

Ristori's doublet is described as 'stamped', which suggests that its decorative pattern was not created during the weaving stage (which required significant time and skilled labour), but rather was applied after the fabric had been taken from the loom and out of the dye vat. Comparatively quick and easy to execute, stamping could be done by hand or using a roller press, using metal tools and heat to impress a pattern into pile weave. Stamped textiles survive in many museum collections, showing that the technique was widely used. The metalworker Dave Budd cast

Figure III.6 Claire Thornton cuts out the drafted doublet pieces from the loom width of mockado, 2021.

two stamps, one with a double 'S' and six-ball flower motif, and the other a four-petalled flower, following a surviving stamped crimson velvet in the SHD collections. The creation and control of heat is often a challenge for historical reconstruction, and in this case, Jordan Colls and Jenny Tiramani proceeded with caution to avoid burning the mockado, using an electric hotplate to heat the tools. Unfortunately, the pile of the mockado was several millimetres higher than the design of the double S stamp, and so it simply impressed a rectangle rather than a legible design into the fabric. Nevertheless, the flower stamp was successful, and working by eye (as many surviving examples seem to suggest), a striking square and bar strapwork pattern transformed the mockado into a more three-dimensional fabric that catches the light (Figure iii.7).

Ristori's doublet was one part of an outfit and could not have been worn alone. Hand-made metal hooks and eyes were sewn in to attach the doublet to hose, and eyelets were used for lacing bands at the shoulders so that sleeves could be attached to the doublet. A pair of hose, sleeves and a shirt were made for this reconstruction, in order that the doublet could be worn attached and over the closest garments worn on Ristori's body, and a full outfit (comprising stockings, shoes, a swordbelt and hat) was borrowed from SHD for dressing and photography. Only when worn over

Figure III.7 The striking square and bar strapwork pattern, created at the London School of Historical Dress by impressing a hot metal stamp into the pile of the mockado, 2021.

a shirt and attached to sleeves and hose do we get a full sense of the range of motion, effect of movement and overall look of the waterseller's doublet (see Figure iv.1). Each time it is worn, we see the stiffness of the garment soften with the heat of the body, and the stamped decoration starting to wear down in areas where the mockado is rubbed (such as at the sides of the body). The doublet, in its material form, is an active object that has life in motion and will age, fade and wear with use (although it is stored in archival boxes and acid-free tissue paper when it is not being worn or on display).

Methodology: the challenges and benefits of imaginative reconstruction

This experiment explores how we might reconstruct an object that once existed but does not remain in the visual or material record. With limited textual record to work with – in this case, only seven words, written by an inventory appraiser – the finished reconstruction is an imaginative possibility. The doublet was made entirely by hand, from fibre to finished garment, by skilled makers using historically appropriate materials. But there are limits to its historicity: we had to make compromises when raw materials were unavailable or prohibitively expensive, it was made using electric lights and heating, and many of the makers – while highly skilled – were attempting some of these techniques for the very first time. It also cost far more than Ristori would have spent on a garment, with most of the cost paying for the labour of makers and researchers rather than the raw materials, which would have accounted for the bulk of the price of any item of clothing in the seventeenth century.

While there is much we can never know about Ristori's doublet, the silences in the archives spurred on research, and encouraged close reading of those sources that were available to us. We had to approach the scant inventory information inventively. From the very start, questions arose that took us in all directions of economic, social, legal and cultural history. For example, when thinking about the fashions of the period, we had to wonder when the doublet was made. To answer this, we wondered how old might Ristori have been when he purchased it. He died in a pandemic year, and left behind young children, so he was possibly in his thirties or forties. But was his doublet in a 'nasty' condition because it was very old, or second-hand? Should we make it in a style from a few decades before the date of the inventory, or was it something that was more recent but had been damaged through repeated daily use and being splashed by water while Ristori worked? This got us thinking about the practical and legal constraints on a waterseller. What would he have been legally allowed to wear? What would have been appropriate workwear? When and why would a waterseller want to look fashionable? Such questions led us back to archives and libraries to research Ristori's neighbourhoods, the labour and the social standing of watersellers, and to scrutinise other objects in his inventory for more clues into the habits and mentalities of a Florentine waterseller with a small but carefully chosen wardrobe and household.

Watersellers had an important role in early modern cities, bringing clean drinking water from rivers, fountains and springs to the urban population.[9] Given that he would have had a highly visible presence on the streets, Ristori's professional reputation would have likely been reliant on his clean and fashionable appearance. In other words, whether he wore the doublet for work (unlikely given its materials) or saved it for church and festival wear, Ristori's doublet likely contributed to his social and professional standing in Florence. He also seems to have appreciated the novel aesthetic and material innovations available in his city, so, while Ristori was not a wealthy citizen, he did participate in the middling version of late Renaissance Florentine culture. His inventory reveals that he decorated his home with paintings of Florentine *belle donna*, terracotta busts and four angels made of *cartapesta* (papier-mâché) and owned forks as well as knives.[10] We will never know the personal or emotional meanings embedded in Ristori's black mockado doublet, or when he might wear it rather than one of his other four doublets (made of white leather, Nîmes wool, black spun silk and black baize). But the inventory, and the questions and possibilities prompted through this reconstruction, transformed the questions we asked of historical sources, as well as informing a material reconstruction that gives a physical presence to an object and its owner who are otherwise absent from the surviving material record.

Conclusion

The material doublet is an explorative reconstruction informed by rigorous archival, visual, material and scientific research. Reconstruction methodology raised new research questions and shifted the scholarly focus to consider overlooked subjects. It also fostered collaboration with makers, scientists, conservators and curators, enabled us to better appreciate the skills of makers past and present, and even helped us to revive and reassess lost techniques. By giving a material presence to the waterseller's doublet, we can call attention to the fashionable aims and innovative techniques used by the artisan classes to participate in Renaissance clothing culture. We restore Ristori's and his fellow artisans' reputations as discerning dressers living, working and shopping in a dynamic urban environment in which novel and carefully chosen materials, creative and skilled makers and a culture of materially literate consumers generated a lively middling material culture of Renaissance fashion.

Notes

The 'Refashioning the Renaissance' project's doublet reconstruction, led by Sophie Pitman in collaboration with the London School of Historical Dress, was funded by the European Research Council (ERC) under the European Union's Horizon 2020 research and innovation programme (grant agreement No. 726195). See www.refashioningrenaissance.eu/seventeenth-century-artisan-male-doublet. We thank our advisory board members Ulinka Rublack, John Styles, Evelyn Welch, Tessa Storey, Flora Dennis, Maria Hayward and Susan North for the original concept, the costume-makers and craft experts who took part in this project, in particular Jenny Tiramani who oversaw all stages of the reconstruction, and Clair Thornton, Melanie Braun, Dave Budd, Jordan Colls, Ruth Gilbert and Karl Robinson, as well as Valerio Zanetti who agreed to be our model in this experiment.

1 For fuller description of this reconstruction, see Sophie Pitman, 'Reconstructing fashion: The mock-velvet doublet of a seventeenth-century Florentine waterseller', in Sophie Pitman and Paula Hohti (eds), *Remaking Dress and Textile History: Applying Reconstruction Methods to Early Modern Textiles and Clothing*, Special Issue, *Textile History*, forthcoming 2024.
2 See, for example, Ulinka Rublack and Maria Hayward (eds), *The First Book of Fashion: The Books of Clothes of Matthaus and Veit Konrad Schwarz of Augsburg* (London: Bloomsbury Academic, 2015); Sarah A. Bendall, 'The case of the "french vardinggale": A methodological approach to reconstructing and understanding ephemeral garments', *Fashion Theory*, 23, 3 (2019), 363–99.
3 www.refashioningrenaissance.eu/database.
4 Inventory of the waterseller Francesco Ristori, 12 September 1631, State archives of Florence (ASF), Magistrato dei pupilli, 2718, 2, fols 190r–194r.

5 Melanie Braun, Luca Costigliolo, Susan North, Claire Thornton and Jenny Tiramani, *17th-Century Men's Dress Patterns 1600–1630* (London: Thames and Hudson, 2016).

6 Phyllis Ackerman, 'A note on suf or camlet', Bulletin of the American Institute for Iranian Art and Archaeology, 5, 3 (1938), 254–6. See also Florence Montgomery, *Textiles in America, 1650–1870* (New York: W. W. Norton, 1984), 295, and Nancy Cox and Karin Dannehl, 'Mocha – mohair yarn', in *Dictionary of Traded Goods and Commodities 1550–1820* (Wolverhampton: University of Wolverhampton, 2007), British History Online, www.british-history.ac.uk/no-series/traded-goods-dictionary/1550–1820, accessed 23 January 2020.

7 For more, see John Styles, Chapter 1 above.

8 Claire Thornton and J. Tiramani, M. Braun, D. Budd, J. Colls and Ruth Gilbert, *The Mockado Doublet Project Report* (unpublished document for the 'Refashioning' project, 2021).

9 See Emanuela Ferretti, *Acquedotti e fontane del Rinascimento in Toscana: acqua, architettura e città al tempo di Cosimo I dei Medici* (Florence: Olschki, 2016); David Gentilcore, '"Cool and tasty waters": Managing Naples's water supply, c. 1500–c. 1750', *Water History*, 11 (2019), 125–51; Katherine Wentworth Rinne, *The Waters of Rome: Aqueducts, Fountains, and the Birth of the Baroque City* (New Haven: Yale University Press, 2010).

10 On belle donne, see Marta Ajmar and Dora Thornton, 'When is a portrait not a portrait? Belle donne on maiolica and the Renaissance praise of local beauties', in Nicholas Mannand and Luke Syson (eds), *The Image of the Individual: Portraits in the Renaissance* (London: British Museum, 1998), 138–53; on cartapesta see Richard Goldthwaite, *The Building of Renaissance Florence* (Baltimore: Johns Hopkins University Press, 1982); on the significance of forks, see Norbert Elias, *The Civilizing Process* (1939; Cambridge, MA: Blackwell, 1994).

PART II

Adornment and display

4

Né vera né falsa: non-elite ownership of pearls in early modern Italy

Michele Nicole Robinson

Introduction

In the early weeks of 1638, a melodramatic scene played out in the gardens of Siena's Ospedale di Santa Maria della Scala. Pasquino Neri, a gardener (*ortolano*) who worked and lived on Ospedale land near the gate of Porta Tufi in Siena, had recently died. His widow, Monica, quickly remarried. Her new husband, Giovanni, was also an Ospedale gardener, and by February of that year the couple were living together in the house that Pasquino had rented from his employer and where his possessions remained. To ensure these goods were passed on to his underage heirs, Pasquino's adult daughter, Margherita, requested that an inventory be drawn up under the care of Siena's Court of Wards (Curia del Placito).[1]

The document suggests that Pasquino's modest house was filled with modest things: a 'nasty' wooden table, a few 'nice' wool garments and piles of fibres to be spun into yarn. Pasquino also had a small collection of jewellery, which he kept inside a round box. This included two necklaces of coral beads and two gold rings with red stones: one with a little horseman and another with a *fede*, or two hands clasping.[2] We might expect to find this kind of jewellery among a gardener's humble possessions, and authors of costume books describe similar items worn by the wives and daughters of men like Pasquino. In his famous costume book, for instance, Cesare Vecellio described the young women bringing fruit to market from Chioggia, just south of Venice, as wearing strings of coral or round silver beads just below a light veil draped around their necks (Figure 4.1).[3]

But alongside these simple pieces, Neri's round box also held a diamond ring and a necklace of little rubies, other red stones, gold beads and misshapen pearls, with a gold enamelled cross attached to it.[4] These precious materials are unexpected in the inventory of a labourer, and scholars

Figure 4.1 Cesare Vecellio, 'Ortolane' in *Habiti antichi, et moderni di tutto il mondo* (Appresso i Sessa: Venice, 1598), fol. 118r. Yale University Library, New Haven.

have long considered costly jewellery and especially pearls to have been the preserve of the wealthy in early modern Italy. However, Pasquino's necklace of pearls is not unique among Sienese gardeners: Angelo di Giulio possessed forty-five misshapen pearls in 1603, and Giovani Cioncolini had a ring with a pearl and a necklace of pearls and coral beads in 1649.[5] Beyond Siena, an *ortolano* nicknamed *Bechino* from the parish of San Zeno in Pisa had 'a necklace of six strands of pearls with an Agnus Dei', as well as strings of coral, a gold *fede* and a rosary of incised red bone, in 1614.[6]

It was not only gardeners who earned pearls in early modern Italy but also dyers, laundresses and shoemakers. Among the 448 inventories of non-elite homes and workshops in Siena, Florence and Venice gathered by the 'Refashioning the Renaissance' project and considered here, there are over two hundred references to pearls, more than any other gemstone or semi-precious material in these documents.[7] This chapter draws attention to the presence of pearls in jewellery boxes belonging to artisans and labourers (and on their necks, wrists and fingers, too). It shows how and why people without a great deal of wealth were able to obtain pearl necklaces, bracelets and rings.

This is important because it counters long-held, erroneous assumptions about who had access to and possessed pearls in this period. Although some scholars have stated outright that artisans and labourers did not own pearls, most who study jewellery and accessories ignore those below merchants on the socio-economic ladder.[8] This is in part due to the absence of evidence around non-elite people's ownership of pearls (and goods in general) in comparison with aristocrats, patricians and merchants; we cannot connect extant pieces of jewellery to butchers, shoemakers or dyers, nor do we have correspondence about the procurement of gems for gardeners, laundresses or innkeepers. But, as this chapter demonstrates, documents like household inventories as well as records from auctions, pawnshops and goldsmiths' workshops show that artisans and labourers bought and sold pearl jewellery from various sellers at a range of prices.[9] Finally, sumptuary laws and criminal records indicate that non-elite people did not simply lock their pearls away but wore them – sometimes illegally – on their bodies. Their desire for pearls was perhaps not so different from that of their social betters: to store wealth, demonstrate status (whether financial, social or marital) and communicate aspects of identity that were recognised and expected across society.

Inside the jewellery box

Scholars have long assumed that pearls were unattainable by all but the elite of early modern Italy, because of generalisations they have made about pearls themselves. Historians of dress and jewellery point to plump, round, white pearls represented in portraits of aristocratic women like Eleonora di Toledo – Duchess of Florence and first wife of Cosimo I de' Medici – as well as images of unnamed patrician brides (Figure 4.2). This ignores the fact that pearls range in shape, size, colour and quality and come from different places; they therefore have (and had) varied social and monetary values.

Unlike today's scholars, authors of early modern lapidaries and treatises on the natural world recognised and painstakingly described these differences within a sort of pearl hierarchy. The humanist Lodovico Dolce, for instance, explained in his *Libri tre ne i quali si tratta delle diverse sorti delle gemme che produce la Natura* (Venice, 1565):

> There are two sorts of pearls: one oriental, whose colour is pure white, like polished silver, with a shining surface. And this is the most perfect. The other sort is Western, which is brought from the English Channel. The colour is dulled with certain whiteness and tends to the colour of gold. The Oriental are the most perfect of all. And they are large and round.[10]

Many early modern writers and translators ignored pearls from the Americas in their texts, despite the fact that Columbus had encountered

Figure 4.2 Studio of Agnolo Bronzino, *Eleonora di Toledo*, c. 1562–72. Wallace Collection, London.

them on his third voyage there in 1498.[11] The so-called 'Pearl Coast' of Venezuela and its neighbouring islands supplied and even overwhelmed the European market in waves over the sixteenth century, its riches celebrated in scenes like that depicted by Jacopo Zucchi (Figure 4.3).[12] According to the Flemish physician Anselm de Boodt – writing in 1609 – pearls from the Americas were 'not so commendable as the Orientall ones'.[13] However, when it came to price, it was not the place of origin but, in de Boodt's words, 'Their own glory, beauty, and excellencie amount their worth … So according to their bigness, weight, roundness, and fairnesse, their price is raised, doubled, and trebled.'[14] The monetary value of a pearl was determined more by its superficial qualities than where it had come from. According to Molly Warsh, 'Oriental' evolved into a term for any pearls that were large, round, and smooth; they did not necessarily have to come from the east.[15]

Given this hierarchy of pearls, it is unsurprising that those at the pinnacle – 'Oriental' pearls – appear only once in the inventories of non-elite homes considered here: a Venetian dyer in 1614 was seemingly in possession of 'a string of 39 Oriental pearls that weighs 165 carats'.[16] This entry is unusual not only for its presumed value but also for the level

Figure 4.3 Jacopo Zucchi, *The Coral Fishers*, 1585. Galleria Borghese, Rome.

of detail given about the jewellery. Descriptions of pearl items in these documents are usually terse: 'a string of pearls', 'a pair of earrings with pearls' or 'a ring with a pearl'. When documents give supplementary information about pearls, it most often relates to their shape and size, not their quality. For example, Domenico Gritti, another Venetian dyer, in 1557 had a necklace of seed pearls, or *perle da onza* – so-called because they were sold by the ounce rather than being individually priced like larger pearls.[17] These tiny pearls appear regularly in the documents considered here, probably because they were relatively inexpensive and, in some cities, were permitted by sumptuary laws, as we shall see.

Scaramazze, or misshapen pearls (what we call 'Baroque' pearls today), also appear frequently in the inventories studied here, especially on strings worn around the neck or wrists.[18] For example, the inventory of the goods of the smith and mason Antonio Marini, drawn up when he died in Siena in 1608, shows he owned a necklace with four strands of seed pearls as well as a necklace of *scaramazze* with ten gold beads.[19] Like seed pearls, *scaramazze* were generally less desirable than Oriental pearls and therefore less costly.[20]

That being said, pearls – even those sold by the ounce – were not cheap. Domenico Gritti's necklace of *perle da onza* was valued in the inventory of his possessions at 4 ducats, the same as an old lady's gown of crimson satin also listed there.[21] The dyer had a second string of ninety-four pearls, which was worth 15 ducats, the same as a crimson damask gown in his possessions.[22] He also owned a string of sixty-two pearls worth 120 ducats, which was worn by his daughter-in-law. This is the most expensive single item in his household inventory; the only thing that approaches and indeed exceeds the value of the necklace is the rock alum used for fixing dye colours in his workshop, which was worth 190 ducats.[23] The price of this costly necklace would have represented about 240 days of work for a Venetian master artisan, suggesting Gritti's elevated financial status (or his level of debt).[24] He was not a prince, diplomat or patrician, but he owned what would have been considered beautiful and valuable pearls.

Further household inventories demonstrate that other gardeners, dyers, smiths and masons owned pearls and pearl jewellery, too. Although most items are described in a way that suggests they were simple strings of seed or Baroque pearls, various sizes and qualities of pearl could be found even within single households. And just as there were different kinds of pearls, there were different ways they could be obtained, and for a range of prices.

Acquiring pearls

Documents related to goldsmiths' and jewellers' shops, auctions and pawnshops reveal that they sold a wide variety of pearls. Their customers were not necessarily restricted to the social and financial elite but included buyers with different needs and budgets. For instance, when a post-mortem inventory of the workshop of the Venetian goldsmith Domenico Redolfi was taken in June 1629, the most expensive finished piece of jewellery listed was a string of forty-seven pearls, weighing 91 carats and worth 350 ducats. In contrast, a set of forty-two unstrung pearls weighing ninety-two carats was worth only 190 ducats.[25] Additionally, the goldsmith had one pearl which weighed 5 carats and was worth 40 ducats, while another, which weighed only slightly less at 4 carats and 3 grains, was worth 60 ducats.[26] There were clearly differences in terms of the appearance, shape and quality of all these pearls not mentioned in the written descriptions.[27]

Redolfi's workshop also had earrings at a range of prices, including a pair made from gold wire and decorated with pearls, valued at 5 lire; four pairs with 'sad little pearls', worth 155 lire; and a pair with 'nasty Scottish pearls', worth 70 lire.[28] The goldsmith seems to have been able to meet demand for both costly and less expensive pearls. But even the latter were still pricy: the cheapest piece of pearl jewellery in the shop, the pair of earrings worth five lire, would have cost a master artisan in the

Venetian building industry in the late 1620s a day and a half's wages, and just over two days' wages for an unskilled labourer. The least expensive pearl necklace in Redolfi's shop was valued at 25 ducats, or the equivalent of around forty-eight days' work for a master builder and around sixty-five for a labourer.[29]

Buying pearls from goldsmiths and jewellers, therefore, could be expensive. But shoppers could also turn to the second-hand market, where pearls could be purchased from auctions and pawnshops, which served diverse groups of buyers and sellers throughout the northern Italian cities.[30] In Siena, for example, auctions usually took place in Piazza Comunale. It was there that the farrier Mario di Paolo Zani's goods were auctioned off on 12 May 1646 in order to help settle the many debts he had left to his underage heirs when he died. Among the jewellery up for sale was 'a necklace of gold beads and little pearls' purchased by a Jewish buyer, Moise di Flaminio Galleni, for 31 lire and 19 soldi, and 'a string of nineteen pearls' sold to a friar for 51 lire and 19 soldi.[31] The latter was one of the most expensive lots, and would have represented nearly a month and a half's earnings for a Sienese barber or a notary at the time, or three months' wages for a porter or guard, and just over that for an unskilled labourer on a building site.[32] But there were also opportunities for poorer individuals to buy pearls auctioned off at much low prices: a smith paid just 6 lire for a little pearl necklace that was sold as part of a Sienese mason's belongings in 1593.[33]

In addition to the possessions of private individuals or families, auctions sometimes sold unredeemed pawns on behalf of brokers and lending institutions like the *monte di pietà*. *Monti* were established in many central and northern Italian cities so that people could pledge goods as collateral for small emergency and subsistence loans. Borrowers were usually given twelve or eighteen months to repay the debt, after which any unredeemed goods could be sold at auction.[34] Moneylenders also accepted jewellery and clothing as pledges, which would be sold if debts were left unpaid.[35] Records show that artisans both pledged and purchased pearl jewellery this way.

In Florence, for instance, Antonio Masini and Domenico Bonini operated The Little Crown (*la Coroncina*) as both a tavern and a pawnshop. The inventory drawn up when Bonini died in 1644 reveals that nearly all the pawned items were jewellery or other objects that had been worked by a gold- or silversmith. Most had been deposited by local people well known to the proprietors. A kitchen servant at the tavern had pledged a pair of gold earrings with pearls for 2 soldi.[36] There were pawns of necklaces, rings and earrings by 'a member of the household', an apprentice at the tavern and the Bonini family's laundress. Even Bonini himself, according to his business partner Masini, had pawned a gold *fede* ring 'that had been kept on his finger' for 7 soldi.[37] Jewellery had also been pledged by a silk weaver, an apothecary and one Baldassare, 'servant to the Princess of Massa'. The proprietors of The Little Crown offered small loans to people from a range of backgrounds. These individuals, from kitchen servants to apothecaries, pledged jewellery – which sometimes featured pearls – to secure their loans. If these were not repaid, the jewellery would be sold, and certainly at a lower price than it would cost to buy new from a goldsmith or jeweller.

However, the inventories of goldsmiths' and jewellers' shops indicate that in some cases they loaned money against pawned jewellery and gems as well and sold them off if their owners were unable to make their repayments on time. The Venetian goldsmith Antonio Albrici, for example, was in possession of several pawned items when he died in 1644. They were kept in separate boxes and envelopes, many of which were labelled with information about the object inside, its value and its owner. His inventory lists 'a little string of seed pearls in a paper with a note that says "pledged for 5 lire"', although in this case their owner's name is unfortunately not stated.[38] People also left goods with Albrici as collateral for purchases of jewellery on credit, like the carpenter named Gasparo who pledged a silver spoon and fork, plus a ring set with a turquoise stone, in return for a plain ring worth 23 lire, according to the receipt kept with the deposited items.[39]

Records like these demonstrate that there were numerous means by which people from across the social spectrum could obtain pearls and other kinds of jewellery at a range of prices. The many and varied opportunities to both buy and sell pearls enabled more than just the wealthy to own them, though they were not cheap. Pearls represented a considerable investment, but one that held its value and could be sold, pawned or traded in times of financial need.

The social significance of pearls

Like other gems and precious materials, pearls clearly functioned as stores of wealth for early modern shoppers; however, they were also desirable for their symbolic and social value. In particular, they have long been associated with marriage, often appearing in written descriptions of wedding gifts and adorning brides in fifteenth- and sixteenth-century wedding portraits. Many of these images depict or relate to members of wealthy and powerful families, thereby furthering the traditional assumption that pearls were the reserve of the elite.[40]

Pearls were beautiful and emphasised the beauty of the bride, but they had symbolic meaning as well. In the early modern period, the colour white was associated with virtue and with cleanliness of body and spirit, and pearls in particular were considered a sign of purity.[41] They were also emblems of chastity, due not only to their colour but also to the way that they were believed to form: oysters rose to the water's surface to sip celestial dew, through which they conceived a pearl. The purity of the dew, the clemency of the weather and the length of gestation were all thought to determine the quality of the pearl, which the oyster would protect by biting off any fingers or hands inserted into their shells.[42] These beliefs are recorded as far back as Pliny's *Natural History* and survived well into the seventeenth century.[43]

The idea that oysters were impregnated by the heavens with something pure was easy to connect with Christian beliefs around the Virgin Mary's immaculate conception. The molluscs' purported chastity was also appealing to early modern writers. Purity before and chastity during marriage were considered important virtues for brides, which the wearing of pearls both signified and reinforced. According to Cleandro Arnobio, a Venetian trader and the author of *Il tesoro delle gioie* (Venice, 1602), a pearl 'renders chaste whoever wears it'.[44]

This symbolism, along with the status of pearls as luxury items, is perhaps why they were a key focus of sumptuary laws in most north Italian cities in the early modern period. In fact, many of these laws started with a blanket ban on the wearing of 'pearls of any sort, real or false' by anyone except, usually, women of the highest rank.[45] This is perhaps why scholars often assume that people from the lower social orders were not legally allowed to own or wear pearls. But closer examination of these laws reveals that pearls were in fact permitted to the wives of artisans on certain stringently defined occasions. Combined with the establishment of distinctions based on different categories of pearls – much like those made by the authors of the lapidaries discussed earlier – it appears that such legislation was more flexible than has generally been understood.

The city of Siena, for instance, took a nuanced approach to controlling the use of pearls. Law-makers recognised that there were different types, of differing worth, and used these criteria – which shifted over time in response to trends in consumption – to determine who could wear them. In the sumptuary laws of 1576, men and women of Siena's lower social orders were allowed to wear only one pearl – real or false – on a single finger ring worth not more than 100 scudi.[46] However, over time, the restrictions around pearls relaxed, so that in the laws of 1594:

> Any sort of person of any age, grade, state, condition … is prohibited from wearing as ornament at home or in the city of Siena or in the state any sort of pearls, real or false, except the little ones, which are at least 300 [pearls] per ounce.[47]

These tiny pearls were allowed only as a single string about the neck, 'and not in the form of a necklace of six or eight strands, in Agnus Deis, in earrings or headdresses', and were permitted to be worn by 'any sort of person' except sex workers and children under seven, for whom pearls were expressly forbidden.[48] Notably, such strings of tiny pearls are frequently mentioned in non-elite inventories, as touched on earlier.

In 1599, the regulations shifted again to allow women from Siena's ruling families (the *Riseduti*) to wear one string of pearls 'at their pleasure', while other women were limited to seed pearls sold by the ounce (*minute a oncia*). By the end of the sixteenth century, therefore, it was not simply the right to wear pearls that distinguished women from different social orders but their number, size and weight. The distinction between the seed pearls allowed to lower-ranking women and the larger pearls permitted to elite women would have been obvious, enabling onlookers to distinguish between women of high and low status at a glance. Inferior pearls were deemed appropriate for socially inferior people perhaps because they were cheaper or less beautiful, or both. Twenty years later, however, the Sienese laws relaxed further, so that all married women except sex workers were also allowed to wear larger pearls weighing up to 2 carats each.

In Venice, as in most other northern Italian cities, sumptuary legislation relating to pearls was more closely tied to socio-economic status than it was in Siena. Once again, though, there were usually exceptions. For instance, a special decree issued in Venice on 8 July 1599 forbade all but the dogaressa, her daughters and daughters-in-law who lived in the ducal palace from wearing pearls. There was an important concession, however: 'any woman, whether of noble birth or a simple citizen' was allowed to wear pearls for 'fifteen years from the day of her first marriage'. After that, 'this string or any other kind of pearls or anything which imitates pearls' was no longer permitted.[49]

In 1609, this grace period was reduced to ten years, because 'the desired end has not been attained, and the extravagance has continued up to the present time and still continues with the gravest injury to private persons'.[50] And, in the decree of 1619, sex workers were also allowed to wear pearls, which further reduced the visual distinctions between Venetian women of different social orders, as highlighted by Giacomo Franco, an engraver and publisher, among others. As Franco's costume book *Habiti delle donne venetiane* (Venice, 1610) explained: 'In Venice, we find four qualities of women who dress almost in the same way. The only difference is how much jewellery they wear.'[51] There were the city's noblewomen, who surpassed all the others with the great price of their pearls; the *cittadine*, who bedecked themselves as if they were gentlewomen, and the merchants' wives, who appeared little inferior; and the wives of artisans, who possessed such 'lovely necklaces of pearls and other jewels' that they appeared to 'adorn themselves like wealthy foreign gentlewomen' (Figure 4.4).[52] Although Franco's description cannot necessarily be taken as fact, the archival evidence discussed earlier in this chapter shows that the jewellery boxes of Venetian artisans certainly did contain

Figure 4.4 Giacomo Franco, Untitled engraving from *Habiti delle donne venetiane intagliate in rame nuovamente* (Venice, c. 1591–1610). Metropolitan Museum of Art, New York.

'lovely necklaces of pearls', which could legally be worn by their wives for at least ten years from their wedding day.

In Florence, recently married women from the lower social orders were also permitted to wear pearls. Although the sumptuary laws of 1638 prohibited 'any woman of any state, grade or condition to use pearls [or] jewels of any sort, real or false', there were exceptions for 'maidens of any condition' who were entitled to trinkets of 'lapis lazuli, agate, coral and other similar things'.[53] Married women – prior to the sixth anniversary of their 'ring day' – could wear 'a pearl necklace of a value up to 1,000 scudi, plus a pair of earrings and a pair of bracelets of pearl or other good gems which do not exceed the total value of 200 scudi'.[54] After this period was over, matrons were permitted 'a necklace of any real gems, but not false, a pair of bracelets and a pair of similar earrings, as long as the value of all three of the aforesaid does not surpass 100 scudi'.[55] These concessions do not appear to have differentiated between women married to nobles, citizens or artisans: it was marital status not social rank that made pearls licit for most women in Florence. Indeed, to all intents and purposes, the wearing of pearls by a Florentine woman could safely have been taken as a sign that she was married.

Although this was the case in Florence itself, matters were different when it came to the surrounding Tuscan countryside. According to the laws of 1638, peasant women (*contadine*) were

> prohibited from wearing pearls and other gems, real or false, also gold and silver, real or false, and silk fabrics, except for their aprons, belts for cinching and hat linings. But they are permitted, at the neck, one string of beads of silver, coral or other [material] which does not exceed the value of 4 scudi, and two rings which do not exceed the value of 3 scudi between them.[56]

Peasant women, however, did not necessarily comply with the letter of the law (compare Chapter 9). The records of the *birri* – the officials who patrolled the streets of Florence enforcing sumptuary regulations – reveal many instances of *contadine*, and indeed women of other ranks, wearing forbidden clothing and jewellery.[57] On 25 September 1639, Costanza, the wife of an *ortolano*, was spotted in Via Vacchereccia wearing 'a necklace of beads and pearls and attached to it a little gold cross with pendant pearls as well as three gold rings with different stones'.[58] According to the *birri*, these items – including the pearls – were forbidden because Costanza's husband was a manual labourer (*manovale*) who 'worked the earth'.[59]

Similarly, the laws banning imitation pearls and gems were also routinely disobeyed (for imitation pearls, see Experiment in focus VI). A complaint filed against the unnamed wife of a Florentine shoemaker who lived in Via de' Pescioni reported that she was seen on the morning of 27 February 1639 passing the Church of San Michelino degli Antinori wearing 'a necklace of black beads and Venetian false pearls', along with other forbidden items.[60] Whether or not the shoemaker's wife was within the first six years of her marriage, according to the most recent sumptuary laws it was forbidden to wear such pearls because they were false.[61]

Although the sumptuary laws of northern Italian cities in this period sought to control the wearing of pearls and often started with a blanket ban on their use, close examination shows that there were opportunities for women from the lower social orders to wear pearls legally. The relatively lenient laws regarding the consumption of pearls in Siena perhaps explain, at least in part, why there are substantially more instances of their ownership here than Florence or even Venice in the 'Refashioning the Renaissance' database. But the various types of pearls, and their range of prices, meant that they could be obtained by families that were not particularly wealthy. Pearls were closely associated with purity, chastity and marriage – as well as social and financial status – so it is unsurprising that the wives and daughters of labourers, craftspeople, shopkeepers and other artisans wanted to – and did – wear them.

Conclusion

Owning and wearing pearls was a sign of economic, social and marital status in early modern Italy. Unsurprisingly, the largest, plumpest and roundest pearls were the preserve of the elite; they were rare, costly and often forbidden to most. But there were also other, less coveted types of pearls that were available at lower prices for buyers without substantial resources. And these could be purchased from jewellers, goldsmiths, pawnshops and auctions, making pearls more widely accessible than scholars have previously acknowledged.

As this chapter has shown, when we take the time and care to examine the jewellery boxes, cupboards and shelves inside the homes and shops of non-elite people, it quickly becomes apparent that they did not lead the dirty, drab and depressing lives often attributed to them by scholars. The wives of gardeners flaunted their pearls in the city streets, kitchen workers traded pearls for small loans and farriers had pearls auctioned to pay off debts. Indeed, as the 'Refashioning the Renaissance' project shows, non-elite people were keen to participate in both fashion trends and longer-held customs to construct and demonstrate their social, financial and marital identities. They at once sought to belong to and stand out in the groups of which they were a part, and pearl jewellery was just one way of achieving this.

Notes

The research leading to these results was done as part of the 'Refashioning the Renaissance' project that has received funding from the European Research Council (ERC) under the European Union's Horizon 2020 research and innovation programme (grant agreement No. 726195).

1 State Archives of Siena (ASS), inventory of the gardener Pasquino, Curia del Placito, 280, 85, 1 March 1638, fol. 119v.
2 'Un vezzo di coralli piccoletti usi bene di numero quarantuno, un altro [vezzo] a collaretta piccola di coralli piccoli, Un vezzo a postine di rubinetti e pietrine rosse

con bottoncini di oro di dodici poste con dodici rochette e alcune perline scaramazze, quali perle disse detto Giovanni esser sue, una Crocettina d'oro e smalto attaccada a detto vezzo, un anello d'oro con cavalierino, un altro [anello] d'oro con pietra rossa, un altro [anello] con pietra rossa, un altro con pietra rossa a ponta d'diamante, disse detto Giovanni esser suo, una fede d'oro', ASS, Curia del Placito, 280, 85, fol. 120v.

3 Cesare Vecellio, *Habiti antichi, et moderni di tutto il mondo* (Venice: Appresso i Sessa, 1598), fol. 118r.

4 The pearls on the necklace, and the diamond ring, are specifically noted as having been the property of Giovanni. The mixing of these goods underlines the need for an inventory. ASS, Curia del Placito, 280, 85, fol. 120v.

5 Inventory of the gardener Angelo di Giulio, 6 April 1603, ASS, Curia del Placito, 275, 1505, fol. 31v; and inventory of Giovani Cioncolini, 30 Septemer 1649, ASS, Curia del Placito, 286, 9, fol. 34v.

6 'Una fede di oro' and 'Una corona grossa di osso rosso intagliata', inventory of the gardener Becchino, 21 October 1614, State Archives of Florence (ASF), Magistrato dei Pupilli, 2717, fol. 73v.

7 www.refashioningrenaissance.eu/database. For pearls among Italian artisans, see also my online blog post 'Luxuries that cost human lives?: Pearls in early modern Italy', at www.refashioningrenaissance.eu/blog.

8 For example, see Eugenia Paulicelli, *Writing Fashion in Early Modern Italy: From Sprezzatura to Satire* (Farnham: Ashgate, 2014), 157. Karen Raber states that pearls 'were deployed more broadly by women of all ranks'; however, she does not look lower on the social scale than women from the merchant classes: Karen Raber, 'Chains of pearls: Gender, property, identity', in Bella Mirabella (ed.), *Ornamentalism: The Art of Renaissance Accessories* (Ann Arbor: University of Michigan Press, 2011), 159–81.

9 I thank the 'Refashioning' team, in particular Stefania Montemezzo, for identifying, transcribing and making available the archival inventories of artisans used in this chapter, available open access at www.refashioningrenaissance.eu/database.

10 Lodovico Dolce, *Libri tre ne i quali si tratta delle diverse sorti delle gemme che produce la Natura …* (Venice: Gio. Battista, Marchiot Sessa, e Fratelli, 1565), 51B. This work was an uncredited translation from Latin to Italian of Camillo Leonardi's *Speculum lapidum* (Venice: Giovanni Battista Sessa, 1502).

11 Molly A. Warsh, *American Baroque: Pearls and the Nature of Empire, 1492–1700* (Chapel Hill: University of North Carolina Press, 2018), 30.

12 Warsh, *American Baroque*, 113.

13 Anselm de Boodt, *Lapidary, or, The History of Pretious Stones*, trans. Thomas Nicols (Cambridge: Thomas Buck, 1652), 78.

14 Anselm de Boodt, *Lapidary*, 80.

15 Warsh, *American Baroque*, 213.

16 'Un fil de perle oriental de n. trentanove fate pesar per il sudetto commissario et uno delli commessarri peso carati 165', inventory of the dyer Rizzo Francesco, 17 December 1614, State Archives of Venice (ASV), Giudici di Petizion, Inventari, 345, 80, fol. 1v.

17 'Item uno fil de perle per la moier de miser jacomo stimade ducati 120 n. 62 … Item uno fil de perle una pizol et una granda stimada n. 94 ducati … Item uno altro fileto (de perle) da onza stimade in tutto ducati 4', inventory of the dyer Domenico Gritti, 24 July 1557, ASV, Cancelleria Inferiore, Miscellanea, 39, 1, fol. 2r.

18 For example, see Elizabeth Rodini, 'Baroque pearls', *Art Institute of Chicago Museum Studies*, 25, 2 (2000), 68–106.

19 Inventory of the smith and mason Antonio Marini, 14 January 1608, ASS, Curia del Placito, 278, 1835, fol. 56r.

20 They were, however, desired for pendants. See Rodini, 'Baroque pearls'.

21 'Item una vestura di raso cremesin trista usada de m.a vechia d. 4', inventory of Domenico Gritti, 24 July 1557, ASV, Cancelleria Inferiore, Miscellanea, 39, 1, fol .7r.
22 'Item una vestura de Damasco cremesin d. 15', ASV, Cancelleria Inferiore, Miscellanea, 39, 1,fol. 3v.
23 ASV, Cancelleria Inferiore, Miscellanea, 39, 1, fol. 11v.
24 Brian Pullan, 'Wage-earners and the Venetian economy, 1550–1630', *Economic History Review*, 16, 3 (1964), 414–15.
25 'Un fillo di perle no 47 K91 350 d' and 'Perle no 42 K92 190 d', inventory of Domenico Redolfi, 26 June 1629, ASV, Giudici di Petizion, Inventari, 351, 62, fol. 12v.
26 'Una perla di K4 grani 3 60 d[;] Una detta [perla] di K5 40 d', inventory of Domenico Redolfi, ASV, Giudici di Petizion, Inventari, 351, 62, fol. 12v.
27 Inventories of Medici jewellery, for instance, show how closely and carefully pearls and gems were evaluated when necessary. Entries in these documents describe in painstaking detail the size, shape, clarity and 'skin' of the many pearls owned by the family. See Maria Sframeli (ed.), *I gioielli dei Medici: dal vero e in ritratto* (Livorno: Sillabe, 2003).
28 'Un paro detti [rechini] di fillo con perle L5[.] Quatro para detto [rechini] con perlette triste L 155[.] Un paro di detti [rechini] con perle cative scozese L 70', inventory of Domenico Redolfi, ASV, Giudici di Petizion, Inventari, 351, 62, fol. 12r.
29 Based on the average wages given for c. 1628–30 in Pullan, 'Wage-earners', 414–15.
30 On auctions, see Paula Hohti, '"Conspicuous" consumption and popular consumers: Material culture and social status in sixteenth-century Siena', *Renaissance Studies*, 24, 5 (2010), 659–60, and Chapter 7 below by Stefania Montemezzo. Also see Patricia Allerston, 'L'abito come articolo di scambio nella società dell'età moderna: alcune implicazioni', in Anna Giulia Cavagna and Grazietta Butazzi (eds), *Le trame della moda* (Rome: Bulzoni, 1994), 109–24; Jack Hinton, 'By sale, by gift: Aspects of the resale and bequest of goods in late-sixteenth-century Venice', *Journal of Design History*, 15, 4 (2002), 245–62; Ann Matchette, 'To have and have not: The disposal of household furnishing in Florence', *Renaissance Studies*, 20, 5 (2006), 701–16; Evelyn Welch, *Shopping in the Renaissance: Consumer Cultures in Italy, 1400–1600* (New Haven: Yale University Press, 2005), 191–203.
31 'Vezzo di rochetti doro e perline peso di 28 L trentuna et 19? Pago Moise di Flaminio Galleni Ebreo L 31 19/18 [...] Vezzo di perle di 19 L cinquantuna et 10 pagò Padre Fra Jacinto di L Girolamo L 51.10', ASS, Curia del Placito, 1339, 12 May 1646, fols 4v–5r. I thank the 'Refashioning' team for this reference.
32 Paula Hohti, 'Material culture, shopkeepers and artisans in sixteenth-century Siena' (PhD thesis, University of Sussex, 2006), 53–4.
33 'Un vezzino di perle in un filo a maestro francesco di Lorenzo fabbro L sei L6', ASS, Curia del Placito, 1328, 29 January 1593, fol. 72r.
34 Maria Giuseppina Muzzarelli, 'From the closet to the wallet: Pawning clothes in Renaissance Italy', *Renaissance and Reformation*, 35, 3 (2012), 25.
35 See Isabella Cecchini, 'A world of small objects: Probate inventories, pawns, and domestic life in early modern Venice', *Renaissance and Reformation*, 35, 3 (2012), 39–61.
36 'Un paro di orecchini d'oro con perle che disse esser di Cintio servo di cucina di bottega per soldi 2', inventory of the innkeeper Domenico Bonini, 6 August 1644, ASF, Magistrato dei Pupilli, 2720, fol. 7r.
37 ASF, Magistrato dei Pupilli, inventory of the innkeeper Domenico Bonini, 2720, fols 7r, 6v–7r.
38 Inventory of the goldsmith Antonio Albrici, 10 March 1644, ASV, Giudici di Petizion, Inventari, 359, 4, fol. 2v.
39 ASV, Giudici di Petizion, Inventari, 359, 4, fol. 2v.

40 See Deborah L. Krohn, 'Rites of passage: Art objects to celebrate betrothal, marriage, and the family', in Andrea Bayer (ed.), *Art and Love in Renaissance Italy* (New York: Metropolitan Museum of Art, 2008), 62.

41 Andrea Bacci, *Le XII pietre pretiose* (Rome: Giovanni Martinelli, 1587), 21. Bacci also noted that pearls possessed contradictory meanings, and in many places, including Venice, they were associated with sex workers. See also Thomas Coryat, *Coryats Crudities* (London, 1611), 266–70.

42 George Frederick Kunz and Charles Hugh Stevenson, *The Book of the Pearl: The History, Art, Science, and Industry of the Queen of Gems* (New York: The Century Co., 1908), 36–7.

43 Pliny the Elder, *Natural History: A Selection*, trans. John F. Healy (London: Penguin, 1991), 109–10. Travel, and European control of oyster-fishing spots, improved understanding of how pearls were formed and harvested. See Girolamo Cardano, *The De Subtilitate of Girolamo Cardano*, ed. and trans. J. M. Forrester (Tempe: Arizona Center for Medieval and Renaissance Studies, 2013), vol. 1, 402; de Boodt, *Lapidary*, 75.

44 Cleandro Arnobio, *Il tesoro delle gioie: trattato marauiglioso in torno alle vertuti e proprieta più rare di tutte le gioie, perle, gemme* (Venice: Appresso Gio. Battista Ciotti, 1602), 236.

45 *Prammatica sopra le perle, gioie, drappi, ricami, et altro, per la città di Pisa e Territorio dello Stato Pisano* (Pisa: Per il Tanagli, & il Dote, 1638), 4.

46 ASS, Balia 830, Libro dei bandi, 1576, fols 246v and 248v.

47 ASS, Regolatori, 767, Bandi, 1594, fol. 99v.

48 ASS, Regolatori, 767, Bandi, 1594, fol. 100r.

49 Parti prese nell'eccellentiss. conseglio di Pregadi. 1599. Adì 8. Luglio. In *Materia di rerle* (Venice: Stampata per Antonio Pinelli, Stampator Ducale, 1599), quoted in Kunz and Stevenson, *The Book of the Pearl*, 26.

50 Parti prese nell'eccellentiss. conseglio di Pregadi. 1599. Adì 8. Luglio, & 1609. 5. Maggio. In *Materia di perle* (Venice: Stampata per Antonio Pinelli, Stampator Ducale, 1609), quoted in Kunz and Stevenson, *The Book of the Pearl*, 26–7.

51 Giacomo Franco, *Habiti delle donne venetiane intagliate in rame nuovamente* [Venice, c. 1610], 3. Translated from Italian to English in Paulicelli, *Writing Fashion*, 157.

52 '[B]elli vezzi di perle & altre gioe', Franco, *Habiti delle donne*, 3. On this passage, see Paulicelli, *Writing Fashion*, 157–8.

53 *Riforma, e Prammatica sopra l'uso delle perle, gioie, vestire, et altro per la Città & Contado di Firenze* (Florence: Massi e Landi, 1638), sig. A3 and B2.

54 *Riforma, e Prammatica sopra l'uso delle perle*, sig. B2.

55 *Riforma, e Prammatica sopra l'uso delle perle*, sig. A2.

56 *Riforma, e Prammatica sopra l'uso delle perle*, sig. A4.

57 Giulia Calvi, 'Abito, genere, cittadinanza nella Toscana moderna (Secoli XVI–XVII)', *Quaderni Storici*, 37, 110 (2002), 492.

58 'Haveva un drappo in capo guarnito di Gillietto nero e di piu un vezzo di bottoni chucitovi da delle Perlle et attachatovi una crocellina doro con perle pendente e di piu anco 3 anella doro Tutt a 3 con pietre diferente', ASF, Pratica Segreta, 176, 26 September 1639, fol. 230r.

59 Calvi, 'Abito, genere, cittadinanza nella Toscana moderna', 494–5.

60 She was also wearing a long head-covering that was trimmed with black silk ribbons and lace which were almost twice the allowed length, as well as cuffs trimmed with gilded lace – 'haveva in capo Un drappo cioe di spumiglia lungho piu et cordinario e Guarnito di Giglietti di seta nero altro piu che la Misura quasi che il doppio e di piu un vezzo di bottoni neri e perle di Venezzia false et un paio di Manichini Guarniti con trine oltrea palchi dorete', ASF, Pratica Segreta, 176, 28 February 1639, fol. 169r.

61 *Riforma, e Prammatica sopra l'uso delle perle*, sig. A2.

5

Adorning the everyday: male artisan jewellery in early modern England

Natasha Awais-Dean

Introduction

Hanging in the galleries of the Victoria and Albert Museum in London is an oil painting by the nineteenth-century artist Charles Robert Leslie, who was known for the accurate renderings of dress in his works referencing historical literary subjects.[1] Titled merely *Autolycus*,[2] it presents a romanticised vision of the 'rogue' and pedlar of that name in William Shakespeare's late romance *The Winter's Tale*.[3] As Autolycus makes his entrance on stage, he advertises the wares he has for sale in a song:

> Lawn as white as driven snow;
> Cyprus black as e'er was crow;
> Gloves as sweet as damask roses;
> Masks for faces and for noses;
> Bugle-bracelet, necklace-amber,
> Perfume for a lady's chamber;
> Golden quoifs and stomachers,
> For my lads to give their dears;
> Pins and poking-sticks[4] of steel,
> What maids lack from head to heel:
> Come buy of me, come; come buy, come buy;
> Buy, lads, or else your lasses cry: Come buy.[5]

Leslie's Autolycus occupies the centre of the canvas, his travelling case hanging at his neck and opened wide to prominently display his goods to the single male and four female inhabitants of the countryside who hover around him, intrigued by what he has to offer.

Literary depictions of this early modern wandering seller appear elsewhere as we see in John Heywood's mid-sixteenth-century work, *A Play Called the Four PP.* Heywood introduces the figure of the Pedlar in an exchange with the character of the Apothecary, who asks what the former carries. The Pedlar replies that 'euery pedler / In euery tryfull must be a medler', alluding to his vital role in allowing rural men and women to purchase contemporary fashionable items.[6] Among the gloves, purses, combs, ribbons and knives that Heywood's Pedlar offers, he also hawks various items which we can classify as jewellery, as they added embellishment or decoration to dress or the body: 'Pomanders, hooks, and lasses knotted / Broches, ryngcs, and all maner bedes'.[7] That the itinerant Pedlar saw fit to stock jewels among his other wares suggests that he would have found willing buyers among lower-class consumers, particularly those unable to access larger, urban centres in England. This chapter explores the ownership and wearing of jewellery by men of the artisanal classes within early modern England.[8] The male focus is a deliberate riposte to the way that traditional scholarship's concentration on women's jewellery has suggested that this was a female preoccupation. As I have discussed elsewhere, jewellery mattered to men from a range of social classes, all of whom 'had the power, money, networks and status to commission, wear, give and bequeath jewels'.[9]

Accessing jewels: people and spaces

As Autolycus's song and the speech of Heywood's Pedlar suggest, featured among the wares of pedlars were often ones designed for the body, as well as ornamenting it. While it is unlikely that a pedlar would be selling very expensive goods, such as gem-set jewellery, that does not necessarily mean what he offered was of inferior quality. The account book of William Wray, a farmer, draper and haberdasher who kept a shop in Ripon, Yorkshire, contains a series of entries in April 1581 recording the sale of various goods – from pepper and saffron to linen cloth and Norwich points (i.e. the tags for laces) – to Thomas Marshall, whom Wray describes as a 'petty chapman'.[10] It is likely that Marshall then sold these goods as he travelled, thereby facilitating access among those who were removed from the wealthy towns and cities of England. While this account provides little evidence of jewellery being purchased by Marshall, what this record does tell us is that the goods offered by pedlars could originate from more established, fixed and legitimate sites of commerce.

In her work *Artisans, Objects and Everyday Life in Renaissance Italy* (2020), Paula Hohti commented on the increased presence of pedlars in urban centres and rural areas across Europe in the later sixteenth century, noting that 'all kinds of inexpensive smaller items … could also be purchased from itinerant pedlars'. Hohti observes that, through a combination of low prices and the availability of a wide range of goods, pedlars were able to provide 'lower-class consumers in particular with easier access to affordable luxuries'.[11] We see some of what a pedlar might have hawked

Figure 5.1 Woodcut book illustration of a pedlar from Jost Amman, *Panoplia omnium illiberalium mechanicarum* ... (Book of Trades), Frankfurt am Main, 1568. British Museum, London, 1904, 0206.103.33.

in one of Jost Amman's woodcuts for Hartmann Schopper's *Book of Trades* (1568) (Figure 5.1, see also 2.3). Although a German example, as we might expect, the pedlar appears in a rural setting, holding a tray that displays various goods, from playing cards and gloves to rosary beads, while in his right hand he holds a mirror, a case with utensils and another string of beads.

In choosing to embody the complexities of the character of Autolycus within the figure of the pedlar, Shakespeare reflected a real contemporary concern (among other) over these itinerant sellers: their mobile nature and the range of wares they sold meant they operated outside the traditional guild system, making them difficult to regulate.[12] Nevertheless, their ability to access customers beyond the reach of fixed-site retailers – 'At wakes and wassails, meetings, markets, fairs', as Shakespeare put it – fulfilled a valuable role in bringing the latest fashions to those who lived outside urban centres.[13] That the pedlar was a controversial figure in the early modern period is undeniable. For fixed-site retailers, whose activities were governed by the various trade guilds operating within England, pedlars were a threat to their business and so they were strongly opposed to any attempts to license (and therefore legitimise) them. An Elizabethan statute from 1597 classified pedlars and petty chapmen alongside rogues, vagabonds and beggars, who should be whipped 'until his or her body be bloudye' and then forced to return to their place of origin.[14] Yet in 1618, James I passed a royal proclamation allowing pedlars and petty chapmen to continue trading in recognition of the benefits such 'industrious and well-disposed' sellers brought to 'our loving subjects dwelling remote from Cities and Market Townes'.[15] However, this licence was revoked only three years later, in 1621, suggesting that pedlars were still considered a threat to guild-regulated trade.[16]

Concerns over pedlars continued to abound throughout the seventeenth century. In 1691, seven years before they were eventually fully licensed to trade by an Act of Parliament in 1698,[17] a proclamation was printed and posted at the Three Pigeons on Cornhill, in the City of London, refuting any arguments in favour of allowing pedlars to operate.[18] Among the anxieties evident within the pamphlet are the notions that pedlars subverted the guild system of training and regulation and undermined long-established retail networks, similar concerns to those that were prevalent in the preceding century. The activity of pedlars, the anonymous author argued:

> hinders and spoyls all ordinary and common Fairs, and impoverishes and ruins all the Markets, and consequently all the Market-Towns in *England*, by taking away from thence the substantial part of Trade, and preventing recourse of People thither, and by turning the Trade out of the right Channel (where apprenticeships have been served).

The paper also accused pedlars of exploiting their mobility to deal in and distribute 'Stolen, Smugled, and Prohibited Goods, such as Linnens, Silks and Spices of all sorts, Tobacco, Brandy, Tea, Coffee, Chocolate; and also *French* Goods'. Taking into consideration the range of wares offered by pedlars, it is clear that jewellery was not merely a concern of the elite and cheaper goods were available, allowing less wealthy citizens to participate in contemporary fashions and own highly personal jewels.

Another perspective on the extent to which men (and women) from the lower classes had access to jewellery is provided by exploring further other spaces from where they could purchase these goods. We have already seen how they might have been able to obtain them from itinerant sellers, but fairs and markets were also important retail settings in early modern England, especially for individuals who normally lived some distance from towns and cities. Each year in the second half of the sixteenth century, there were more than eight hundred individual fairs in England.[19]

Their scale and importance were remarked upon by the historian and topographer William Harrison (1535–93) in his panoramic *Description of England* (1577):

> There are (as I take it) few great towns in England that have not their weekly markets, one or more granted from the prince, in which all manner of provision for household is to be bought and sold for ease and benefit of the country round about ... as there are no great towns without one weekly market at the least, so there are very few of them that have not one or two fairs or more within the compass of the year.[20]

That jewellery was commonly bought and sold at fairs and markets is clear from the records of the Court of Wardens of the Worshipful Company of Goldsmiths in London. The wardens were responsible for regulating the craft of goldsmithing, thereby maintaining its legitimacy throughout England. As early as the fourteenth century, provincial craftsmen were required to attend Goldsmiths' Hall in London so that they could familiarise themselves with the guild's hallmarks, while officials from the metropolis visited provincial shops and fairs in order to inspect and assay wares, ensuring that the silver and gold adhered to the prescribed standards of purity.[21] The Court of Wardens imposed fines on those sellers whose goods failed the guild's inspection, with the minutes of its proceedings recording the nature of the infraction and the resulting punishment, thereby enabling us to gain some sense of the extent to which customers might be tricked with substandard goods.[22] Thus, for example, at Our Lady's Fair in Southwark in 1568, an individual referred to as a pilgrim named Arnold was fined 2 s as a consequence of the impurity of the silver in eighteen gilt rings.[23] And the following year, at the fairs held in Bury in Lancashire, Harleston in Norfolk and Woodbridge in Suffolk, there were widespread attempts to foist inferior 'claspes without hooke, claspes for cloke, earepykers, pynnes, whistells, paire of gyltehooke, and paire of eyes and claspes' on unsuspecting shoppers.[24] The court's records are full of references to small-scale and inexpensive dress accessories and jewellery, suggesting that fairs and markets were popular and legitimate places for citizens of the lower and middling classes to purchase such items of goldsmiths' work, despite the occurrence of illegitimate practices.

Shifting perceptions of the ownership of jewels through material evidence

Our perceptions of who owned and wore items of jewellery have been skewed by the sources that conventionally have been considered, whether material, visual or archival. However, in recent years, there has been a shift in our understanding of the material culture of those living beyond what we might view as the more elite centres of the court and urban spaces. In his seminal work *The Dress of the People* (2007), John Styles explored the idea that participating in new fashions of dress was not limited to elite society in eighteenth-century England, allowing us to refocus our view on the consumption habits of those from the lower classes, whom we might deem ordinary people. Paula Hohti's most recent work reflects on this idea of non-elite fashions, with its investigation of the material culture of those of lower social standing in Renaissance Italy. Hohti defines the artisanal class as those who had an economic and social position between that of professionals, such as merchants or notaries, and workmen. She notes that artisans are what we would consider craftspeople, shopkeepers and local tradespeople with small commercial outlets.[25] Similarly then, when examining the wearing and ownership of jewellery in the early modern period, it is now possible to shift our perspective away from the higher levels of society in order to gain a broader understanding of male artisans' ownership of jewellery in early modern England.

It is true that the material evidence – the jewels themselves – seems to favour the survival of the sorts of high-status pieces we commonly see in museum displays. Though in spite of the high material worth of such jewels, these, and items of jewellery across all levels of society were often imbued with layered narratives that gave them intangible emotional value. Even those pieces of seemingly lesser worth, such as a simple silver-gilt clasp, were made from materials with an intrinsic fiscal worth, no matter how small. This is important to remember when we consider evidence for artisan ownership, with the ability to bequeath a jewel indicating relatively stable wealth, but conversely the paucity of surviving examples suggests that such items may have needed to be pawned or sold.

Nowhere is this monetary value of jewellery more obvious than in the use of gold chains as payment for mariners, revealing that, at the most fundamental level, jewellery often remained nothing more than a wearable and portable store of wealth. An evocative narrative letter sent to the Spanish king by Francisco de Cuéllar, a captain shipwrecked off the Irish coast during the failed 1588 Spanish Armada campaign, explains how he was stripped of his clothes to reveal a gold chain hanging at his neck. The chain, he writes, was 'worth more than a thousand reals', adding that 'I was only a poor soldier and this money was what I had earned on board ship'.[26] Gold chains recovered from the 1622 wreck of the Spanish guard ship, the *Nuestra Señora de Atocha*, personal possessions of the forty-eight male passengers or the 220 crew members, provide another case in point.[27] The weight of the links corresponded with the contemporary Spanish escudo coin, and the malleability of the gold made removing them from the chain simple, suggesting that these items of jewellery were considered a form of currency. This is perhaps not altogether surprising given that ships journeying between the New World and Europe were subject to looting by

Dutch and English privateers, and a gold chain could be concealed about the body more easily than a purse full of coins.

As styles changed, jewels were broken up in order to be refashioned and updated. Sixteenth-century jewels were particularly susceptible to this fate, as the seventeenth century brought with it a shift in aesthetics away from favouring the figurative art of the goldsmith and enameller and towards a preference for an abundance of gemstones.[28] So those jewels that have been preserved are remarkable survivals and probably remained intact precisely because of their significance to their multiple owners. The four 'Hunsdon jewels', given by Elizabeth I to her cousin Henry Carey, first Baron Hunsdon (1526–96), illustrate this. Comprising an enamelled gold ship pendant, a rock-crystal bracelet, an onyx cameo and a tiny enamel and gold girdle prayer-book, they have survived because of their association with the queen.[29] That is not to say that jewels owned by the less wealthy meant less to them, but the financial value of a single silver clasp would have represented a greater proportion of their wealth, so a jewel such as this was more susceptible to destruction, thereby allowing an owner to benefit from the intrinsic value of the raw material.

Equally, the survival of jewels can equally be the result of chance, as is certainly the case with the cache of late sixteenth- and early seventeenth-century jewels and loose gemstones that form part of the 'Cheapside Hoard'. These items are believed to have been part of the stock-in-trade of an as yet unidentified Jacobean goldsmith. Deliberately hidden sometime after 1640 or 1641 likely as a result of the outbreak of Civil War, the hoard was discovered only in 1912 by workmen excavating a site on Cheapside in London.[30]

However, it is chance survivals of another nature that have improved our understanding of the ownership and use of jewellery by men and women from across the entire social spectrum. Since the implementation in England, Wales and Northern Ireland of the Treasure Act in September 1997, all finds of objects with at least 10 per cent gold or silver content that are more than three hundred years old must be legally reported to the district Coroner via a local Finds Liaison Officer. The discovery of objects whose value is solely archaeological is subject to voluntary registration with the Portable Antiquities Scheme (PAS) administered under the aegis of the British Museum.[31] The publicly accessible PAS database of finds has transformed the scholarly narrative so that we no longer perceive jewels as the preserve of the social elite. Because the majority of the early modern pieces of jewellery recorded in it are not Treasure, it is likely that they were once owned by members of the middling and lower levels of society. The range of items – all presumably lost by their owners, on the basis of how they were discovered in the ground – has broadened our understanding of the prevalence of jewels, and their different types, among non-elite English men and women.[32]

As well as demonstrating that early modern individuals from all walks of life owned and wore jewellery, the PAS database also reveals the affinities these objects share with those valuable enough to be classified as Treasure. Take, for example, a ring found in Oxfordshire in December 2018 (Figure 5.2).[33] Thought to date from some time in the early modern period (between 1450 and 1700), it is made of copper-alloy and clear traces of gilding remain to its surface. Its oval-shaped bezel, the shoulders of which taper to a narrow hoop, was designed as the setting for some now-lost decoration, such as a stone. It is not hard to imagine that, when gilded and with a cheap coloured stone or paste in place, the ring would have appeared very similar to a gold ring set with a precious or semi-precious stone. A bezel and remnants of the hoop of another copper-alloy ring, probably dating from the sixteenth century, were found in Pembrokeshire in 2003.[34] The bezel takes the form of a heart, from the top of which sprout two flowers, which are surmounted by a crown. The heart is held between two hands, each with cuffs, which form the shoulders of the missing hoop. There is significant gilding or silvering of the surface, again inviting comparisons with higher-value gold or silver pieces of similar design. It seems likely that a ring with this iconography would have been a token of love or affection, possibly in the context of a betrothal or marriage. Such base-metal rings would have enabled those of particularly limited means to adorn themselves fashionably at least in some small way.

Figure 5.2 Cast copper-alloy finger ring with gilding, England, c. 1450–1700, found in Merton, Oxfordshire, England. Portable Antiquities Scheme NMS-88E563.

The definition of jewellery in the early modern period encompasses items that we might now regard as dress embellishments, such as buttons. Although buttons are ostensibly functional objects used to fasten the fabric of clothing together, the manner in which they were often decorated and the materials from which they were made in the early modern period suggest that they should be regarded as forms of jewellery. There is clear evidence for this in the PAS database of finds. Although the gold buttons recorded there are not numerous, they all bear decorative features of varying types.[35] Silver buttons are far more common, and the majority of these are engraved or stamped with decorative features.[36] One popular style features a heart and crown motif commonly associated with the marriage of Charles II and Catherine of Braganza in 1662. There are variations in the design, including one with two hearts beneath a crown, and one with a flaming heart pierced by two crossed arrows.[37] While silver versions of these buttons – some of which were gilded – were probably worn by men of the middling sort among whom we might count male artisans, there are also examples made from copper-alloy and other base metals such as a seventeenth-century disk (presumed to be a sleeve button) found at Cundall with Leckby in North Yorkshire. This has stamped decoration of two hearts conjoined beneath a crown.[38] These base-metal objects were almost certainly worn by men of more limited means, providing evidence both for their participation in fashion and for the sharing of fashionable jewellery designs across the social spectrum.[39]

Much the same can be seen with a particular type of jewellery, which was worn exclusively by men in the sixteenth century – the hat ornament. The fashion spread north from the Italian peninsula from military origins to a more secular context across Europe to England. Portraits of elite European men with these large, brooch-like, often iconographic jewels abound, with Tudor men more accustomed to being depicted with jewels that are not emblematic in nature.[40] Nevertheless, the fashion for substantial jewels on the cap remained popular in sixteenth-century England, with a preparatory drawing by Hans Holbein of William Parr suggesting that emblematic hat ornaments were adopted by Tudor men.[41] Earlier scholarship has proposed that these were often worn as a mark of distinction by the elite, noting specifically that 'a man of standing might wear a badge on his hat or cap as a status symbol'.[42] However, there is strong material evidence that these adornments were popular among men across the social spectrum, with even those of lesser means participating in the fashion, though documentary evidence for such use in England is limited.

A discrete group of copper-alloy roundels, termed 'plaquettes',[43] was donated to the British Museum in 1915.[44] Those that we can safely identify as hat ornaments are circular in form and have holes pierced at their edges to enable them to be sewn on to a cap. They resemble in style and form the gold and enamelled versions that were evidently in fashion among the

Figure 5.3 Embossed bronze hat ornament with traces of gilding and remains of green, red and blue enamel depicting a scene from the tale of Pyramus and Thisbe, which is enclosed within a garlanded border, sixteenth century. Diameter 5 cm, weight 23 g. British Museum, London, 1915,1216.133.

wealthy. Indeed, the remnants of gilding on most of them, and the presence of enamelling on one in particular, indicate that they may well have been intended to echo costlier examples (Figure 5.3). The same can also be assumed from the way that the decorative schemes on the plaquettes are derived from classical iconography, the particular choice of subject matter no doubt fulfilling some emblematic function.

These ornaments were not only produced from cheaper materials but were also manufactured using a more economical process. Instead of being unique commissions, often embossed and chased, they were cast in one piece, allowing multiple copies to be produced at relatively low cost once the mould had been made. This method enabled men of lower economic standing to participate also in the trend for emblematic hat ornaments as a mark of distinction. Indeed, the speed with which the fashion took off, along with its geographical spread and its century-long duration, suggest that artisanal engagement with it was to a significant degree independent of elite involvement.[45]

Using documentary sources to understand artisanal engagement with jewels

One disadvantage that finds of Treasure and the objects recorded by the Portable Antiquities Scheme share with much early modern material evidence is that they are often impossible to connect with their original owners.[46] Instead, to gain a real insight into the ownership of jewellery by men from the lower and middling classes, it is necessary to resort to documentary evidence, and especially to that provided by wills and inventories. For the following discussion, I have relied on two published sets of probate inventories from the sixteenth and seventeenth centuries for men we might deem artisans (from Ipswich and Bristol), and a published series of wills of men from London, with artisans numbering among them. A singular will from Lincolnshire is also examined, for the richness of evidence it provides.

The inventorial records of the inhabitants of Ipswich allow us to consider ownership within the context of a busy early modern trading and commercial town in the provinces, where, because of its location, considerable wealth was generated.[47] In all, the published set comprises seventy-two inventories dating between 1583 and 1681, of which fifty-nine belong to men. Of these, eighteen do not list the occupation of their subject; the remaining forty-one represent a total of twenty-five different occupations, many of which are artisanal, such as blacksmiths, butchers, tailors or linen weavers.

Jewellery is present in only five of the male inventories. The contents of the earliest, however – that of John Seely, dated 8 September 1584 – suggest he may have been a tailor and so the items listed do not appear to be of personal use.[48] The sailor Edward Barnes, whose inventory is dated 14 March 1590, owned two dozen silver buttons, which are listed in such a way as to suggest that they were two separate sets that could be moved between items of clothing.[49] The possessions of another mariner, Matthew Nicholas, inventoried on 25 April 1599, appear relatively modest: some simple furniture, two candlesticks, pewter dishes to dine off, and limited clothing.[50] Nevertheless, he also had 'a silver whistell and silver chaine' together valued at £1 1 s, which seem to be the most valuable goods he owned. It is likely that these items were related to his profession; as such, they would not have been deemed frivolous jewels, but rather a symbol of office. The inventory of a second tailor, Simon Isam, taken on 22 March 1618, records 'a whistle and pick' (probably an ear and toothpick), both made of silver, along with a silver spoon. The combined value of the three items is relatively low, at only 6 s.[51] The final man whose inventory includes jewellery is the clerk Richard Rainsford and this dates to 14 June 1631. His occupation would seem to preclude him from being considered a member of the artisan class, and the goods listed in his inventory reflect both his relative wealth and learned status, so we read of his 'Lybrarye of Bookes' valued at £10, as well as noting that his pewter dishes and clothing are of a greater value than the men we have just seen.[52] The jewels he owns are only two rings and together they are valued at £1 10 s.

The second group of published inventories comes from Bristol, another large urban centre in early modern England, and covers the period from 1542 to 1650.[53] In all, the inventories number 108 but in the period up to 1635, in total only eight men whom we might consider to be artisans appear from their inventories to have owned jewels.[54] The inventories of two of them include jewelled goods that appear indicative of stock-in-trade rather than personal possessions – for example, the 'ij gould bands' and 'vij copper bands' recorded in the inventory of the haberdasher Robert Clement, dated to 19 March 1589 or the 'thrid, buttens, pins, Laces, needles' that are listed in John Noble's 1625 inventory.[55] Of the six men who owned jewellery in a personal capacity, five possessed only rings. According to his inventory of 16 October 1634, the shoemaker John Shipway had 'two gold rings with a gilt gimmall Ringe'.[56] Three of the others – the clothworker Francis Baylie (4 July 1620), the hosier Michael Threkelle (14 July 1623) and the haulier John Davies (1635) – owned a single gold ring each; while another haulier, Nathaniel Wright (inventory dated 1620/1), owned a single ring of unspecified material.[57] The surgeon Richard Woodson, whose inventory is dated 11 February 1623, possessed instruments for his trade of silver and trimmed with silver. These plus his silver toothpicker were valued at £3 18 s 4 p.[58]

It is highly likely that, apart from Shipway's gimmel ring (a form of love token), the rings owned by these men were signet rings engraved with their marks to legitimise their business transactions. Whether used to stamp the wax seals on documents or to prominently display as a mark of identity, these were important personal possessions crucial also in establishing trust.[59] An example of a ring bearing a mark likely to represent an individual involved in the trade of wheat is the fragment of a copper-alloy ring discovered in 2018 in Somerset. The flat, oval-shaped bezel is engraved with a wheatsheaf with seven pellets flanking the decoration. The design is encircled with a corded border (Figure 5.4).

The one minor exception to the general paucity of jewellery in the Bristol artisans' inventories is provided by the plumer Richard Saunders. Saunders's original inventory is dated 6 July 1629, with two further addenda recorded noting his gold ring and the following items: 'one dozen of old silke, silver and gould poynts', worth 18 d; 'one silver bodkin', worth 12 d; 'one old silke and silver hat band and one old small twist silke and silver hatt band', together worth 2 s 6 d; and 'one seale of silver with a boaning handle', worth 12 d.[60] This last item would have been used to authenticate documents, and so would have been essential for the effective conduct of Saunders's trade, just like his signet ring. By way of contrast to artisan ownership, the inventory of the merchant Nathaniel Butcher dated to 25 November 1628 reveals he owned four rings – two signet rings, a ring set with a blue sapphire and one described only as 'a little hooped ringe'. The fifth jewel recorded within the inventory is recorded only as 'an old Jewell'.[61]

Figure 5.4 Fragment of a copper-alloy signet ring engraved with a wheat sheaf encircled by a corded border, sixteenth–seventeenth century. Found in Whitelackington, Somerset. Portable Antiquities Scheme DEV-416E32.

Inventories are able to provide only a snapshot of what goods an individual had in their possession at one particular moment in time. Nevertheless, they do serve to highlight what might have been owned by men from a range of social backgrounds. They have shown how, among the artisan class, rings were the most common single item of male jewellery and were probably used for proving one's identity. However, given that jewels were highly personal objects, their owners might choose to pass them on to family or friends after their death, and, in such instances, it is wills that often enable us to get a sense of what particular jewels might have meant to a man, as he made the deliberate choice to bestow them as marks of remembrance.

A published collection of 245 wills from the diocese of London, dating to the period 1507–47, provides an insight into the nature of the goods owned by the middling and wealthier artisan classes.[62] Of these, ninety-five relate to religious figures and so these can be discounted since they are out of the scope of this chapter. Of the remaining 150 wills, only twenty-two contain any evidence of jewellery within them, of which only twelve were made by male testators. This might suggest that ownership of jewels was not especially common – or, more accurately, that the bequeathing of jewels was not commonly practised even among the citizens of by far the largest and wealthiest city in England. In ten of the twelve wills, there is language such as 'I have putto [*sic*] my seale', which is indicative of the testator using – and therefore owning – a personal sealing device, whether a signet ring or a handheld seal such as the one described in the plumer Richard Saunders's inventory. But only two of these self-sealed wills specify actual bequests of jewels. The haberdasher William Turke, in his will of 14 August 1541, left a 'ryng of golde of the value of 20*s*.', along with a black gown, to his executor.[63] The inclusion of the monetary value of the ring may indicate that Turke wished to emphasise the extent of his gratitude for the role undertaken, though in another context it may also be necessary to indicate the value that was to be spent on a mourning ring. The skinner Wylliam Chambarlayn, in his will of 20 May 1542, bequeathed two rings – one 'a byge houpe off gollde' and the other 'a golde rynge with a dyamond' – in a manner that suggests they were treasured possessions.[64] The remaining two wills do not refer to any form of sealing device but do include bequests of jewels. Wyllam Symons is described as a merchant tailor and on 28 December 1538 he left 'bedes of corall' and 'black bedes of get', both of which were 'gawded with sylver', to his two nieces.[65] The surgeon Antony Copage, on 14 December 1537, left two rings. The one he describes as 'mye rynge' is for the brother of Doctor Laye. Is it possible that this was a form of a signet ring he was bestowing on a fellow medical practitioner? The second he describes as 'the ryng that is a ponne my fynger' and this he leaves to the wife of a one Thomas Austyn, which suggests a level of intimacy between testator and recipient; but, as a member of one of the professions, he cannot be classed as an artisan.[66]

Moving away from London, the will of the mercer John Leek, from Boston in Lincolnshire, and dated 19 August 1527 includes a number of bequests which reveal the items of jewellery that he owned.[67] As one might expect, rings predominate, with six being bequeathed. Three of these rings are described as a 'gymmowe' or 'gymmow', meaning 'gimmel'. Most often associated with love and marriage, gimmel rings take their name from the diminutive of the Latin word for 'twin', *gemellus*, since they are often formed of two interlinked hoops. Each hoop bears its own inscription or stone, essentially forming half a ring; so that when the two hoops are joined, the ring is complete.[68] An example of this is a ring believed to be the wedding ring of the London merchant Thomas Gresham, with the two hoops each inscribed with one half of a fuller inscription and the bezels set with a ruby and diamond, united as the hoops come together. Leek bequeathed William Pakker a gold gimmel ring, and a silver one to Pakker's wife. John Neyll's wife also received 'a lyttyll gymmowe off sylver', while her husband was left a ring of 'basse golde', which might be a reference to an alloy. John Neyll also received a 'sylver hernest gyrdyll' which he had once given to Leek. The fact that Leek bequeathed his wife's wedding ring to one Alice Arley suggests that she was probably a close kinswoman. Such an item of jewellery would have been particularly symbolic, representing at once Leek, his wife and their union. Arley was left a significant number of other items, mostly furniture, but she also received a silver buckle for a girdle and a pendant, the form and material of which is not specified. Isabel Arley received 'a lyttyll golde ryng' and a 'tryangle off sylver' (likely a small triangular-shaped ornament, possibly a dress accessory), while John Arley was left an Agnus Dei pendant, a talismanic jewel of some form depicting the Lamb of God. The final gift of jewellery was the 'crystall stone closyd in sylver' bestowed on John Smyth, though the lack of any further description leaves the precise form of this object unclear. Although it is evident from his will that Leek was considerably wealthy for a mercer, it is not possible to determine which, if any, of the few pieces of jewellery mentioned in it he might have worn himself.

The documentary evidence from inventories and wills allows us to connect male artisans with the types of jewelled items that they owned. What is clear is that we cannot consider these artisans as a homogeneous group, with varying degrees of wealth manifest, and so what these men owned differs. There is limited jewellery within the written sources, contrasting with what material survivals indicate,[69] but we do see that of all items the singularly most important piece was the signet ring. And this is not surprising given that these men demanded trust, authenticity and identification daily within their respective trades.

Conclusion

Men from the artisan class were among those from almost every social and occupational background who wore and owned jewellery in early modern England. The material evidence that we now have access to illustrates the extent to which the consumption of jewels was not restricted to those from wealthy urban or courtly circles. However, the documentary evidence from both wills and inventories – which allows us a glimpse of behaviours at an individual level – is perhaps more nuanced and nor does it reflect what we now understand from objects uncovered as Treasure or through the Portable Antiquities Scheme. However, here in these written sources it seems that a single signet ring may have been the item of jewellery most commonly owned by a male artisan, as it was the one most necessary to his occupation. That is not to say that these men did not own or wear other jewellery; it is just that we cannot see the evidence for this as clearly in the archival records. No doubt this is in part because they owned fewer jewels than their wealthier counterparts, but it may well also be the case that these objects succumbed to the whims of fashion or times of financial hardship without leaving a trace or being found among records of pawnbrokers. Jewellery, therefore, was not merely an elite concern, and nor was it a female preoccupation. Men bought, wore and exchanged jewelled goods, participating in the latest fashions as they did.

Notes

1 Dorinda Evans, 'Leslie, Charles Robert (1794–1859), literary genre painter and author', *Oxford Dictionary of National Biography* (2004), https://doi.org/10.1093/ref:odnb/16485, accessed 30 October 2022.
2 Victoria and Albert Museum, London, inv. No. VAM.FA.115[O], https://collections.vam.ac.uk/item/O80881, accessed 30 October 2022.
3 Autolycus is listed as a 'rogue' in the dramatis personae of William Shakespeare, *The Winter's Tale* in The *Complete Works of William Shakespeare*, Shakespeare Head Press Edition (Ware: Wordsworth Editions, 1999), 1101.
4 Poking sticks were small rods used to stiffen the pleats of ruffs.
5 Shakespeare, *The Winter's Tale*, Act IV, scene iv, ll. 219–30, *Complete Works*, 1121.
6 John Heywood, *The Play Called the Four PP*, ed. John S. Farmer (London: Tudor Facsimile Texts, 1908), sig. A4v.
7 Heywood, The Play Called the Four PP, sig. B1r.
8 Portions of the research for this chapter were originally funded as part of an Arts and Humanities Research Council Collaborative Doctoral Award held jointly with Queen Mary, University of London, and the British Museum between 2009 and 2012, and supervised by Professor Evelyn Welch and Dr Dora Thornton. This resulted in Natasha Awais-Dean, 'Bejewelled: The male body and adornment in early modern England' (PhD thesis, Queen Mary University of London, 2012), which then formed the basis for Natasha Awais-Dean, *Bejewelled: Men and Jewellery in Tudor and Jacobean England* (London: The British Museum, 2017). This chapter incorporates material from both projects, as well as new research conducted since then.
9 Awais-Dean, *Bejewelled*, 3.
10 Joseph T. Fowler (ed.), 'The account-book of William Wray', *The Antiquary*, 32 (1896), 117.
11 Paula Hohti Erichsen, Artisans, Objects and Everyday Life in Renaissance Italy: The Material Culture of the Middling Class (Amsterdam: Amsterdam University Press, 2020), 178.
12 Nancy Cox and Karin Dannehl, *Perceptions of Retailing in Early Modern England* (Aldershot: Ashgate, 2007), 49.
13 *Love's Labour's Lost*, Act 5, scene ii, ll. 317–18, *Complete Works*, 227.
14 Cox and Dannehl, *Perceptions of Retailing*, 50.
15 Cox and Dannehl, *Perceptions of Retailing*, 50.
16 Margaret Spufford, The Great Reclothing of Rural England: Petty Chapmen and Their Wares in the Seventeenth Century (London: Hambledon, 1984), 8.
17 Cox and Dannehl, *Perceptions of Retailing*, 50.
18 The Pedlars Case Stated, Or, Some Remarks upon the Pedlars, and All Their Carrying of Goods Abroad to Proffer to Sale in All Cities, Towns, and Places throughout England and Wales, &c., in Order to the Prevention Thereof (London, 1691).
19 Margaret T. Hodgen, 'Fairs of Elizabethan England', *Economic Geography*, 18, 4 (1942), 391.
20 William Harrison, *The Description of England: The Classic Contemporary Account of Tudor Social Life*, ed. Georges Edelen (Washington, DC, and New York: Folger Shakespeare Library and Dover Publications, 1994), 246, 253.
21 John Cherry, *Goldsmiths* (London: British Museum Press, 1992), 57.
22 Awais-Dean, *Bejewelled*, 14–15. *The Art of Glass* by Antonio Neri (1612) includes recipes for the making of pastes but these were not seen as deceptive. See *L'arte vetraria*, trans. Paul Engle (Heiden & Engle, 2005). For imitation gold, gems and pearls, see also Chapter 3 below by Sophie Pitman and Experiment in focus VI by Michele Robinson.

23 'Wardens' Accounts and Court Minutes', Goldsmiths' Company Library, London, 24 September 1568, 420.
24 'Wardens' Accounts and Court Minutes', 424.
25 Hohti Erichsen, Artisans, *Objects and Everyday Life*, 32.
26 Evelyn Hardy, *Survivors of the Armada* (London: Constable, 1966), 67.
27 This ship was part of a Spanish treasure fleet that sank off the Florida Keys. The *Atocha* was recovered in 1973. See R. Duncan Mathewson III, *Treasure of the Atocha: Sixteen Dramatic Years in Search of the Historic Wreck* (London: Sidgwick & Jackson, 1986), 59–62, and Awais-Dean, 'Bejewelled', 20–2.
28 Awais-Dean, *Bejewelled*, 114.
29 These four objects are on loan from Berkeley Castle, Gloucestershire, to the Victoria and Albert Museum, London, where they are on display in the British Galleries. The Hunsdon Onyx, Victoria & Albert Museum London, LOAN:MET ANON.2:5–1998, collections.vam.ac.uk/item/O124999/cameo-masnago-alessandro; pendant ship, Victoria & Albert Museum London, LOAN:MET ANON.2:3–1998; collections.vam.ac.uk/item/O124996/pendant-ship/ bracelet, Victoria & Albert Museum London, LOAN:MET ANON.2:2–1998, collections.vam.ac.uk/item/O78342/bracelet/; and prayer book, Victoria & Albert Museum London, LOAN:MET ANON.2:4–1998, collections.vam.ac.uk/item/O124997/prayer-book/. For a discussion of their bequest by the second Baron Hunsdon, George Carey (1549–1603), see Richard Edgcumbe, 'O, those jewels! The pride and glory of this kingdom!', in Olga Dmitrieva and Tessa Murdoch (eds), *Treasures of the Royal Court: Tudors, Stuarts and the Russian Tsars* (London: V&A Publishing, 2013), 146–7.
30 The location of this hoard is significant, as Cheapside was known for its proliferation of goldsmiths' shops, particularly on a section known as Goldsmiths' Row, which is described by John Stow in his Survey. See John Stow, *A Survey of London, Reprinted from the Text of 1603*, ed. Charles Lethbridge Kingsford (Oxford: Clarendon Press, 1971), vol. I, 81 and 345. For further discussion of the Cheapside Hoard (now mostly within the collections of the Museum of London), including possible reasons for its deposit, see Hazel Forsyth, *London's Lost Jewels: The Cheapside Hoard* (London: Philip Wilson Publishers, 2013).
31 The laws on portable antiquities are different in Scotland, where all archeological objects must be reported under Treasure Trove. For further information, see treasuretrovescotland.co.uk, accessed 12 November 2022.
32 The Portable Antiquities Scheme database is available at https://finds.org.uk, accessed 12 November 2022. It incorporates Treasure finds, allowing searches to identify similar finds made from a variety of materials.
33 Post-medieval finger ring, Portable Antiquities Scheme ID NMS-88E563, finds.org.uk/database/artefacts/record/id/1029887, accessed 12 November 2021.
34 Post-medieval finger ring, Portable Antiquities Scheme ID NMGW-C8DB58, finds.org.uk/database/artefacts/record/id/31255, accessed 12 November 2021.
35 A restricted search for 'post medieval / button / gold' on the PAS database yielded only eight results, with one of these likely to be modern, date of search 13 November 2021.
36 A restricted search for 'post medieval / button / silver' on the PAS database yielded 180 results, date of search 13 November 2021.
37 See Awais-Dean, *Bejewelled*, 161, for further discussion of the various designs on surviving examples of these buttons.
38 Post-medieval button, Portable Antiquities Scheme ID NCL-4E3126, https://finds.org.uk/database/artefacts/record/id/60289, accessed 5 January 2024. When discovered in 2003, it was originally thought to be made of silver and reported under the requirements of the Treasure Act. Later analysis at the British Museum confirmed it is produced from a base-metal alloy and so it was deemed not Treasure and returned to the finder.

39 Astrid Wendel-Hansen shows in Chapter 8 below that gold buttons were rather exclusive and rare. These were forbidden by sumptuary laws to artisans in early modern Tallinn, and artisan inventories in the period included only silver and tin buttons.

40 For the wearing of non-emblematic hat jewels in England, see the portrait of William Cecil, National Portrait Gallery, NPG 362 or King Henry VIII, National Portrait Gallery, NPG 4690. These can be contrasted with Bartolomeo Veneto, *Portrait of a Gentleman*, c. 1512, Galleria Nazionale d'Arte Antica di Palazzo Barberini, Rome, inv. Prov. Coll. Torlonia 1892 or Francesco Mazzola detto il Parmigianino, *Gian Galeazzo Sanvitale*, 1524, Museo di Capodimonte, Naples, Q111.

41 Hans Holbein the Younger, *William Parr, Marquess of Northampton*, c. 1538–42, The Royal Collection, RL 12231.

42 Quoted in 'A mark of distinction', *Bonhams Magazine*, 21 (Winter 2009), 8. The idea that hat ornaments were the preserve of elite men is influenced by Yvonne Hackenbroch, *Enseignes: Renaissance Hat Jewels* (Florence: Studio per edizioni scelte, 1996), which provides a good survey of extant examples, although some have since been identified as of nineteenth-century manufacture.

43 For a general discussion of the potential uses of plaquettes of all shapes, see Marika A. Leino, 'Italian Renaissance plaquettes in context' (PhD thesis, University of London, 2003), in particular chapter 6 where it is proposed that surviving plaquettes may have been used as hat ornaments.

44 The British Museum's collection was donated by the collector Thomas Whitcombe Greene in 1915. See George F. Hill, 'The Whitcombe Greene plaquettes', *The Burlington Magazine for Connoisseurs*, 30, 168 (1917), 103.

45 The taste and fashion for hat ornaments in the Italian artisan context is seen in the 'Refashioning the Renaissance' database, which includes many examples of hat badges (centigli), gilded medallions (medaglia) or gold studs (borchie). These were usually attached to straw or felt hats and worn by both artisan men and women. See, for example, 'Un cappello di paglia nuovo con medaglia indorata e piume bianche', inventory of the baker Bartolomeo di Andrera, 2 March 1591, State Archives of Siena (ASS), Curia del Placito, 266, 624, fol. 111r; and other examples at www.refashioningrenaissance.eu/database.

46 One notable exception is a gold mourning ring found in Cobham, Surrey, in 2008, Portable Antiquities Scheme ID SUR-676831, finds.org.uk/database/artefacts/record/id/231085, accessed 20 July 2021. The inside of the hoop is inscribed 'FV Ob: 16 May 70', denoting the initials of the deceased and the date on which he died. Convention tells us that the year represented on the ring is 1670. Local parish registers (see PAS database record for a report written by David Taylor on the link to these parish registers) record the death of a Sir Francis Vincent on 16 May 1670, and his will mentions certain bequests of mourning rings. These involved leaving money to relations, friends and associates of the deceased for the purchase of rings in their memory, and so it is more than likely that the ring found in 2008 belonged to one of the individuals named in Sir Francis's will (see 'Will of Sir Francis Vincent of Stoke D'Abernon, Surrey', 29 June 1670, TNA, PROB 11/333/253, fols 67r–69r, fol. 68v).

47 Michael Reed (ed.), *The Ipswich Probate Inventories, 1583–1631* (Woodbridge: Boydell Press for the Suffolk Records Society, 1981), 1.

48 Diocesan archives in Norfolk and Norwich Record Office (NRO) INV/2, No. 37, published in Reed (ed.), *The Ipswich Probate Inventories*, 22–3.

49 Archdeaconry archives in Ipswich branch of Suffolk Record Office (SRO) (I) FE1/2, No. 104, published in Reed (ed.), *The Ipswich Probate Inventories*, 30–1.

50 NRO INV/16, No. 135, published in Reed (ed.), *The Ipswich Probate Inventories*, 51–2.

51 NRO INV/29, No. 29, published in Reed (ed.), *The Ipswich Probate Inventories*, 93–7.

52 NRO INV/37, Box 47, No. 106, published in Reed (ed.), *The Ipswich Probate Inventories*, 110–11.
53 Edwin George and Stella George (eds), *Bristol Probate Inventories, Part 1: 1542–1650* (Bristol: Bristol Record Society, 2002). The scope of my research has meant that only those inventories up to and including the date 1635 have been analysed.
54 This excludes three men who owned swords or daggers: although such weapons could be elaborately jewelled objects, it is not clear from the inventories whether these ones were ornamented in any way or were purely functional. See the inventories of the wire-drawer Jonas Seldon (16 October 1617), the baker John Gibbons (17 September 1623) and the soap-maker John Bittfield (12 May 1624): George and George (eds), *Bristol Probate Inventories*, 25–6, 43–5, 50–1.
55 Bristol Probate Inventories, 6.
56 Bristol Probate Inventories, 87–9.
57 *Bristol Probate Inventories*, 31–3, 36–8, 94–6, 33–4.
58 Bristol Probate Inventories, 36.
59 For more on the role of signets in the early modern period, see Awais-Dean, *Bejewelled*, chapter 9.
60 Bristol Probate Inventories, 69–71.
61 Bristol Probate Inventories, 64.
62 Ida Darlington (ed.), *London Consistory Court Wills, 1492–1547* (London: London Record Society, 1967).
63 London Consistory, 82.
64 *London Consistory*, 143. It is the wording of the bequests, such as noting 'to my spesyall frend, Syr Rouland Hylle', that alludes to these being objects of particular emotional importance.
65 London Consistory, 66.
66 London Consistory, 61.
67 Charles W. Foster (ed.), *Lincoln Wills*, vol 2, 1505–1530 (London: British Record Society, 1918), 35–58, www.british-history.ac.uk/lincoln-wills/vol2/pp35–58#h3–0010, accessed 31 October 2022.
68 For a fuller discussion of gimmel rings, see Awais-Dean, *Bejewelled*, 83–4.
69 Italian archival sources, by contrast, contain a wealth of indications to the ownership of jewellery among artisan families. The 448 Italian inventories included in the 'Refashioning the Renaissance' database records in total over 5,800 items of jewellery, including brooches, necklaces, earrings, rings, crosses, rosaries, medallions, ornaments and rosettes. In addition, circa 1,130 buttons are recorded, many of which were of gold or silver. See www.refashioningrenaissance.eu/database.

6

Dressed to kill: arms, armour and protective attire in Renaissance men's middle- and lower-class dress

Victoria Bartels

Introduction

In sixteenth-century Italy, bearing arms was no longer the preserve of princes, lords, knights or upper-class citizens. We know that weapons were owned and worn – both licitly and illicitly – by men from all social classes, including well-to-do merchants, members of the middling classes such as artisans, and even humble farmers. The diplomat and writer on manners Stefano Guazzo (1530–93) observed in 1574 how 'peasants dare to compete in their clothing with artisans, and artisans with merchants, and merchants with noblemen, so much so that once a grocer has taken up the habit of carrying arms and wearing the clothing of a noble, you cannot tell who he is until you see him in his shop selling his wares'.[1] But just how widespread were weapons and armour in the middle and lower classes really? Although the answer is by no means a simple one, weapon ownership still predominantly appears to have been an upper-class affair. Nonetheless, weapons were also owned by a sizable minority of the popular classes, and when middle- and lower-class men armed themselves, they did so to enhance their personal and public identities, not unlike their elite counterparts. Consequently, these objects served to boost a wearer's sense of security, status and manliness, no matter which rung of society they occupied.

Traditionally, donning weapons was a privilege reserved for upper-class men. Weapons were considered fashionable accessories for the elite wardrobe and, as a result, were considered crucial components of male dress. Countless portraits from the period emphasise this notion. In Giovanni Battista Moroni's *A Knight with His Jousting Helmet*, for instance, the sitter dons a leather jerkin with mail sleeves over a black silk doublet and breeches (Figure 6.1). One arm rests upon a burgonet with an

Figure 6.1 Giovanni Battista Moroni, *Portrait of a Gentleman with Armour (A Knight with His Jousting Helmet)*, 1554–58. National Gallery, London.

elaborate feathered headpiece, complete with matching gold rivets and buckles, and black fabric fastenings. His side sword's steel hilt shines brilliantly against his luxurious black outfit, and the corresponding sheath that encapsulates its blade seems to diagonally cut across the composition's lower half. Disembodied pieces of armour, including a greave, gauntlet and gorget, are scattered along the foreground. The message is clear: martial ability is a manly virtue.

The practice of adding martial touches to portraits was not reserved for men with chivalric titles, however, as the artist and writer Giovanni Paolo Lomazzo quipped in 1584 that 'merchants and bankers, who have never seen an unsheathed sword, to whom one would properly expect a pen behind the ear with a full gown, and a book in front [of them], have themselves painted in armour with batons in hand like generals, [a] thing truly ridiculous, and [which] manifestly displays such little sense and judgment, both in the painted and the painter'.[2]

In most early modern Italian cities, citizenship was a pivotal component required for the privilege of arms-bearing, and men who held upper-class professions (such as the merchants and bankers whom Lomazzo references) were often eligible for citizenship. Although a well-known custom in Florence, for instance, the documentary evidence confirming this dispensation is scant: the only place I have found where this privilege is explicitly stated is a Florentine sumptuary law of 1562, which noted that male citizens could wear a sword and dagger at their belt, and could possess gold, silver or gilt knives and horse trappings.[3] The fact that this information is contained in sumptuary legislation instead of arms regulations indicates how commonplace the practice of donning arms was for elite men. For full Florentine citizenship was not a distinction granted lightly. To qualify, a man had to have resided in the city for thirty years, own and pay taxes on property of a certain value and be approved by the Council of Two Hundred or by the duke himself.[4]

Even with these laws and customs firmly in place, we know that non-elites also kept arms and armour. Paula Hohti's recent study on the possessions of early modern Sienese artisans, for instance, recorded various arms, including swords, daggers and polearms in some of the Sienese inventories she examined. In 1551, a second-hand dealer by the name of Vincenzo di Matteo, for instance, owned 'three swords, one with a sword belt' as well as a dagger with a gilded, velvet sheath with a sword belt or strap.[5] Inventories pose their own challenges when considering weapons, however, as these objects were technically illegal for lower-class inhabitants to own. Circumstances relating to eligibility varied and depended on many factors, including current state policies, the organisation of its militia and the geographical location of where inhabitants resided. Further complicating matters was the practice of awarding arms licences to individuals through special dispensation.

In order to explore these issues comprehensively, this chapter is split into three sections, each of which relies on different types of archival sources.[6] Using contemporary legislation and the Medici ducal collection of letters, the regulation of arms, armour and other sorts of protective clothing is explored first. The large dataset of Italian inventories collected by the 'Refashioning the Renaissance' project is then interpreted and contextualised under this lens. The final section explores petitions asking for permission to wear arms and armour by middling members of society from the records of the Otto di Guardia e Balia, the Florentine magistracy responsible for law enforcement and criminal affairs. By examining an assortment of sources, I hope to shed light on the role that arms, armour and protective attire played in the lives of the popular classes in early modern Italy.

The regulation of arms and armour

Even though weapons were commonplace in early modern society, most Italian city states banned or heavily restricted their usage. Indeed, the right of common individuals to bear arms was often seen as a major threat to government security.[7] One of the earliest acts of the Florentine Duke, Cosimo I de' Medici (1519–74), after assuming power in Florence in 1537 – like that of many other new rulers – was to prohibit the possession of weapons by his subjects.[8] The ban was first referred to in a public notice (*bando*) that year, but it was not until 28 May 1539 that a comprehensive edict dedicated entirely to the question of weapons appeared, under the aegis of the magistracy of the Otto di Guardia e Balia.[9] The edict opened with a long-winded justification for the action being taken:

> The noble and most worthy lords of the Otto di Guardia e Balia of the city of Florence, hoping to provide for the tranquillity and general well-being of this city with every appropriate remedy, and considering the many excesses, scandals, injuries and killings caused by the carrying of weapons, which undoubtedly would not occur if men were stripped of them, and understanding that recently some have been bearing the aforesaid arms about the city and other prohibited places and also keep and retain them within their own homes despite the prohibitions and bans which were previously decreed by the said noble lords of the Otto di Guardia, and knowing this could cause many disorders and inconveniences should no new remedy be provided – therefore they publicly proclaim, notify and command that each and every person of whatever station, rank, quality or condition, whether ecclesiastical or secular, [to declare and surrender any weapon kept in the home or worn] in any place, house or shop in Florence and within eight miles of the said city.[10]

In order to avoid any possible confusion, all major offensive and defensive weapons (as well as their accoutrements) were listed by name. Firearms, both old and new models, along with gunpowder and other crucial accessories, occupy some of the first positions on the list. Other weapons included in the ban appear more appropriate for the battlefield than the urban centre. For instance, the ban restricted seven different types of polearms, explicitly listing each model by name. Although ostensibly cumbersome for urban use, it appears the need to ban such items was warranted. In a 1568 letter to Medici majordomo Matteo Bartoli, the Duke regent Francesco I de' Medici (1541–87) requested that actions be taken to stop priests in Castrocaro from carrying and using polearms, an order seconded by Pope Pius IV.[11]

Even previously legal arms were now strictly prohibited, and the penalty for not surrendering one's weapons within ten days of the edict's issue were pulls on the *strappato* (a form of corporal punishment in which a victim's arms were secured behind his back and tied to a hook or a pulley that would hoist him up in the air and 'drop' him) and monetary fines, some as high as 300 scudi. The populace was encouraged to turn in offenders with the promise of receiving one-quarter of any fine as a reward. These were substantial sums: 300 scudi was roughly nine times the annual earnings of a builder's labourer.[12]

More common weapons, such as 'swords, daggers, knives, pointed instruments, stones, or clubs, or other similar sorts of offensive arms' were also prohibited in the city and within the surrounding three miles, instead of the eight miles set out for the first group of weapons.[13] This would appear to signify that these weapons were more widespread outside immediate city limits and, therefore, slightly more tolerated. Defensive armour and other types of protective clothing were another heavily policed item. Reinforced garments could be fashioned from a wide variety of materials, including iron or steel – in the form of mail and plate armour – leather, wood, bone and fabric.[14] Protective elements could be worn under – or even stitched into – articles of dress. In the *cassone* of Lorenzo de' Medici, for instance, a 'damask fabric doublet lined in chain mail, for use in times of danger' was recorded.[15] Mail and/or steel plates were also added to berets or other types of felt hats, often called *segrete*.

In his combat treatise, master-of-arms Pietro Monte explained to readers that defensive clothing was being made throughout Europe and new modes of making were being 'invented daily'.[16] He noted:

> these days they are made in various manners, sometimes of cord in the manner of mail, sometimes like the stitching of shoes, sometimes just of mail, and sometimes doublets are made with tinned iron. A rather good form of light armour involves taking iron mail, section by section, spreading it over a loom, and stitching it to the fabric with strong, waxed cord, every stitch securing it along the line of mail rings. These days in various places better doublets are made from tinned steel that commonly comes from old swords.[17]

In response to the growing trend for creating armoured clothing, another Florentine ban was published in July 1570.[18] The proclamation outlawed 'doublets, or jerkins, as one says, reinforced, padded, or made with artifice of any strength of small mail, or *agore*, or thick rings made with the force of waxed cord, or other mixture'.[19] If caught with reinforced clothing in the city, or within the surrounding eight miles, perpetrators could be subjected to the penalties for carrying defensive armour, 'namely, penalty of life, and confiscation of goods', although this strict level of punishment was rarely enforced.[20]

Producing such illicit garb was also prohibited, and it was forbidden to ask a tailor or 'any other person, either male or female, from working on them in any way'.[21] This was not surprising, as tailors, seamstresses, shoemakers and shop assistants could also be punished for failing to adhere to contemporary sumptuary legislation. Per a 1546 law on outlawed clothing, for instance, if tailors, seamstresses, shoemakers or their assistants turned in clients who commissioned illegal garments, those in the trade would be pardoned for their part in any wrongdoing.[22] Analogously, if patrons denounced the workshops that agreed to fashion their forbidden vestments, they, too, were immune from being prosecuted. Grace was thus given to the party who denounced the other first, offering an incentive for guilty parties to betray accomplices.

Another law on protective clothing was issued in January 1585 and barred the wearing of *colletti di dante*.[23] These sturdy jerkins made from fallow-deer [*dante*] hide, as the edict explained, were 'new types of leather doublets ... heavy, out of the ordinary'.[24] They probably resembled or were inspired by brigandines: doublets strongly reinforced with iron plates, mail, bone, leather or even toughened cardboard, originally worn by 'brigands' – that is, by light-armed or irregular infantry such as archers, and later by pikemen and musketeers.[25] As such, they were 'really made for the security of the person, and not for ordinary wear' – hence the Florentine authorities' strong aversion to them.[26] The 'Refashioning the Renaissance' database inventories record three *colletti di dante* jerkins, presumably of the type which the Florentine authorities banned in 1585, among the possessions of a carpenter in Siena, and a clothes-seller and an innkeeper in Venice.[27]

Weapons and the urban middle and lower classes

In the sample of 448 post-mortem inventories of urban middle- and lower-class households collated by the 'Refashioning' project, 122 (around 27 per cent) mention at least one weapon or piece of armour. In total 196 weapons were identified. Of the 196 items documented, seven were present in Venice, 80 in Florence and 109 in Siena. But since arms were illegal for non-citizens – such as most trade and craftsmen – without special dispensation, it is hard to know how far the inventories accurately reflect either the number of non-elite owners or the number of weapons owned by them. Given the harsh penalties in place for lawbreakers, it is possible that any illicitly held items were given away or otherwise disposed of before the notaries were called in to conduct their surveys. If illicit objects were found, they were to be turned over to the authorities at once.

Another incentive for getting rid of forbidden items was the fact that charges could be pressed for solely having knowledge of an illegal weapon that was kept in the household. In a Florentine firearms ban from June 1547, the comprehensive law outlawed wheellock arquebuses and *schioppi*, in addition to matchlock [*da corda*] and what appears to be early flintlock [*da fucile*, *da pietra*, and *da acciaiuolo*] styles that were smaller than one braccio and a half.[28] If the informed party did not denounce the perpetrator(s) within three days' time of gaining such knowledge, he or she would be charged 150 scudi, half of the fee charged to persons illegally caught with firearms.

Geographical location, especially in terms of its proximity to the city centre, additionally affected arms legislation efforts. The 'Refashioning the Renaissance' inventories were sourced from households located in the urban centres of Florence, Siena and Venice, which possessed much stricter laws compared to rural areas, the countryside and towns included in a state's dominion. In the Medici-governed Sienese territories in 1560, for instance, anyone who lived at least eight miles from the centre could keep weapons in their home and carry on their person a sword and dagger, while anyone who lived even further afield was permitted to keep arms at home and carry whatever offensive or defensive weapon they wished.[29] The manner in which the restrictions became looser the further one ventured from the city centre reflected the more lawless nature of the countryside and the limitations of the state's authority: in some instances, entire rural locales were allowed to carry arms.[30] But it also demonstrates the priority given to keeping the area within the city walls as free from weapons as possible – even to the extent of stationing several companies of Duke Cosimo I's militia outside the city in various parts of the Florentine domain.

One's city of residence similarly plays an important role in the 'Refashioning the Renaissance' data. Of the arms and armour recorded, 75 per cent of weapon owners were Sienese inhabitants. This is most certainly a reflection of the looser regulations in force there than of any sort of behavioural tendency. When Duke Cosimo I acquired Siena in 1557, increasing his domains by a third, the restrictions imposed on the city's inhabitants appear to have been laxer than Florence's. In a *bando* of June 1560 prohibiting *archibusetti* (small, hand-held arquebuses), for instance, it appears that the Florentine government continued to recognise the rights to bear arms which some Sienese had possessed before their city's annexation.[31] In fact, all of the intact firearms – ten in total – found in the 'Refashioning the Renaissance' sample were owned by a trumpet player, a bookseller, an oil producer and a blacksmith, all located in Siena.[32] The home of the bookseller Giovanni Tantucci contained four guns in total. His comprehensive collection included two wheellocks, an early flintlock model and a matchlock, along with some additional accessories.[33]

The most commonly owned weapons were undoubtedly swords and daggers, making up roughly 84 per cent of the arms and armour recorded (Figure 6.2, Figure 6.3). Despite the popularity of long and short bladed weapons in this period, and the variety of types, the documentary sources often describe them quite vaguely with the single term *spada*. A mason's inventory in Venice, for example, simply records the presence of 'una spada' that was found, along with various items of clothing and shoes, inside a big, wooden chest that was unlocked with a key.[34] Only if the weapon was unusual in some way, or its fittings were noteworthy or valuable, was more detail recorded. Thus, for example, a Venetian innkeeper's inventory from 1612 boasted 'two scimitars in the Turkish style, with their sword belts' ('due samitere alla turchesca con suoi centuroni'), while a Florentine arms-maker by the name of Niccolo di Antonio had 'one sword

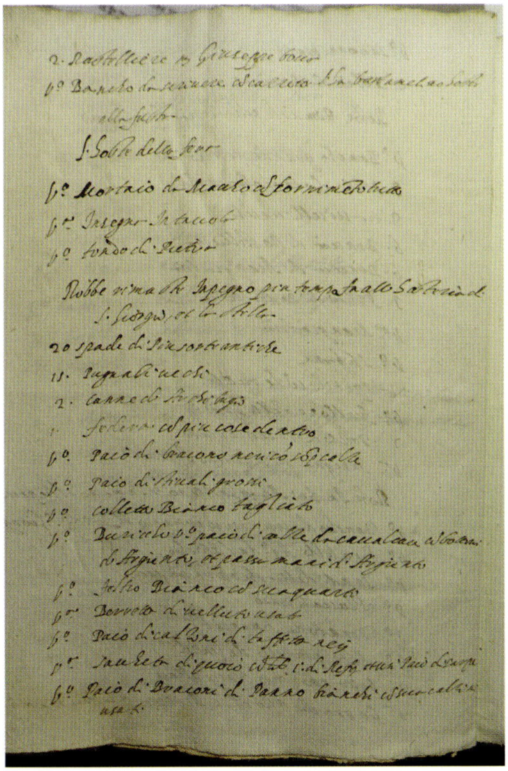

Figure 6.2 Arms and armour recorded in an inventory of a Florentine innkeeper, 1571. Archivio di stato, Florence, Magistrato dei Pupilli, 2653, fol. 119v, 1571.

with a silver hilt' ('1 spada con manico di argento') in his sitting room.[35] Damaged or timeworn elements were similarly noted, as was the case for a tailor in Siena who possessed 'a blade of a sword without finishings' ('una lama da spada senza finimenti') and an innkeeper in Florence with 'a sword without a sheath and pommel' ('una spada senza fodero e senza pome') that was found 'in the room above the living room' ('nella stanza sopra la sala').[36] Individuals with the financial means could have the blade of their weapon etched with ornate patterns, or its hilt and sheath adorned with precious metals or even jewels. Even if the matching pair of sword and dagger hilts set with 680 diamonds by the Medici goldsmith Giacomo Biliverti in 1601 was beyond one's reach, it was still possible for someone on a modest income to add a touch of luxury, such as the dagger and pair of swords boasting silver hilts owned by a Sienese candle-maker in 1595, or the 'dagger with its little belt and silver hilt' ('un pugnale con suo cinturino e manica d'argento') belonging to a Florentine fishmonger (Figure 6.4).[37]

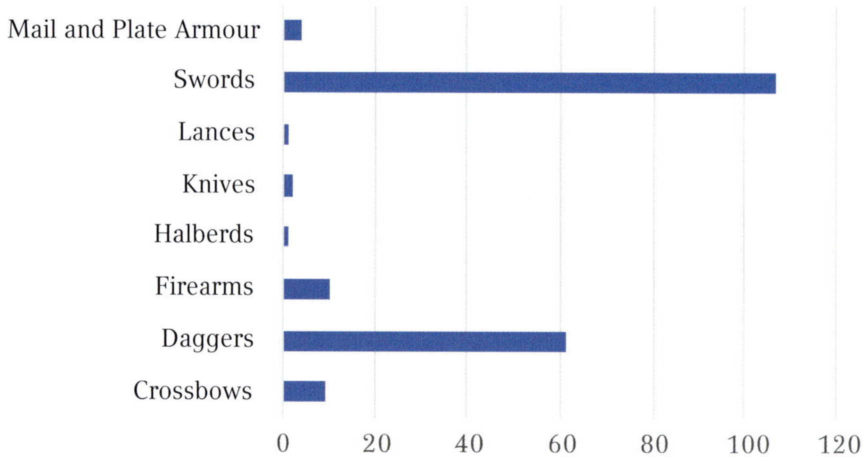

Figure 6.3 Arms and armour recorded in artisan inventories in Florence, Siena and Venice, 1550–1650.

Figure 6.4 An example of a dagger and scabbard with silver embellishments. Sword maker Wolf Paller, German, *Dagger with Scabbard*, c. 1575. Steel, silver, ray skin, leather, wood; length 38.4 cm. Metropolitan Museum of Art, New York, Bashford Dean Memorial Collection.

Interestingly, the inventories of innkeepers often appeared to be chock-full of weapons. This almost certainly reflects the fact that innkeepers often acted as pawnbrokers, lending customers money against pledges of goods. Like clothing and jewellery, weapons were objects that could retain value, and as a result they were frequently pawned – especially by soldiers, given their itinerant and unpredictable lifestyle.[38] In 1533, the inventory of Marchione di Paulo, innkeeper in the town of Mulazzo in north-western Tuscany, catalogued six swords (including one 'belonging to the house'), four pikes, two spears, three sword belts, one dagger in the Spanish style and another with silver furnishings, and one pistol. A further ten swords, twelve daggers, one white sword belt and a silver hilt were noted as being out at pawn.[39] In 1571, the household inventory of the innkeeper Nicco di Michielino recorded that twenty antique swords of various types ('spade di piu sorte antiche'), fifteen daggers categorised as 'old', and two arquebus barrels had similarly been pawned.[40]

Nine crossbows were also recorded in the inventories. In Siena, they belonged to an oil producer, linen weaver, sausage-maker, weaver and three different greengrocers.[41] The oil producer Michele Biancardi's inventory records a crossbow, as well as a sword and dagger, kept in a chest with various types of women's items and fabrics.[42] Also found inside a chest was a Florentine miller's *balestro da pallottole*, a type of crossbow invented and popularised in Tuscany that used stone balls for ammunition.[43] These weapons were a popular choice for hunting, but were also used in acts of assault. In March 1562, for instance, an arquebusier named Cencio di Pierantonio received a 25 fiorini fine for having shot a certain Giovanni di Rimedio in Carmignano with the weapon the preceding November.[44]

Defensive wear was also present, although uncommon. One such case was a 'rusty' and 'damaged' armour with a morion (an open-faced helmet typically fashioned with a crest) recorded in a list of items that had been pawned by a Florentine innkeeper.[45] Protective jacks of mail were owned by a dyer, soap-maker and baker in Venice, the latter of whom pawned his for 60 lire.[46] The soap-maker Zuan Francesco Fondi safeguarded his two protective garments, a jack and shirt of mail ('doi zachi, cioe una camisiola et un zaho'), in a chest kept in one of the rooms in his house. Three *colletti di dante*, as mentioned above, were also recorded.[47]

In contrast to the verbose legislation issued during the same period, the 'Refashioning the Renaissance' database demonstrates that as many as a quarter of inventory owners possessed arms and armour in their households. As mentioned earlier, however, it is difficult to ascertain whether these numbers accurately reflect arms possession in these social groups, as family members could be prosecuted if they knew such illegal objects were present in the household. In summary, the majority of the arms that were recorded were owned by Sienese men. This makes sense given the looser regulations in place there, as the city was absorbed by the Medici duchy in the mid-sixteenth century.

Arms licences and the lower and middle classes

Despite all the restrictions imposed by state authorities on the private possession of weapons, members of the city's middling and lower classes were still able to petition the state for the right to bear arms. As with other types of personal entreaty and appeal to the government, these pleas were made via formal written petitions called 'supplications' (*suppliche*). In Florence, they could be submitted by dropping the documents through an aperture known as the *buca de suppliche*, located on Via Lambertesca, near the Piazzale degli Uffizi (Figure 6.5).[48] This was intended to provide Florentines with direct and discreet access to their duke – a 'good custom', as Duke Cosimo acknowledged in 1568, explaining that 'any kind of person, for comfort and facility in negotiating, could write to us and have the letter arrive in our hands'.[49]

The task of reviewing petitions relating to arms in Medici territories fell to the Otto di Guardia e Balia. Many petitioners requested licences or renewals of previous privileges, as seen in the case of the Flemish artist

Figure 6.5 An aperture for submitting supplications in Florence. Also known as the '*buca de suppliche*'. Located on via Lambertesca, near the Piazzale degli Uffizi, Florence.

Giovanni Stradano (or Johannes Stradanus; 1523–1605). On 29 October 1565, while Stradano was busy creating decorations for the imminent wedding of Duke Cosimo's heir, the future Grand Duke Francesco I, the court artist petitioned the duke for 'a pardon to own [and] to carry offensive and defensive arms, namely a sword, a dagger, and a jack of mail, such as your most illustrious excellency had granted [for] many years'.[50] A copy of the original licence that Stradano was seeking to renew was attached to his supplication (Figure 6.6 and 6.7). Dated 14 March 1558, and probably written by the then secretary of the Otto di Guardia e Balia, Francesco Borghini, it declared:

> Master Giovanni di Giovanni della Strada, Flemish Painter of the age of 35 years, black hair, very bushy beard, white in the face, average stature, has the ability to own and carry in the domain of his most illustrious excellency, and by his grace, [a] jack of mail, sword and dagger, as the son-in-law of Master Giovanni Rosti, as [noted] in the file of supplications under number 320.[51]

'Number 320' refers to a past supplication submitted by the noted tapestry-maker Jan Rost dated 8 January 1558, which asked permission to equip with swords, daggers and jacks of mail four members of his household (*domestici*): his brothers Pieter and Jacob; his assistant Baltasar, 'who

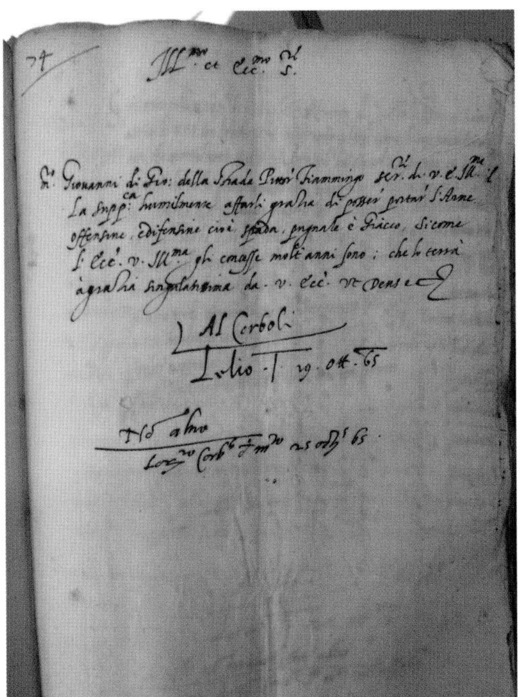

Figure 6.6 Giovanni Stradano's supplication to wear offensive and defensive arms, Archivio di Stato, Florence, Otto di Guardia e Balia del Principato, vol. 2247, fol. 74v.

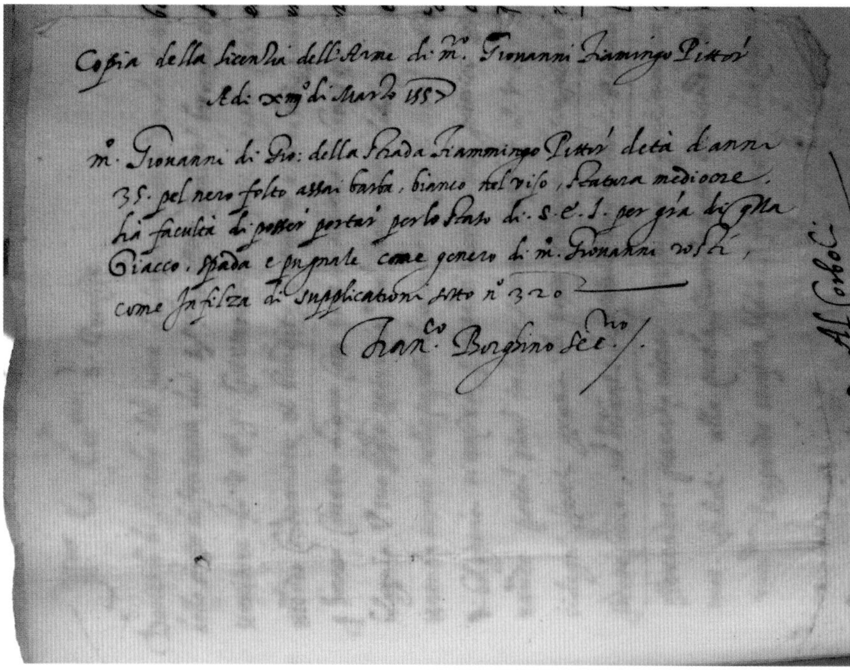
Copia della licentia dell'Arme di m.o Giovanni Fiamingo Pittore
Adì xiiij° di Marzo 1557
m.o Giovanni di Gio: della Strada Fiammingo Pittore d'età d'anni
35. pel nero folto assai barba, bianco del viso, statura mediocre,
ha facultà di potter portar per lo Stato di S. E. I. per già di q(ue)lla
Giacco, spada e pugnale come genero di m. Giovanni Rost,
come in filza di supplicationi sotto n° 320
Fran.co Borghino Sec.rio

Figure 6.7 Copy of Giovanni Stradano's arms licence issued in 1557. Archivio di Stato, Florence, Otto di Guardia e Balia del Principato, vol. 2247, unnumbered, between folios 74 and 75. Archivio di Stato di Firenze.

live[d] above his workshop'; and Giovanni Stradano.[52] Rost noted that all four men were Flemish and, somewhat curiously, that they would if necessary take up their arms 'in service of your illustrious excellency'.[53] Rost's request was granted on 13 February, five weeks after its submission; but whether his workshop actually needed protecting, or whether the permits were sought purely to demonstrate the status of Rost and his household, is unknown.[54] Interestingly, Stradano's supplication to renew his licence was submitted in the year of Rost's death, suggesting that he needed to transfer the authorisation into his own name. The outcome of his application, though, remains uncertain.[55]

High-profile artists at court appeared to be likely candidates for arms licences. Another such case was the flamboyant Florentine goldsmith Benvenuto Cellini (1500–71). After he returned to Florence in 1545, Cellini cherished his licence to bear arms, acknowledging the favour shown him by Duke Cosimo when he declared in a supplication in 1562 that 'just as you had done for all your good servants, you granted me the right to carry weapons'.[56] As is evident from the goldsmith's many other supplications, he took full advantage of his privilege to maintain a well-stocked arsenal. According to an inventory from 1570, he was keeping in the antechamber to his Florentine lodgings 'two *pistolesi* daggers, a knife in the Turkish style, a sword, a dagger with silver mounts, a hand-and-a-half sword and a Moorish spear' – although a note on the inventory indicated that Cellini did not have permission to own the last two items and they would be confiscated by the state, as a result.[57]

The duty of reviewing such cases required much time and effort, and it was not uncommon for the duke to intervene personally in tricky cases, such as the supplication submitted in 1560 by a *contadino* (peasant) named Tonino Frosini from Meleto Valdarno, a hamlet in the countryside south-east of Florence.[58] As the 'capo della casa' (head of the household), Tonino explained, he had to attend several markets each day for work. This meant he often returned home late in the evening, making him susceptible to attacks from his enemies. He therefore sought permission to equip himself with a sleeved jack of mail, a sword and a dagger. As well as attesting that these weapons were to be worn ten miles outside the city, Tonino rather curiously admitted that he was not capable of using them – he wished only *to be seen* wearing arms, in order to give the impression that he was the sort of man who would defend himself at a moment's notice. He did not intend to 'offend anyone', merely avoid further 'insults and injuries in order to attend more securely to his affairs and to his wife and children'.[59]

Forwarding Tonino's supplication for Duke Cosimo's attention, the secretary of the Otto di Guardia e Balia, Lorenzo Corboli, noted that many weapons were already worn by 'would-be bravoes and bullies', and that, if licences to bear arms were granted on the grounds put forward by Tonino, there would be 'infinite [men] more frightened than him' demanding the same right.[60] Allowing Tonino to carry weapons could only make the situation worse, the *contadino* 'not being capable, as he says', and so he recommended rejecting the supplication.[61] The duke, however, chose to ignore Corboli's advice, and granted the petition, for the supplicant's safety and security.

Although protection and self-defence were the reasons most often cited for carrying weapons, their presence in early modern Italian society went far beyond their utilitarian function. Indeed, the Venetian humanist Andrea Navagero (1483–1529) went so far as to confess in his poem *De imagine sui armata* that he wore weapons in his portrait only because everyone else did so, 'not because [he was] skilled in fighting'.[62] One of the best ways to embody a manly persona, therefore, was literally to dress the part. But it would be an oversimplification to suggest that arms and armour were solely theatrical props that allowed members of civic society to pretend to be knights or warriors. Rather, as elements of fashionable dress, they were central to contemporary notions of masculinity and social status, giving expression to them in a multitude of ways.

One of the most significant features of this episode is the way it illustrates how the laws relating to arms-bearing, and its traditional associations with martial, patriarchal ideologies, were being reinterpreted to fit profoundly non-elite contexts. Tonino's enemies were only briefly mentioned in his petition. Instead, he emphasised factors relating to his trade and to his masculine role as head of the household. This was perpetuated in the gendered language and assumptions of the state's administrators, who stressed the peasant Tonino's status as 'capo della casa' and 'padre di famiglia' and his role as protector and provider: in order to safeguard his wife and son, who depended on him for their welfare, he must first safeguard himself. And even though Tonino lacked the skill to defend himself with arms, it was understood just how powerful weapons could be symbolically, as merely the sight of them could guard men from harm.

The Otto di Guardia e Balia's records demonstrate that a wide variety of residents submitted petitions to bear arms. In 1551, for instance, a thirty-year-old Portuguese priest by the name of Francesco Barradas asked to carry weapons on an upcoming trip to Florence.[63] Francesco noted that he had already received permission to bear arms in his parish but required the extension of this authorisation because he had to visit the archbishop in Florence for work. He stated that a rowdy group of men from his parish – who previously menaced him, almost taking his life – befriended a group of young citizens who resided in Florence. Because he lived in fear of his enemy's Florentine allies, Francesco asked to carry his concealed dagger and have his manservant armed with his sword during his visit to the city centre. He demonstrated the urgency of this request by explaining that his enemies had been given permission from his Excellency to carry arms. It appears this reasoning fell short of convincing either the duke or the Otto di Guardia e Balia's secretary, as the Portuguese priest's application to carry arms in Florence appears to have been denied.

Conclusion

What is clear from all this is that, as items of male dress, weapons were objects which embodied bravery in the fuller sense of the word – that is, courage, strength and beauty – and that these attributes were transferrable to those who wore them. An essential part of martial, civic and patriarchal ideologies, the practice of bearing arms appeared to be an extension of a man's personal and public identity, fundamentally shaping the privileges individuals received from state authorities. Yet, as we have seen, there were legal and illegal ways of navigating these strata. By understanding the social and cultural context in which these objects existed, we shed light on the complex role played by weapons, protective materials and masculine dress in early modern Italy.

Notes

The research leading to these results was done as part of the 'Refashioning the Renaissance' project that has received funding from the European Research Council (ERC) under the European Union's Horizon 2020 research and innovation programme (grant agreement No. 726195).

1 Stefano Guazzo, *La civil conversatione*, ed. Amedeo Quondam, vol. I (Ferrara: Franco Cosimo Panini, 1993), 140. Quoted and translated in Elizabeth Currie, *Fashion and Masculinity in Renaissance Florence* (London: Bloomsbury Academic, 2016), 107.

2 'Per incontro poi i mercanti e banchieri, che non mai videro spade ignuda, ai quali propriamente si aspetta la pena nell'orecchia con la gonnella intorno, ed il giornale davanti, si ritraggono armati con bastoni in mano da generali, cosa veramente ridicola, e manifestamente accusa il poco senno e giudizio, così del dipinto, come del dipintore', Giovanni Paolo Lomazzo, *Trattato dell'arte de la pittura di Gio. Paolo Lomazzo ... diviso in sette libri* (Rome: Presso S. Del-Monte, 1844), vol. 2, 374. Also cited by Carolyn Springer, *Armour and Masculinity in the Italian Renaissance* (Toronto: University of Toronto Press, 2010), 162. For more on the link between arms, dress, and masculinity, also see Victoria Bartels and Katy Bond, "Dress & Material Culture," in *A Companion to Late Medieval and Early Modern Augsburg*, edited by B. Ann Tlusty and Mark Haberlein (Leiden: Brill, 2020).

3 Lorenzo Cantini (ed.), *Legislazione toscana: raccolta e illustrata dal dottore Lorenzo Cantini*, 32 vols (Florence: Albizziniana da Santa Maria in Campo per Pietro Fantosini e figlio, 1800), vol. 4, 402–10.

4 Robert Burr Litchfield, *Emergence of a Bureaucracy: The Florentine Patricians, 1530–1790* (Princeton: Princeton University Press, 2014), 46.

5 State Archives of Siena (ASS), Curia del Placito, 746, 457, in Paula Hohti Erichsen, *Artisans, Objects and Everyday Life in Renaissance Italy: The Material Culture of the Middling Class* (Amsterdam: Amsterdam University Press, 2020), appendix 6. For more examples of weapons owned by craftsmen, see Paula Hohti, 'Material culture, shopkeepers and artisans in sixteenth-century Siena' (PhD thesis, University of Sussex, 2006).

6 I thank the 'Refashioning' team, in particular Stefania Montemezzo, for identifying, transcribing and making available the archival inventories of artisans used in this chapter, available open access at www.refashioningrenaissance.eu/database.

7 In some other regions of early modern Europe with citizen militias, however, cities and states actually required men to own weapons. For more on this topic, see B. Ann Tlusty *The Martial Ethic in Early Modern Germany: Civic Duty and the Right of Arms* (Basingstoke: Palgrave Macmillan, 2011).

8 John K. Brackett, *Criminal Justice and Crime in Late Renaissance Florence, 1537–1609* (Cambridge: Cambridge University Press, 1992), 103. It is unclear precisely when the first arms ban was issued in Florence, although evidence suggests the fifteenth century, if not earlier: Brackett, *Criminal Justice and Crime*, 123 and n. 153.

9 The Florentine year traditionally began on 25 March; where necessary in this chapter, the year has been adjusted to accord with the modern calendar's new year on 1 January. Statutes issued as bandi were written in everyday Italian, and then printed so that they could be read aloud and posted in prominent public places within the city, and circulated to Florentine magistrates, commissars, vicars and captains in the regions. Brackett, *Criminal Justice and Crime*, 80–1.

10 'Li Spettabili, & Dignissimi Signori Otto di Guardia, & Balìa della Città di Fiorenza, desiderando con ogni remedio opportuno provedere alla quiete, & bene universale di questa Città, & considerando di quanti eccessi, & scandali, ferite, & homicidii sia causa il portare l'arme, le quali senza dubbio alcuno non resultano, quando li

huomini si trovono di quelle spogliati, & intendendo come alcuni da pochi giorni in quà portano l'arme predette per la Città, & altri luoghi prohibiti, & etiam ne hanno, & ritengono in le loro case contro le prohibitioni, & Bandi altra volta per detti Spettabili Signori Otto mandati, & conoscendo tal cosa potere causare molti disordini, & inconvenienti se con nuovo rimedio non si provedessi. Però fanno publicamente Bandire, notificare, & expressamente comandare a ogni, & qualunque persona di qualunche Stato, grado, qualità, o conditione si sia così ecclesiastica, come secolare'; 'in qualsivolgia luogo, casa, o bottega in Fiorenza, & appresso a essa Città a miglia otto', Cantini (ed.), *Legislazione toscana*, vol. 1, 183–6.

11 State Archives of Florence (ASF), Mediceo del Principato, 229, fol. 255v. Medici Archive Project, Doc ID 19206. Special thanks to Maurizio Arfaioli for bringing this document to my attention.

12 Currie, *Fashion and Masculinity*, 19.

13 'spade, pugnali, coltelli, appuntati, sassi, o bastoni, o altre simili sorti d'arme offendibili', Cantini (ed.), *Legislazione toscana*, vol. 1, 185.

14 Lionello G. Boccia, *Armi difensive dal Medioevo all'Eta Moderna* (Florence: Centro Di, 1980), 17.

15 Translated and quoted by Richard Stapleford, *Lorenzo de Medici at Home: The Inventory of the Palazzo Medici in 1492* (University Park: Penn State University Press, 2013), 28.

16 Jeffrey L. Forgeng (trans.), *Pietro Monte's Collectanea. The Arms, Armour and Fighting Techniques of a Fifteenth-Century Soldier* (Woodbridge: Boydell Press, 2018), 169.

17 Forgeng (trans.), *Pietro Monte's Collectanea*, 105. The use of tinned iron in defensive clothing likely protected the plates from rusting, especially considering the perspiration one accumulated while fighting. Special thanks to Chassica Kirchhoff for pointing this out to me.

18 Cantini (ed.), *Legislazione toscana*, vol. 7, 227–30.

19 'giubboni, o imbusti, come si dice, rinforzati, imbottiti, o fatti con artificio di alcuna forte di magliette, o agore, o buochi spessi farti con forza di spaghi incerati, o altra mistura', Cantini (ed.), *Legislazione toscana*, vol. 7, 227–30.

20 'cioè pena della vita, & confiscatione de' Beni', Cantini (ed.), *Legislazione toscana*, Vol. 7, 227–30. Modern year given in text to adjust for the Florentine calendar. For more, see Brackett, *Criminal Justice and Crime*, 103, n. 19.

21 'o qualsi voglia altra persona così mastio come femina lavorarle in alcun modo', Cantini (ed.), *Legislazione toscana*, vol. 7, 227–30.

22 Cantini (ed.), *Legislazione toscana*, vol. 1, 323–4.

23 Cantini (ed.), *Legislazione toscana*, vol. 11, 354–6.

24 'nuove invenzioni con Giubboni di cuoio … gravi, fuori dell'ordinario', Cantini (ed.), *Legislazione toscana*, vol. 11, 355.

25 Stephen V. Grancsay, *Arms and Armor: Essays from the Metropolitan Museum of Art Bulletin, 1920–1964* (New York: The Metropolitan Museum of Art, 1986), 282.

26 'fatto veramente per sicurtà della persona, è non per vestimento ordinario', Cantini (ed.), *Legislazione toscana*, vol. 11, 355.

27 In the 'Refashioning the Renaissance' database, colletti di dante are categorised as clothing and have not, therefore, been added into the total of arms and armour recorded here. For examples, see the inventory of the wood-carver Gaspare Radi, 24 October 1636, State archives of Siena (ASS), Curia del Placito, 279, 12, fol. 66v; the inventory of the innkeeper Giovanni Suster, 11 September 1646, State Archives of Venice (ASV), Giudice di Petizion, Inventari, 360, 25, fol. 11v; the inventory of the cloth-sellers Isach, Consseglio and Mazo, brothers of Udine Valvasoni, 9 March 1634, ASV, Giudice di Petizion, Inventari, 354, 37, fol. 8r.

28 Cantini (ed.), *Legislazione toscana*, vol. 1, 358–62. One braccio equalled approximately 58 cm, meaning that the prohibition outlawed firearms shorter than 85 cm

total. Firearm size was another factor that mattered greatly to Florentine legislation. Small pistols were considered extremely dangerous, as they were easily concealed, could be worn in multiples and were primarily used for harming others, unlike larger guns that could also be used for hunting.

29 Cantini (ed.), *Legislazione toscana*, vol. 4, 30–2.
30 Brackett, *Criminal Justice and Crime*, 103.
31 Cantini (ed.), *Legislazione toscana*, vol. 4, 30–2.
32 See the inventories at ASS, Curia del Placito: inventory of the trumpet player Paolo Rimbombi, 30 September 1597, 270, 1052, fol. 96v; inventory of the bookseller Giovanni Tantucci, 12 September, 1636, 279, 5, fol. 26v; inventory of the oil producer Leonardo de Lullis, 21 August 1642, 282, 177, fol. 142v; inventory of the blacksmith Giovanni Battista Bonavoglia (date n/a), 1647, 283, 297, fol. 203v.
33 Inventory of the bookseller Giovanni Tantucci, 12 September 1636, ASS, Curia del Placito, 279, 5, fol. 26v.
34 Inventory of the mason Maphei de Zazziis ASV, 16 February 1554, Cancelleria inferiore, Miscellanea, 39, 62, fol. 1r.
35 Inventory of the innkeeper Zuanne Aider, 13 August 1612, ASV, Giudice di Petizion, Inventari, 345, 12, fol. 5r; inventory of the arms-maker Niccolo di Antonio, 17 February 1572, ASF, Magistrato dei Pupilli, 2653, fol. 693r.
36 Inventory of the tailor Vincenzo di Giovanni (date n/a) 1591, ASS, Curia del Placito, 266, 562, fol. 16v; inventory of the innkeeper Oratio di Iacomo di Anton di Piero Franceschini, 7 August 1617.
37 Inventory of the candle-maker Pellegrino Peruzzi (date n/a) 1595, ASS, Curia del Placito, 269, 924, fol. 76v; inventory of the fishmonger Tommaso di Salvadore Mariti (date n/a) 1620, ASF, Magistrato dei Pupilli, 2717, fol. 284r. For Biliverti's hilts, see Angus Patterson, *Fashion and Armour in Renaissance Europe: Proud Lookes and Brave Attire* (London: V&A Publishing, 2009), 60.
38 William P. Caferro, 'Warfare and economy in Renaissance Italy, 1350–1450', *Journal of Interdisciplinary History*, 39, 2 (2008), 191.
39 Inventory of the innkeeper Marchione di Paulo, 1533, ASS, Curia del Placito, 687, 20, in Hohti Erichsen, *Artisans, Objects and Everyday Life*, appendix 4.
40 Inventory of the innkeeper Nicco di Michielino, 2 May 1571, ASF, Magistrato dei Pupilli, 2653, fol. 119v.
41 Inventory of the oil producer Michele Biancardi, 9 July 1587, ASS, Curia del Placito, 262, 106, fol. 147r; inventory of the linen weaver Giacomo Alessandrini (date n/a) 1595, ASS, Curia del Placito, 269, 938, fol. 98r–98v; inventory of the sausage-maker Lazzaro Tartagli, 18 November 1600, ASS, Curia del Placito, 273, 1314, fol. 56r; inventory of the weaver Antonio del Serva, 12 September 1638, ASS, Curia del Placito, 280, 73, fol. 45r; inventory of the grocer Giovambattista di Piersega, 16 November 1600, ASS, Curia del Placito, 274, 1398, fol. 21v; inventory of the grocer Angelo di Giulio, 6 April 1600, ASS, Curia del Placito, 275, 1505, fol. 31r; inventory of the grocer Gismondo di Agnolo, 1 June 1603, ASS, Curia del Placito, 275, 1503, fol. 24r.
42 Inventory of the oil producer Michele Biancardi, 9 July 1587, ASS, Curia del Placito, 262, 106, fol. 147r.
43 Inventory of the miller Giovanbattista di Paulo Pini, 5 September 1565, ASF, Magistrato dei Pupilli, 2708, fol. 171v.
44 ASF, Otto di Guardia e Balia del Principato, 2239, fol. 11.
45 Inventory of the innkeeper Nicolo di Michielino, 28 May 1571, ASF, Magistrato dei Pupilli, 2709, fol. 78r.
46 Inventory of the dyers Marci e Joanni Jacobi Bartholomei, 18 April 1577, ASV, Cancelleria inferiore, Miscellanea, 42, 7, fol. 5r; inventory of the soap-maker Zuan Francesco Fondi, 26 January 1642, ASV, Giudice di Petizion, Inventari, 357, 91, fol. 1v; inventory of the baker Foresto Foresti, 16 October 1636, ASV, Giudice di Petizion, Inventari, 355, 33, fol. 13v.

47 Inventory of the wood-carver Gaspare Radi, 24 October 1636, ASS, Curia del Placito, 279, 12, fol. 66v; inventory of the innkeeper Giovanni Suster, 11 September 1646, ASV, Giudice di Petizion, Inventari, 360, 25, fol. 11v; inventory of the cloth-sellers Isach, Consseglio e Mazo brothers of Udine Valvasoni, 9 March 1634, ASV, Giudice di Petizion, Inventari, 354, 37, fol. 8r.

48 In 1576, the celebrated architect Bernardo Buontalenti (1531–1608) constructed the eye-catching Porta delle Suppliche (Door of the Supplications), located roughly two metres to the buca's left. The Mannerist portal (which now welcomes guests into the Uffizi Museum's ticket office) boasts a dramatic split pediment with a bust of Grand Duke Francesco I de' Medici, who was rumoured to have spied on petitioners submitting their letters in the space between the pediment's two halves. See Christopher Hibbert, *Florence: The Biography of a City* (London: Penguin, 2004), 357.

49 Quoted and translated in James E. Shaw, 'Writing to the prince: Supplications, equity and absolutism in sixteenth-century Tuscany', *Past & Present*, 215 (2012), 57.

50 'Mastro Giovanni di Giovanni: della Strada Pittore Fiammingo servitore di vostra eccellentia illustrissima La supplica humilmente affarli gratia di possere portare l'arme offensive, e difensive cioè spada, pugnale e giacco, sicome li eccellentia vostra illustrissima gli concesse molt'anni sono', ASF, Otto di Guardia e Balia del Principato, 2247, fol. 74v. All of the city's most eminent artists were tasked with creating elaborate street decorations for the official entry into Florence of Francesco's bride, Joanna of Austria, which took place in December 1565. Stradano's supplication may have been handled by Francesco, as he had begun acting as his father's regent in 1564, especially for administrative tasks.

51 'Mastro Giovanni di Giovanni: della strada Fiammingo Pittore detà d'anni 35. pel nero folto assai barba, bianco nel viso, statura mediocre, ha facultà di possere portare per lo stato di sua eccellentia illustrissima per gratia di quella Giacco, spada e pugnale come genero di Mastro Giovanni Rosti, come in filza di supplicatione sotto numero 320', ASF, Otto di Guardia e Balia del Principato, 2247, unnumbered, between folios 74 and 75. This licence is the only evidence for the hitherto unknown fact that Stradano was Rost's son-in-law, although the practice of intermarrying between craftsmen and the family members of artisans in the same or similar trades was quite common in many early modern European cities. Messer Francesco di Raffaello di Bernardo Borghini was secretary of the Otto di Guardia e Balia from September 1547 to October 1559. For more on the magistracy's personnel, see Brackett, *Criminal Justice and Crime*, appendix I ('Names of other officers of the Otto di Guardia e Balìa, 1537–1609, as We Have them'). For more on this particular request, see Victoria Bartels, 'Masculinity, Arms and Armour, and the Culture of Warfare in Sixteenth-Century Florence' (PhD Thesis, University of Cambridge, 2020), 79–90.

52 'sopra state della soa [sua] bottega', ASF, Otto di Guardia e Balia del Principato, 2233, fol. 320.

53 'in servizio di vostra illustrissima eccellentia', ASF, Otto di Guardia e Balia del Principato, 2233, fol. 320.

54 Rost's elevated standing is evident in the witnesses he selected to vouch for the members of his household: the chaplain of the Medici guard, Hadriano Elie de Candidis; and the ducal printer, Lorenzo Torrentino, who produced some of the first editions of Giorgio Vasari's *Lives of the Artists*. Both had Flemish ancestry, as well as senior positions in the Medici court.

55 The ruling written at the bottom of Stradano's supplication was 'non altro', short for 'non occore altro', or 'nothing else needed'. It probably denoted something like 'case closed', meaning that no further action would be taken. Although somewhat ambiguous, this phrasing typically indicated a petition's refusal. However,

it seems surprising that Stradano's request for a renewal would not have been approved.

56 Transcribed and translated in Margaret A. Gallucci, *Benvenuto Cellini: Sexuality, Masculinity, and Artistic Identity in Renaissance Italy* (New York: Palgrave Macmillan, 2003), appendix 3D.

57 'Dua pistolesi. Una coltella alla turchescha. Una spada. Uno pugniale con fornimenti argentati. Una spada a una mano e mezzo. Una zagaglia: quale due arme dissono si havevano a mandare alli Octo', ASF, 'Filza legata in carta pecora bianca, intitotlata: Filza d' Inventari pupillari di Firenze dal 1570 al 1572', No. 2653, transcribed by Eugène Plon, *Benvenuto Cellini, orfèvre, medailleur, sculpteur: recherches sur savie, sur son oeuvre et sur les pièces qui lui sont attribuées* (Paris: Plon, 1883), 381.

58 ASF, Otto di Guardia e Balia del Principato, 2235, unnumbered, immediately after folio 23. For more, see Victoria Bartels, 'Masculinity, Arms and Armour, and the Culture of Warfare in Sixteenth-Century Florence' (PhD Thesis, University of Cambridge, 2020), 95–101.

59 'offendere nessuno'; 'insulto et aciaccho et accio possa piu sicuramente provedere a sua affari et sua donna et figluoli', ASF, Otto di Guardia e Balia del Principato, 2235, unnumbered, immediately after folio 23.

60 'da bravi o da belli in piazza'; 'nello stato ne saranno infiniti che stanno con piu sospetto di lui', ASF, Otto di Guardia e Balia del Principato, 2235, unnumbered, immediately after folio 23.

61 'non essendo habile come dice', ASF, Otto di Guardia e Balia del Principato, 2235, unnumbered, immediately after folio 23.

62 Andrea Navagero, *Opera omnia*, ed. Giovanni Antonio Volpi and Gaetano Volpi (Venetiis: Ex Typographia Remondiniana, 1754), 193. Quoted and translated in Tobias Capwell and Sydney Anglo, *The Noble Art of the Sword: Fashion and Fencing in Renaissance Europe, 1520–1630* (London: The Wallace Collection, 2012), 31.

63 ASF, Otto di Guardia e Balia del Principato, 2228, fol. 537.

EXPERIMENT IN FOCUS IV: Embodied experience of a tailor-made doublet

Valerio Zanetti

Introduction

As a historian of the body and a dress scholar, I am used to relying on a variety of textual, material and visual sources in order to recapture the life of absent and long-lost bodies. Collaborating with the 'Refashioning the Renaissance' project, I braced myself for a methodological challenge: exploring the heuristic potential of my own embodied experiences as a research tool. During six sessions at the School of Historical Dress in London, I was measured, fitted and finally wore the tailor-made doublet modelled on the *giubbone* of black stamped mockado (mock velvet) owned by the Florentine waterseller Francesco Ristori, who died in 1631 (See Experiment in focus III) (Figure iv.1).

The experiment

It is undeniable that my body differs in many ways from that of the waterseller Ristori. Moreover, the conditions of the fittings at the School of Historical Dress were certainly not meant to replicate the atmosphere of an early modern tailor's workshop. The experiences I discuss, therefore, have no pretension to perfect authenticity. After each fitting I recorded the bodily sensations felt and the thoughts they generated. I then reflected on how my own experiences could be used to shed light on those of Ristori and countless other early modern men and women who underwent a similar process to have their garments made to fit.

In the first session I was measured by the School Principal Jenny Tiramani, who started by tying a tape around my torso to mark the waist. She immediately noted that my waistline was quite high, adding that this would suit the intended shape of the doublet. As the meticulous measuring

Figure IV.1 The outfit of reconstructed doublet worn over a shirt with hat, sword belt, hose, stockings, garters and shoes.

progressed, she then observed that one of my shoulders was higher than the other, a fact that would have to be corrected with some light padding. This discovery sparked a brief discussion about the possible cause of this physical irregularity, until we agreed that it was probably due to the long practice of a unilateral racket sport. Thus, the measuring turned into an occasion to reflect on specific features of my body, either inherent or acquired through habit, some desirable and others implicitly undesirable since they would have to be artificially remedied. Early modern 'artisans of the body' possessed a considerable degree of medical knowledge in addition to the skills specific to their trade.[1] The experience of being measured provided me with a space to reflect on my lifestyle while also gaining new insights about my health and anatomy. This made me consider that tailors too performed a medical function in so far as they possessed the ability to identify and qualify salient corporeal features which they could, if necessary, enhance, conceal or correct. Like a patient's consultation with a physician, this knowledge was generated through conversation between customer and artisan. Yet there was no doubt that the latter wielded more authority.

The same power dynamic persisted during the fittings that followed, when a team of makers progressively assessed how the doublet looked and felt on my body. The successful fitting normally developed through a flow of questions and answers going both ways, while the experience remained overall marked by a fundamental disparity. As the wearer, I had to provide crucial information on how clothing felt, articulating physical sensations as clearly as possible.

However, it was not uncommon for this information to be reinterpreted by the makers present, who would also offer advice on how I should stand and move to be more comfortable. An additional challenge was posed by the fact that I had to rely almost entirely on the people around me to form an impression of how the doublet looked on my body. People of Ristori's standing almost certainly did not have access to a full-length mirror, so I too decided to experience the fitting with a limited visual perspective. Since I had to rely on external eyes to evaluate the success of the doublet and the outfit as a whole, I came to appreciate the high level of trust customers had to place on the tailor's aesthetic judgement.

Once I moved past the initial frustration, I also found that the absence of a mirror helped me to better appraise how the garment felt. It must be said that being aware of physical sensations during fittings was not always a matter of choice. Standing still for up to an hour did take its toll on the body: my legs started to hurt and lose sensitivity while my hands invariably turned purple when I left them dangling at my side for too long. Once again, I had to rely on the tailors' expertise in picking up on the signs of physical discomfort and suggesting ways to cope with it. Upon Tiramani's suggestion, I began holding one arm akimbo and shifted my body weight from one leg to the other, assuming a stance that can be admired in many Renaissance portraits. Few artisans could aspire to having their likeness painted, yet when being fitted they too needed to develop the physical skills necessary to hold a pose.

So far, I have emphasised the tailor's agency in shaping my carriage and embodied sense of self. However, the most important influence was exercised by the garment itself as it progressively came together. During the first fitting, I tried on a canvas model of the doublet without sleeves, fixed with only a few safety pins at the front (Figure iv.2a and iv.2b). Despite the lightness of the material at that stage, the perfect cut of the model was sufficient to instantly change my posture. My shoulders were pulled back and I was forced to keep straight. I felt the garment's full force when I first tried on its main body, still sleeveless but cut from actual fabric and provided with buttons. That day, my neck and back hurt from exercising and I experienced discomfort when standing straight. Wearing the doublet immediately made it easier for me to keep an upright posture and I gradually felt more comfortable. When I expressed my relief, I was told that the remarkable effect was achieved through the insertion of thin baleen strips and light fustian padding at the front. I realised the extent of the aid provided by the doublet only once I had to take it off half an hour later and I felt the pain creeping up again. It may be easy to imagine that a garment of this type – tightly cut, padded and reinforced – would feel constrictive, especially on a body used to an active lifestyle or even physical labour. Many articles of historical clothing worn by both men and women have traditionally been deemed uncomfortable and cumbersome on the basis of visual and textual sources. Pioneering experiments of

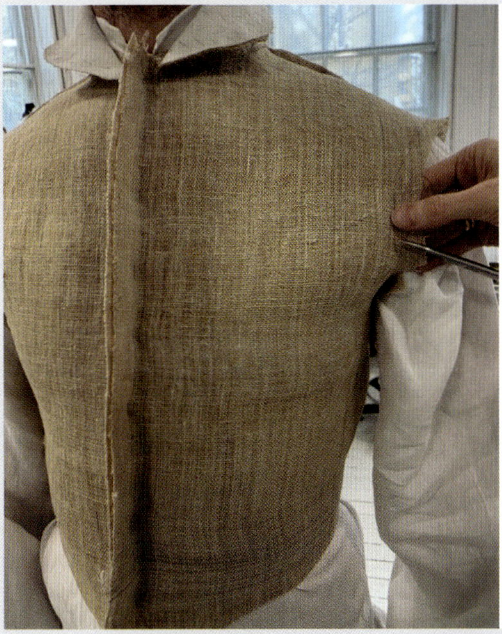

Figure IV.2a Front view of the fitting for the hemp interlining.

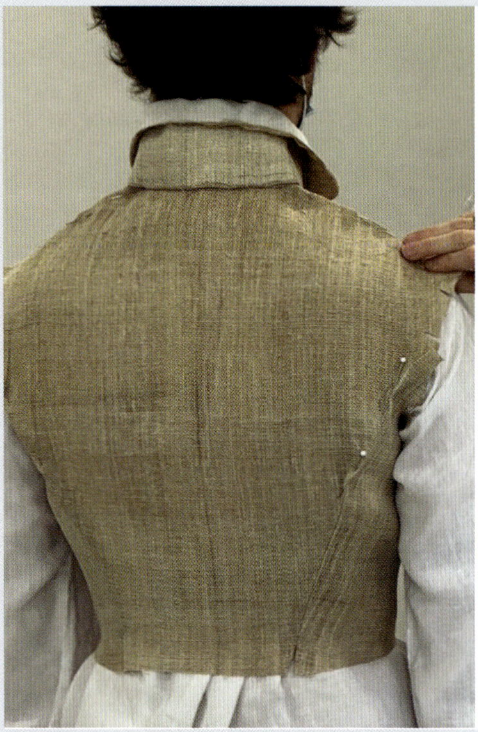

Figure IV.2b Back view of the fitting for the hemp interlining.

scholarly re-enactment, however, are starting to reassess such assumptions in favour of more balanced accounts.[2] My own experience of wearing the doublet similarly attests that its perfect fit translated into greater support (Figure iv.3).

In combination with the tight cut, what rendered the interaction between clothing and my body more tangible during the fittings was the changing sensation of the fabric. At first the doublet did feel a little rigid and stuffy but, as Tiramani put it, the fabric just needed a little time to adjust to the temperature of my body and vice versa. As promised, the garment felt rapidly suppler without losing its ability to hold up my torso. It also managed to regulate my body temperature according to the changing environment, with the help of the underlying linen shirt. Renaissance people believed in clothing's power to affect their health through body contact.[3] Seen as animate entities, garments and accessories were also feared for their potential to overcome their wearers. Feeling the doublet coming alive, I realised to what extent this cultural discourse was shaped by first-hand experiences of the osmotic relationship between dress and the body.

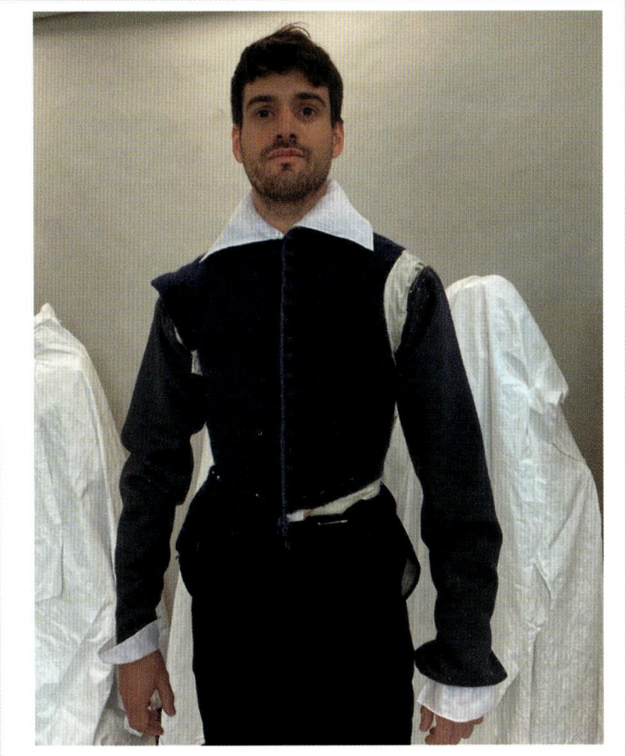

Figure IV.3 Front view of the second fitting in mockado.

Conclusion

What did the experiment teach us about early modern embodied experiences of fashion? Being dressed up in the doublet provided me with a deeper understanding of how early modern sartorial practices were the product of constant negotiation between human and material agents. To achieve a successful fit, wearers such as Ristori and myself had to acquire specific corporeal skills which have left no trace in the records. Re-enacting the dressing process in all its phases revealed that these represented crucial occasions for people to rediscover and interpret their body in light of unique sensorial and cognitive experiences. This experiment represents a first attempt to explore the tacit skills and embodied knowledge gained by men and women of the middling sort through their experience of clothing. By drawing our attention to the role of ordinary wearers in the process of sartorial creation, it opens a new path towards ever more nuanced and inclusive accounts of the non-textual world of artisanal practice.

Notes

This experiment was part of the 'Refashioning the Renaissance' project's reconstruction of an early modern male doublet, funded by European Research Council (ERC) under the European Union's Horizon 2020 research and innovation programme (grant agreement No. 726195). See www.refashioningrenaissance.eu/reconstructions and Experiment in focus III. I would like to thank Sophie Pitman and all costume and craft experts who took part in the fitting and the construction of the doublet, in particular Jenny Tiramani, Claire Thornton, Melanie Braun.

1 Sandra Cavallo, *Artisans of the Body in Early Modern Italy: Identities, Families and Masculinities* (Manchester: Manchester University Press, 2007).
2 Sarah A. Bendall, 'The case of the "french vardinggale": A methodological approach to reconstructing and understanding ephemeral garments', *Fashion Theory*, 23, 3 (2019), 363–99.
3 Elizabeth Currie, 'Health hazards: Clothing's impact on the body in Italy and England, 1550–1650', *Bulletin of the John Rylands Library*, 95, 2 (2019), 115–33.

EXPERIMENT IN FOCUS V: Digital doublet

Maarit Kalmakurki

Introduction

Doublets were complex upper-body garments that were commonly made by skilled tailors and consisted of several layers of fabrics and supportive materials to create sculpted shapes. The way in which the linings, interlinings, smaller pieces of linen and the main fabric are layered together, however, are hidden, and we cannot see the layers because of 'indirect contact with an object'.[1] The 'Refashioning the Renaissance' project carried out a digital reconstruction experiment to explore how new technological digital garment-making tools can be used to render the layers of materials visible and explore period garment-making. This was achieved by digitally making an animation of the seventeenth-century male doublet that used to belong to a Florentine waterseller, Francesco Ristori (see Experiment in focus III).[2]

The experiment

The doublet animation is based on several different source materials from the 'Refashioning' doublet project.[3] The realisation of this doublet used notes on historical pattern and tailoring books and the patterns and materials of the physical doublet that was made at the School of Historical Dress in London. In addition, my personal skills in pattern cutting and garment making as well as dimensions of the digital space are implemented in building the digitally animated doublet. The practice-led experiment investigated the opportunities and limitations in recreating a seventeenth-century male doublet via digital tools. The focus was to advance knowledge within the practice, and practice was an integral part of this experiment. Creating the digital animation facilitates a more thorough understanding of the doublet structures via embodied experience.

The digital animation of the garment was created with Clo3D, a program that is originally designed for the fashion industry. The program was selected as it allows the construction of garments with existing patterns, easy alteration of patterns and the creation of new ones, which proved to be useful in replicating the shape of the historical doublet. Patterns were the starting point of the digital garment reconstruction because the doublet's shape and fit are dependent on the shape of the pattern, as in the making of garments by hand with physical materials. I digitised the 'Refashioning' doublet patterns created at the School of Historical Dress with Gerber Accumark pattern program and imported them to Clo3D. These patterns were then multiplied to replicate each of the materials typically used in a physical doublet. One of the important benefits of digital animation is being able to illustrate the vast number of material pieces that are included in a historical doublet. In addition, the Clo3D program enables illustration of the pieces next to the avatar body, which in our case helped to visualise the pattern positions on a human body (Figure v.1). Another beneficial aspect of Clo3D is that the program allows alteration of patterns and shows the alterations simultaneously on the avatar body. Easy pattern alteration is beneficial for dress history research because historical patterns often differ from contemporary shapes. This feature facilitated the finalised pattern shape, which enabled visualisation of the doublet's sculpted shapes.

Once imported into the program, the patterns fly in the digital space and need to be assembled on the avatar body. This process is comparable to that in real life: the many layers of different fabrics that are included in

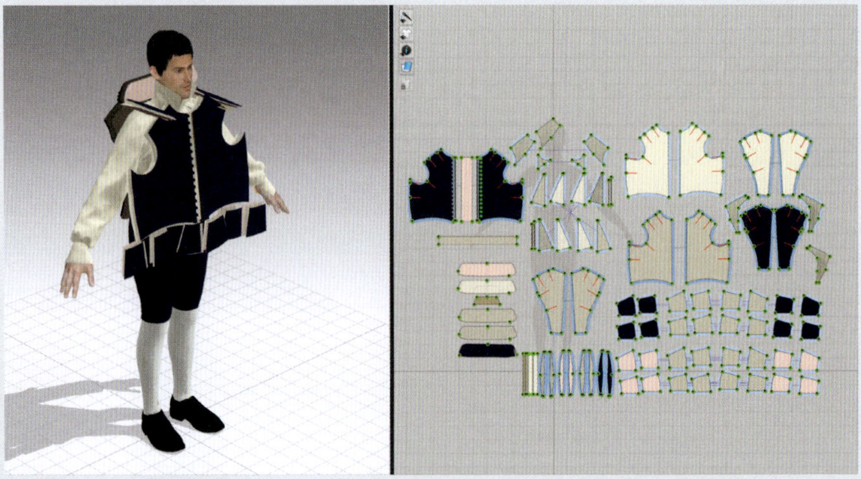

Figure V.1 Doublet pattern pieces on the right with all patterns laid over an avatar body on the left.

a doublet are placed on a body in the same order in the digital space. This is one of the important findings of this animation project: it can render visually all the different layers of fabrics and supports that are not visible in the physical object. For example, typically in a Renaissance doublet, the collar is constructed with multiple layers of various fabrics and supportive materials such as baleen or bent grass. In the 'Refashioning' animation we were able to demonstrate the ways in which the silk lining, interlinings from hemp, linen and a smaller piece of coarse linen, and the main wool fabric (mockado) are layered together (Figure v.2).

Additionally, the Clo3D program can mimic some of the stitching types used in the assembly of the physical doublet. This is helpful to demonstrate the supportive materials and stitchings that are positioned beneath the doublet's main fabric layer, such as the different wool fabrics on the front and back of the shoulder. Often in physical doublets, a specific shape of wool was pad-stitched to the supportive linen between the lining and main fabric. The purpose of these materials in the physical doublet is to create softness and thickness, and pad-stitching creates round forms for the shoulder area.[4] The animation can illustrate these different layers and stitching types, as well as the pad-stitching in the collar of the physical doublet that forms its round shape.

Sewing in the digital space functions somewhat as in real life. The seams that join are connected; however, the digital garment does not include any seam allowances, which are necessary for sewing physical materials. Material qualities, such as yarn density, fabric weight and thickness, can also be modified at any point in creating a digital garment. This feature is beneficial for research purpose as it allows one to test different

Figure V.2 The complex layers of doublet collar can be visualised via animation.

Figure V.3 Final digital reconstruction of a sleeveless, seventeenth-century male doublet.

material qualities on the avatar. In animation, garment simulation is the phase when the flat patterns transform into three-dimensional, finished digital garments (Figure v.3). The texture of the garment is created by inserting any digital image of a material on top of a pattern where texture is required. Any kind of digital image can be imported to the program, which is beneficial, as this enables existing historical fabrics to be replicated on digital garments. The colour and shine of the texture can also be adjusted, which is a useful tool for testing different colour options based on historical sources such as dye recipes. The 'Refashioning' doublet surface is a combination of the material characteristics of the physical doublet made at the School of Historical Dress. The texture and colour are slightly altered with the Procreate digital drawing program. The doublet's ribbon is a combination of an image of an extant material, which was reworked with Adobe Photoshop before importing to Clo3D.

Another valuable feature of digital animation is that, after the simulation, the garment and patterns can still be adjusted to finalise a perfect shape. The sculpted form of a historical doublet that requires layers of different materials and specific sewing methods executed by skilled tailors in real life is, however, created differently in the animation. All layers between the main fabric and lining have to be removed before the digital simulation process, as they cause collision effects. The volume is created digitally by adding pressure underneath the garment, inserting 'bond' to the pattern pieces, which should look more solid or in contemporary terms 'fused'. One of the outcomes of this experiment was the discovery that the software lacks trimmings that resemble historical ones. Specifically designed buttons, buckles or ribbons must be created beforehand with a digital sculpting program[5] and then imported to Clo3D.

Conclusion

The results of this experiment stem from studying Clo3D and finding ways to make the most of the program features to represent the historical doublet. The digital reconstruction is not a copy of the physical doublet but a digital creation showing aspects of the doublet that cannot be seen in real life. The digital reconstruction does not directly copy physical making because the digital space has different applications, dimensions and rules. However, it offers many beneficial tools to experiment, analyse and visualise historical garments for the purpose of demonstrating their making so that we can better understand unique craftsmanship of the skilled Renaissance tailors.

Notes

The 'Refashioning the Renaissance' digital reconstruction project, co-ordinated by Sophie Pitman and carried out by Maarit Kalmakurki, was funded by the European Research Council (ERC) under the European Union's Horizon 2020 research and innovation programme (grant agreement No. 726195). See www.refashioningrenaissance.eu/reconstructions.

1 Maria Cybulska, 'To see the unseen: Computer graphics in visualisation and reconstruction of archaeological and historical textiles', in Nobuhiko Mukai (ed.), *Computer Graphics* (Rijeka: InTech, 2012), 213.
2 Digital reconstruction is emerging in historical research prompted by the possibility of recreating objects that are damaged or no longer exist. See, for example, studies by Aleksei Moskvin et al., 'Digital replicas of historical skirts', *The Journal of the Textile Institute*, 110, 12 (2019), 1810–26; Viktor Kuzmichev et al., 'Computer reconstruction of 19th century trousers', *International Journal of Clothing Science and Technology*, 29, 4 (2017), 594–606.
3 The initial design idea for the garment stems from one of the descriptions of a doublet marked in the post-morterm inventory of the waterseller Francesco Ristori, 12 September 1631, State archives of Florence (ASF), Magistrato dei pupilli, 2718, 2, fols 190r–194r. See Experiment in focus III for the description of the historical reconstruction of this garment.
4 Melanie Braun, Luca Costigliolo, Susan North, Claire Thornton and Jenny Tiramani, *17th-Century Men's Dress Patterns 1600–1630* (London: Thames and Hudson, 2016), 16.
5 In the frame of the 'Refashioning' project, creation of these details were omitted since the aim was to investigate the possibilities of the software.

EXPERIMENT IN FOCUS VI: Imitation of amber and pearls

Michele Nicole Robinson

Introduction

False. Counterfeit. Imitation. There were – and are – many ways of talking about goods created to replicate or mimic costly and rare materials. And in the early modern period these terms carried connotations ranging from derision to admiration. The wearing of false gems, for instance, could be seen as an attempt at social climbing (and a violation of sumptuary laws in many northern Italian regions). But it could also be a means of reinforcing status or displaying one's interest in artistic and technological innovation.[1] There were thus many reasons people in early modern Italy purchased and wore imitation stones and metals.

We know much more about the false gems sought out by aristocrats, diplomats and merchants; however, the 'Refashioning the Renaissance' database shows that shopkeepers, cutlery-makers and shoemakers appear to have had some interest in imitation gems. Were these meant to pass for richer materials? Were they markers of an interest in innovative processes and production practices? Or was there simply a desire for products with the same look and feel as the materials they referenced (and those who wore them)? And what was that look and feel?

With these questions in mind, the 'Refashioning the Renaissance' project organised a workshop in March 2020 to re-create recipes for imitations of amber, pearls and other luxury goods. The workshop, designed by Sophie Pitman, offered insight into processes of making and a unique sensory experience with the resulting products, which showed that some imitation goods were more attractive and appropriate alternatives to the real thing than others.

The experiment

Amber was valued for its beauty, rarity and scent in the early modern period. It was burned as incense, but also made into paste for perfuming ointments, pomanders and gloves. Amber also releases a pleasant smell when warmed through handling, which made it a desirable material for rosaries (*corone*).[2] Along with its pleasing sensory properties, because amber was imported into Italy from the Baltics, it was expensive. Consequently, it is unsurprising to find amber simulants among the possessions of Italian artisans, though examples of them are rare. The database listing artisans' possessions includes two examples of imitation amber, both from Siena: Bernardino Ciampi, a cutlery-maker, owned 'a rosary of false amber with two crosses and silver medals' when he died in 1646;[3] and the shopkeeper Lorenzo Brogi had 'a rosary of amber of glass' in 1650.[4]

Glass-making was beyond the scope of this workshop, so Sophie Pitman, who designed the experiments, selected two recipes for imitation amber. These were from the best-selling *Secreti del reverendo donno Alessio Piemontese* (Venice, 1555), attributed to Girolamo Ruscelli:

> To make cleere stones of Amber:
> Seeth Turpentine in a pan leaded, with a little cotton, stirring it until it be as thick as paste, and then poure it into what you will, and set it in the sunne eight dayes, / and it will be cleaer and hard inough. You may make of this little balles, haftes for knives, and manie other things.
>
> [Untitled recipe for imitation amber]:
> Take the yelkes of sixteene egges, and beat them well with a spoone: then take two ounces of Arabicke, an ounce of the gumme of Cherrie Trees: make these gummes into a powder, and mire them with the yelkes of the egges, let the Gummes melt well, and poure them into a pot well leaded. This done, set them six daies in the sunne, and they will become hard, and shine like glasse, and when you rub them, they will take up a straw unto them, as other amber stones doe.[5]

We were more successful with the first recipe, where both Canada balsam and thicker and darker Venetian turpentine were warmed separately, poured into silicone moulds with differing amounts of raw cotton added to each sample; this created the striations and cloudiness observed in real amber (Figure vi.1). Importantly, the turpentine released a strong pine smell during preparation of the recipe and with handling of the final product, like real amber. It remains to be seen if the simulants can be shaped into beads like those found on extant rosaries and necklaces, but the experiment clearly showed that not only the colour but also the smell that made amber so desirable can be fabricated.

The second experiment focused on false pearls. If there are few instances of imitation amber listed among the possessions of non-elite

Figure VI.1 Imitation amber made from turpentine, 2020. Aalto University, Espoo.

families, there are fewer still of false pearls. In fact, among the nearly three hundred references to pearls in the inventories that compose the database, there is just one that mentions false pearls, which appears in a Sienese oil-maker's home in 1642: 'a pair of bracelets with little coral [beads] with baroque pearls of glass, but the coral is good'.[6] That being said, there are references in other sources to false pearls being worn by non-elite women. For instance, a complaint was filed against the unnamed wife of a Florentine shoemaker who lived in Via de' Pescioni in 1639 for wearing, 'a necklace of black beads and Venetian false pearls', among other forbidden items.[7]

Whether or not artisans owned and wore false pearls, contemporary recipes for creating them are numerous; in the Italian context alone, Paola Venturelli has identified fourteen different sources with recipes for making pearls from the thirteenth to the seventeenth century.[8] Most common are recipes that call for shell, pearl or mother of pearl to be softened or dissolved (usually with an acid) and made into a paste and then formed into larger pearls.

For the workshop, we recreated this type of recipe from a popular German text, *Allerley Mackel* (1532).[9] The recipe calls for snail shells to be softened, ground and made into a paste with egg whites. Beads were then

Figure VI.2 Imitation pearls made from crushed snail shells (top) and covered with gold and silver leaf (bottom), 2020. Aalto University, Espoo.

made from the paste and, once dry, boiled in linseed oil. This resulted in lumpy, gravelly beads not at all resembling pearls (Figure vi.2).

We also recreated a recipe of another common type, which was based on clay rather than nacreous paste from another popular text, Isabella Cortese's *I secreti* (1584).[10] This required baked clay beads to be coated with Armenian bole mixed with egg whites; once dry, a layer of gold or silver leaf was applied, and the bead dipped in parchment glue and left to set (Figure vi.2). Up close, these looked like large gold and silver beads, but, when viewed from a distance of 2 m or more, they could more plausibly be taken for real pearls (Figure vi.3).

That neither recipe produced what could be easily mistaken for a genuine pearl is in part because we did not have the necessary technical skills or tools to accurately follow the instructions. The German recipe called for quicksilver (mercury) to coat the beads, which was too dangerous for use in the facilities available for the workshop. Additionally, without proper grinding tools and experience, we were unable to crush the shells into a powder fine enough to make a smooth paste. In fact, most recipes for making false pearls call for shells or smaller pearls to be well ground, suggesting this skill was critical to the manufacture of persuasive pearls; it was this type of imitation that contemporary writers cited as able to fool

Figure VI.3 Imitation pearls seen at a distance, 2020. Aalto University, Espoo.

even jewellers.[11] Others considered glass pearls to be the most convincing simulants, and difficult to detect when set into rings.[12]

If the best imitations were those made from smaller pearls or by expert glass bead-workers, perhaps they went undetected in the probate inventories that compose the 'Refashioning the Renaissance' database.[13] It is also possible that those manufactured from shell or smaller pearls were large, spherical and costly, and others made by glass-workers were too expensive for most buyers.

Conclusion

The amber and pearls created in our workshop show that there was considerable technical skill required to produce these materials. It would have taken time, practice and materials to produce convincing pearls and amber pieces. This, in turn, reinforces the idea put forth by scholars such as Timothy McCall and Marlise Rijks that an interest in producing, purchasing and wearing well-made false gemstones was an elite pursuit related to status and an interest in innovation, rather than simply a desire by artisans and others with limited financial means to wear cheaper imitative products.[14] This especially seems to be the case with false pearls. The most convincing imitation pearls were large, round and expensive; out of reach for most.[15]

However, the experiment also showed that the sensory experience of some materials – like amber – could be recreated in ways that were perhaps more accessible in terms of availability and cost. The false amber beads on the cutlery-maker's rosary were perhaps not intended to stand in for amber, but rather to produce the same smell and warmth when touched, which supported religious practices and protected bodily health. It was amber's transformative effect that was sought, rather than the material itself. Thus, this research experiment combining written evidence with hands-on experimentation shows that in the early modern period different kinds of imitations may have been appropriate for different social and economic groups for various purposes.

Notes

The 'Refashioning the Renaissance' project's workshop 'Imitation in early modern clothes and accessories', organised by Sophie Pitman and held at Aalto University on 10–12 March 2020, was funded by the European Research Council (ERC) under the European Union's Horizon 2020 research and innovation programme (grant agreement No. 726195). See www.refashioningrenaissance.eu/experiments/imitation. We thank our collaborator Tim McCall and the 'Refashioning the Renaissance' team members for participation, presentations and discussions, and the expertise they brought into the workshop.

1 See Timothy McCall, 'Material fictions of luxury in Sforza Milan', in Catherine Kovesi (ed.), *Luxury and the Ethics of Greed in Early Modern Italy* (Turnhout: Brepols, 2018), 239–76; and Marlise Rijks, 'Gems and counterfeited gems in early modern Antwerp: From workshops to collections', in Michael Bycroft and Sven Dupré (eds), *Gems in the Early Modern World* (Cham: Springer International Publishing, 2019), 309–42.
2 Rachel King, '"The beads with which we pray are made from it": Devotional ambers in early modern Italy', in Wietse de Boer and Christine Göttler (eds), *Religion and the Senses in Early Modern Europe* (Leiden: Brill, 2013), 153–76.
3 Inventory of the cutlery-maker Bernardino Ciampi, 1 March 1646, State Archives of Siena (ASS), Curia del Placito, 283, 273, fol. 134v.
4 Inventory of the shopkeeper Lorenzo Brogi, 2 April 1650, ASS, Curia del Placito, 286, 18, fol. 69v.
5 Girolamo Ruscelli, *The Secrets of Alexis*, part III, trans. William Ward (London, 1595), fols 251v–252r. On this text see William Eamon, *Science and the Secrets of Nature: Books of Secrets in Medieval and Early Modern Culture* (Princeton: Princeton University Press, 1994), 134–67. See Sophie Pitman, 'Una corona di ambra falsa: Imitating amber using early modern recipes', www.refashioningrenaissance.eu/imitating-amber.
6 Inventory of the oil-maker Leonardo di Michele, 12 August, 1642, ASS, Curia del Placito, 282, 177, fol. 141v.
7 State Archives of Florence (ASF), Pratica segreta, 176, 28 February 1639, fol. 169r.
8 Paola Venturelli, 'Segreti di Leonardo da Vinci per ottenere "perle grosse"', *Arte Lombarda*, 132, 2 (2001), 42–7.
9 *Allerley Mackel* (Mainz: Peter Jordanim, 1532), translated from German to English in Sidney M. Edelstein, 'The Allerley Matkel (1532): Facsimile text, translation, and critical study of the earliest printed book on spot removing and dyeing', *Technology and Culture*, 5, 3 (1964), 317.

10 Isabella Cortese, *I secreti* (Venice, 1584), 205.
11 Venturelli, 'Segreti', 42–7, and Girolamo Cardano, *The De Subtilitate of Girolamo Cardano*, ed. and trans. J. M. Forrester, vol. I (Tempe: Arizona Center for Medieval and Renaissance Studies, 2013), 431.
12 Anselm De Boodt, *Lapidary or, The History of Pretious Stones*, trans. Thomas Nicols (Cambridge: Thomas Buck, 1652), 76.
13 Paolo Zecchin, 'La nascita delle conterie veneziane', *Journal of Glass Studies*, 47 (2005), 77–92.
14 McCall, 'Material fictions', and Rijks, 'Gems'.
15 See my chapter (Chapter 4 above) on the types of pearls owned by non-elite people.

PART III

Status and credibility

7

The dissemination of fashion: consumption habits and non-essential textile goods in early modern Italian artisan inventories

Stefania Montemezzo

Introduction

In the mid-sixteenth century, the Tuscan writer Giovanni della Casa in his courtesy book *Galateo overo de' costumi* notes that 'everyone must dress well according to his status and age', and should avoid using inappropriate or extravagant styles, as it would 'disdain' others. Writing on men's appearance, he also adds that 'not only should clothing be of fine material but a man must also try to adapt himself as much as he can to the sartorial style of other citizens and let custom guide him'. He further adds that every man should follow these guidelines 'even though it may seem to him to be less comfortable and attractive than previous fashions'.[1] Even if uncomfortable, adapting to new fashions and using suitable-quality materials seems to play an essential role in della Casa's idea of good manners that every Renaissance man should possess. Della Casa's prescriptions focus on appearance and social acceptability while disregarding the needs of those groups that needed comfortable and suitable clothing for their daily activities, such as artisans and the working classes. And yet, did the artisans of Florence, Siena and Venice dress in a manner appropriate to their status, as della Casa would have us believe? What kinds of objects, textiles and colours were favoured by the lower social classes? How did fashion vary between cities and what characterised the varying dress and consumption patterns?

This chapter addresses these questions, using a sample of post-mortem inventories of Florentine, Sienese and Venetian artisans and locally based shopkeepers between the sixteenth and seventeenth centuries, collected in the 'Refashioning the Renaissance' database.[2] By looking into their wardrobes and establishing the types of textile materials, colours and clothing styles they had access to, it aims at giving a more coherent picture of the consumption habits of non-essential textile goods and fashion items at lower social levels of society in early modern Italy. First, the chapter looks at inventories as a source, assessing some of the factors that may influence the compilation and content of inventories. These include local institutions, social dynamics and regulations, and economic factors. Second, it analyses the data from the inventories, focusing on fabrics, the colours, the styles and the origins of the garments that belonged to this social class and were found in the household inventories.

The chapter argues that there are two main reasons why it is crucial to analyse the dress, fashion and consumption habits of the artisan classes. First, artisans formed a large segment of the urban population. For this reason, analysing their consumption of fashionable items provides insight into how consumerism emerged in Europe.[3] Second, only recently has historiography focused on the social and cultural aspects of the life of these groups, such as their relationships, gender relations, living standards and everyday life.[4] The lack of comprehensive research is linked to the difficulty of studying heterogeneous social groups such as the artisans.[5] Differences in wealth and craft specialisations suggest a pronounced segmentation and stratification within the artisan order.[6]

Fashion and institutions

The overall picture of artisans' wardrobes that we obtain from post-mortem inventories is of relative richness and variety, despite being conditioned by exogenous factors, such as regulatory limitations. The increasing availability of luxury goods and new fashion manufactures that characterised dress and appearance in early modern Italian cities, as John Styles, Andrea Caracausi and Sophie Pitman have shown (Chapters 1, 2 and 3 above), such as new types of mixed fabrics, knitwear and ribbons, favoured consumption by the middle classes, challenging the 'hierarchy of appearance'.[7]

Within this evolving framework, many elements could impact the consumption of fashion items. The first was related to the availability of certain goods on the market, often linked to local manufacturers and locations, concerning regional and international trade flows (Figure 7.1) The second was linked to institutions (such as governments, guilds and confraternities) that could play a role in the consumption of luxury goods and thus influence the use of certain types of clothing and fashion items. This section aims to contextualise the sample provided by the 'Refashioning the Renaissance' database in terms of the procedures used to compile inventories, the role of the institutions that produced them and the social and institutional factors, such as sumptuary laws or guilds, that may have influenced the consumption of fashion items.

Post-mortem inventories of possessions provide a valuable source of information for investigating material wealth and consumption in

Figure 7.1 Marketplace at the Piazza di Porta Ravegnana, Bologna. A tailor's workshop and a tailor taking measures of a client can be seen in the lower part of the image. *Matricula societatis draperiorum civitatis Bononiae*, ms. 641, fol. 1r, No. 93 (1411). Museo Civico Medievale di Bologna.

pre-industrial societies.[8] This is especially the case for the northern Italian states, where comprehensive collections are often preserved in family and children's court records, such as the Venetian Giudici di Petizion, the Florentine Magistrato dei Pupilli or the Sienese Curia del Placito.[9]

However, the range of individuals represented in these inventory records is somewhat patchy. The records are dominated by individuals who were young enough to leave children under the age of majority, and who died intestate, without having appointed guardians. They therefore exclude whole sections of the population, such as the elderly, single women and the childless. Furthermore, the nature and interpretation of the information in these inventories needs to be considered in light of the differences in the local institutions that were responsible for drafting them.[10] Despite these problems, these legal documents provide a unique insight into the dress and material culture of the lower orders of society, which, like all subaltern groups, presents significant difficulties for historical reconstruction.[11]

Other elements, such as sumptuary laws and guild regulations, complement this composite institutional landscape. The significance of these elements for dress at the lower social levels is difficult to assess, as social constraints generally are in the past. Yet it is essential to be aware of their existence during the analysis. The wardrobes of artisans, for example, must be considered in the light of the owner's social status and profession. It is not only the presence of clothing suitable for carrying out the typical tasks of each profession that is important, but also the presence of clothing indicating membership of a guild, a brotherhood, a particular social group, as well as age and social status (married, young, old, widow etc.). The rules of professional and religious institutions were part of a context in which status and appearance were closely linked, making clothing a public expression of membership. Sixteenth-century manuals and books, such as that of Cesare Vecellio, bear witness to these differences. In the pages of his treatise, Vecellio discusses the differences that characterised the clothing of nobles, foreigners, officials, merchants and shopkeepers. He highlights the role of clothing in the social representation of different social groups by comparing the clothing styles, noting their similarities and differences, and who usually wore one style rather than another. For merchants and shopkeepers, for example, while describing their headpieces, the author says:

> And these wear high berets, which they call 'tozzo', and some also wear it with chopping-board style, with a very narrow fold and with a veil around it: but this [latter] beret is more for the sheltered and mature; and that [tozzo] for younger people.[12]

Vecellio also adds that their usual elbow-length sleeves are the same as those of the aristocracy, showing how social distinction was marked by the use of certain styles or garments, and how the wardrobes and fashion choices of different social groups and the artisans within them were influenced by unwritten social constraints.[13]

Sumptuary laws could influence the content of artisan wardrobes and their post-mortem inventories by imposing limitations to the consumption of luxury goods. Public institutions sought to limit the consumerist tensions especially of the lower orders by enacting a range of sumptuary laws in an attempt to maintain the division between social classes.[14] The actual impact of the sumptuary laws on the consumption of the lower orders, although present in all Renaissance Italian states, remains doubtful and to be determined. For example, in a city like Siena, those not part of the ruling class were not formally allowed to wear silk or velvet since medieval times.[15] Yet post-mortem inventories show that even ordinary artisans possessed items of silk. In the Tuscan city between 1550 and 1650, altogether 926 garments or trimmings listed in 187 artisans' home inventories were made of silk, making an average of about five items of silk per household.[16] These included textile objects of different sizes and value, from small ribbons, scarves and caps that required just a little amount of silk to garments such as jackets and gowns that were made of or trimmed with silk. For example, of the sixty-nine dress items recorded in the inventory of Carlo Bertuzzi, a Sienese miller, four garments (cloaks, gowns) were decorated with silk or velvet, while eight of the dress accessories were made entirely of silk. These included a fine headscarf decorated with gold, two veils, three pairs of colourful tights, one made of regular silk and two others made of silk scraps (*filaticcio*), and two pairs of colourful cuffs made of silk, gold and silver. Although small accessories such as these that did not require lengths of silk, it is important to note that some of these were decorated with gold and silver thread.

As objects of value, such objects were considered valuable family treasures and given as wedding gifts or passed on in inheritances. This is probably why the miller's wife Triburzia was keen to keep the accessories and clothes of silk at the time of Carlo's death. When questioned by the Magistrati dei Pupilli, she explained that the silk items, along with some jewellery and silverware, had been given to her by her family on the occasion of her wedding.[17] This information provides a fascinating glimpse not only into the fashion and material wealth of the artisans and the circulation of precious textiles within the lower class but also into the efforts of women in the line of succession to claim what was rightfully theirs.[18]

The unique path of each city's economic, institutional and social development could produce distinct consumption patterns, especially concerning different colours, materials and styles and the extent of their dissemination among the lower orders. The patterns were influenced by non-standard commercial means, such as bartering, credits and second-hand commerce, allowing for the informal circulation of expensive fashion items among lower-income groups. The manufacturing world also began to offer cheaper new products and fabrics. As we will see below, these variations reflected cultural preferences and differences in trading networks and local manufactures, encouraging the desire of the middling classes to challenge the 'hierarchy of appearances'.[19]

The colours of dress

Intense and spectacularly dyed textiles and clothing appear in many inventories of artisans and shopkeepers, suggesting that even families of modest means appreciated vibrant colours.[20] This was the case, for example, with the wife of the ordinary Venetian shoemaker Dal Paon, deceased in 1627. In the tiny two-room house, the estimators found several garments in mixed fabrics with fascinating workmanship and colouring. Besides a golden robe made of mixed fabric (*ferrandina*, a light cloth made of wool and silk) and two black gowns (one of coarse low-cost silk, *terzanella*, and the other of *ferrandina*), the inventory mentions three brightly coloured skirts. The first of these, in wool, was red; the second, also in wool, was red with a green satin edging; while the last, in *buratto* (a light woollen cloth with a silk warp), was in a whole rainbow of colours including black, golden yellow, green and white.[21] Some of these colours, such as bright reds dyed with expensive insect dyes or imitated through using cheaper dyestuffs such as brazilwood, have been made physically and materially visible by the 'Refashioning the Renaissance' project through colour reconstruction (Experiment in focus VIII).

Garments' colour vibrancy and sumptuous effects were often achieved through edgings, borders, ribbons and linings made of more expensive textiles. It is no coincidence that silk fabrics and furs are the materials that are most often described as multi-coloured (23 and 22 per cent of cases), followed by gold and silver objects – jewels, brooches or buttons – and wool (16 and 14 per cent, respectively). Detachable sleeves, too – a fundamental part to vary the styles of the clothes – are among the garments that were often made of or decorated with fine fabrics and dyed in bright colours (Figure 7.2). Tommaso Maridi, a Florentine fishmonger from San Frediano, had a collection of cuffs ranging from black wool to red and green mocaiardos (woollen velvet) and white and red *drappo*, a mixtures of wool or silk.[22] The use of small quantities of colourful and more expensive fabrics to embellish outfits made primarily of cheaper textiles illustrates the ingenuity of artisan fashion.

With fashions and preferences changing over the decades, the use of monochrome colours was relatively common. Despite the fact that black was the colour of high fashion of the ruling classes since the sixteenth century, Italian artisans showed a fluctuating interest in black throughout the century.[23] This is confirmed by the archival evidence collected by the 'Refashioning the Renaissance' project. The 'Refashioning the Renaissance' database shows that in 1550s, about 43 per cent of the clothing items were black; it reached over 50 per cent in 1560s, with a gradual decrease over the next decades. Between 1550 and 1650, the percentage of colours such as white, red, brown and purple remained almost unchanged, but green gradually increased. The most common green-coloured items of clothing were gowns, petticoats and hose, such as *zimarre* or *turche* made of wool

Figure 7.2 Domenico Ghirlandaio, *Announcement of Death to St Fina* (detail), 1475. The painting displays garments often found in artisans' inventories, such as a *gamurra* and colourful sleeves. Chapel of Santa Fina, Collegiata, San Gimignano.

from Perpignan or Valencia, or skirts made of silk waste (*filaticcio*), which rose from an average of 6–7 per cent in the 1550s to almost 13 per cent in the 1610s. The colour combinations obtained by juxtaposing different elements (from linings to trimmings) also increased.[24]

Yet the use of colours varied from city to city. The choice of high-contrast colours for combinations unites Florence, Siena and Venice, albeit with a different styles between the three cities.[25] Siena and Venice share a preference for the combination of red and yellow – the colours of the Serenissima's coat of arms, in 18 per cent of all combinations – followed by red and white, and black and white – the last being the colours of the Sienese banner.[26]

A different approach to colour between cities also applies to plain colours. In that respect, the Sienese artisans' wardrobes are the most colourful. On average, only one out of three items of clothing was black.[27] The Sienese outfits differed greatly from those of the Florentines and Venetians also by their fancy combinations of bright colours. Multi-coloured, green and pink items were widespread in Siena. Of the city's artisans' clothing and other textile articles 73 per cent were in colours other than black (compared with 62 per cent in Venice and 58 per cent in Florence). Of all the colours, pink is the one that stands out: more than one out of ten items of clothing were in the spectrum between flesh colour and purple, unlike in Florence and Venice (Figure 7.3).

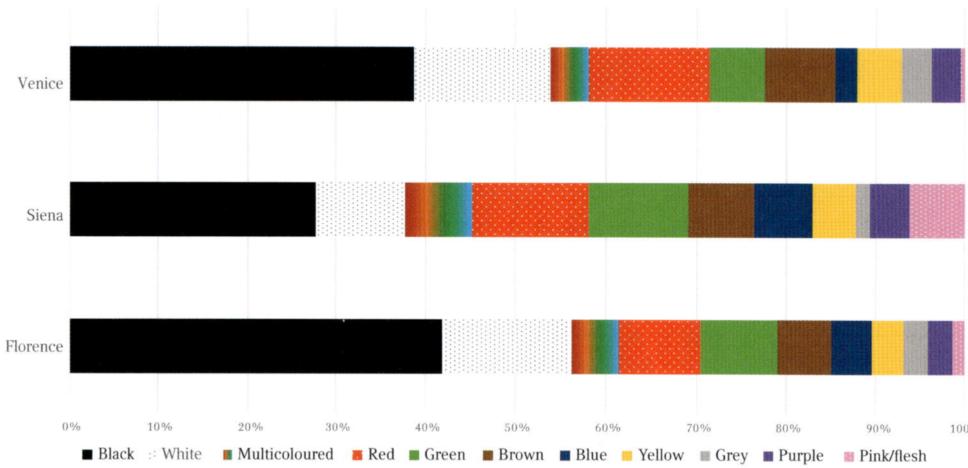

Figure 7.3 The colours of artisans' garments in Florence, Siena and Venice, 1550–1650. The statistics are based on all the inventories of the sample. Raw materials, metallic accessories, grooming items and garments recorded in workshops are excluded.

Some garments were more vivacious than others. The wife of the Venetian shoemaker Dal Paon and her colourful green and red skirts, referred to above, are not precisely an exception in the Italian panorama of the time. In fact, skirts seem to be one of the elements that gave liveliness to clothing. A detailed examination of the colours associated with different garments in the artisan's wardrobes shows that skirts and petticoats, aprons, shirts and body linen were the focal points in terms of colours. For example, a colourful green skirt ornamented with golden flowers that belonged to Anzola, the wife of the Venetian weaver Andrea Padovani, stands out from the family's inventory. Scrolling through the items recorded in the document, it is possible to imagine how Anzola's outfits were to be composed. To begin, a pair of red stockings and a white shirt (perhaps a clean one, instead of the 'sweated' mentioned in the inventory), a black gown made of mixed fabric, *terzanella*, completed with a pair of orange sleeves and a white veil on the shoulders, and a round apron to protect the beautiful skirt.[28] The painted scene from a kitchen of an upper-middle-class household shows what such garments worn in everyday life at artisan levels may have looked like. The woman in the front, plucking a chicken, wears a veil on the shoulders; another is grating cheese wearing a rolled-up apron strapped twice on the waist on an orange *gamurra* with sleeves. Lastly, the woman in the background with a green dress has her pink sleeves pinned in the back to have them out of the way while working (Figure 7.4).

The colourful garments and accessories were often combined with overgarments such as capes, cloaks and tabards to protect the wearer from the cold or bad weather. These are rarely in any colour other than black

Figure 7.4 Vincenzo Campi, *The Kitchen*, 1580s. Pinacoteca di Brera, Milan.

and are usually made of wool or mixed materials (between 72 per cent and 76 per cent of cases). This was due to the nature of these garments. Cloaks, capes and coats were designed for public occasions rather than for domestic use. Indeed, overgarments provided not only protection from the cold and dust of the city streets of the day (which were often unpaved) but also from the prying eyes of neighbours, who might sneakily complain to the city authorities about artisans who broke sumptuary laws.[29]

New textiles and international styles

Besides differences in colours and colour combinations, inventories of Tuscan and Venetian artisans and shopkeepers reveal some clear variations in the consumption patterns between each city also in terms of the textiles materials and styles present in artisans' wardrobes. These reflect differences in both trading networks and local manufactures. Woollen textiles are proportionally more prevalent in Florence, a city renowned for wool production since the Middle Ages, whereas leather and furs, often imported from northern and eastern Europe, are more common in Venice, which had well-established trading connections with those regions (Figure 7.5).[30]

Since the fifteenth century, the textile industries of several northern Italian regions had copied and adapted foreign production processes to manufacture fabrics that looked luxurious, using cheaper, locally

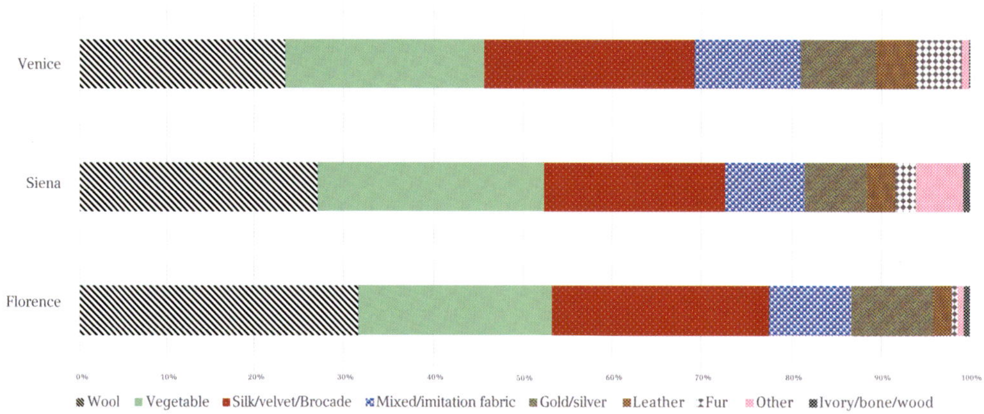

Figure 7.5 Artisans' garments materials in Florence, Siena and Venice, 1550–1650. The statistics are based on all the inventories of the sample. Raw materials, metallic accessories, grooming items and garments recorded in workshops are excluded.

available raw materials that imitated expensive foreign textiles such as the 'Oriental' silks worn by the nobility and upper classes.[31] Employing these foreign inspired imitative textiles allowed eager artisans to indulge their fashionable aspirations. One of such aspiring artisans was the Venetian feather dealer Antonio de Baratis. Antonio owned several items made of mixed textiles, such as a green and gold gown made from brocatel (a mixed fabric usually woven from silk and linen but, in this instance, from wool) or a white woollen velvet (*mocaiardo*) gown, enriched with decorated red velvet.[32] In addition to local production, some foreign textiles were also imported. Clothes from the Serenissima's *Stato da mar*, other Italian states, Spain, Flanders and the Ottoman Empire are on record. The mixture of foreign textiles and fashions that characterised the city's cultural trends is illustrated by items such as the 'two Spanish-style tabards of mixed fabric, one lined with velvet and the other with satin, both new', which belonged to the cloth dealer Antonio Rossati.[33]

Local and imported textiles were used to produce not only 'newness', but also 'historic' and foreign-inspired clothing, such as garments made in *all'antica* style. The *all'antica* style is the subject of much debate in the field of art, especially in works on art and architecture, but it is still ambiguous in the field of clothing.[34] The types of garments in *all'antica* style are varied, ranging from hats to breeches, from jewels to buttons. The only thing they seem to have in common is the fact that many of these, especially hats and sleeves, were often decorated with silk trimmings and ribbons. This was the case, for example, of the sleeves owned by the linen-seller Filippo Guerrieri, made of leather and embroidered with silk ribbons.[35] The French style (*alla francese*) is also mentioned in the inventories of artisans in all three cities, especially in connection with sleeves, gowns, bodices, doublets, breeches, hats and cuffs. Again, the precise definition of the 'French style' is unclear, although Elisabeth Currie explains that these garments might have 'included substantial lengths of fabric to make the kind of voluminous silhouette' shown in contemporary portraits, 'or heavy surface decorations, including pearls or other jewels'.[36]

The habit of owning voluminous clothing seems to concern especially Venetian artisans. Both artisan men and women owned gowns in the *dogalina* style. These ample gowns, usually lined, were characterised by wide sleeves that touched the ground at the bottom and puffed up and tied at the shoulders at the top. Although the taste for *dogalina* style gowns continued throughout the early modern period, it became increasingly rare at the end of the sixteenth century. Even Cesare Vecellio included it among the 'old' Venetian styles.

The Venetian artisan also owned garments and accessories in the maritime style, called *da matelo* in the inventories – another style that fell out of fashion in the Doge's city. This style is more difficult to define, as it was not clearly assigned to any genre in the inventories and the colours varied from blue and red to yellow and green. On the other hand, 'maritime'-style gowns (*muda*) were usually made of satin, velvet or damask.[37] All the garments associated with this style in the 'Refashioning the Renaissance' database are found in the house of the cloth dealer Antonio Rossati, probably waiting to be altered or shredded to reuse the fibres.[38]

The strong relationship of Venice with the Middle Eastern world is testified by textiles and clothing styles from the territories of the Ottoman Empire (Morocco, Egypt, Syria and Turkey) that were also present in the artisans' wardrobes. An example is an Ottoman-inspired jacket, the dolman, appearing in several Venice inventories. This short jacket, worn by both men and women and closed by frogs at the front, was normally made of wool or camlet (mixed fabric of wool and silk) and dyed black – at least in half of the cases. These come in many forms, more or less rich, and range from the old sleeveless dolman in black wool that belonged to Antonia Baldigara, widow of cap-maker Giovanni Marco, to the dolman in red damask, lined with red pine marten fur, found among the possession of the cloth dealer Antonio Rossati, mentioned above.[39] The 'Refashioning the Renaissance' database shows the fate of this garment very clearly. Often mentioned in the 1550s and 1560s, it gradually disappears from Venetian wardrobes so much so that it appears more and more rarely during the 1570s, with the last mention in 1578 in the inventory of the boat-builder Francesco Brazza.[40] Conversely, jackets, mantles and overcoats of different types, such as the overgarment known as 'Turcha', survive in the inventories.[41] 'Turche' generally referred to long overgarments, either sleeveless or with narrow sleeves, which were open in the front. The garment seems to be in vogue especially in Siena during the sixteenth century, only to gradually disappear at the beginning of the next.

For what concerns Florence, its economic and cultural interests decidedly oriented fashions towards western Europe, with most textiles coming from Flanders, Spain, England and France.

Discerning whether descriptions in inventories refer to the style of items, their place of origin or their counterfeiting is often difficult. In some cases, the preposition *alla*, accompanied by a feminine noun clearly defines the style of a garment (*alla veneziana*, *alla fiorentina*, *alla turchescha* etc.), while origin is usually attributed by a masculine adjective (*veneziano*, *fiorentino*, *turchesco*). However, the distinction is sometimes random since vernacular languages were not fully codified or grammatically fixed. Officials, estimators and notaries often expressed themselves idiosyncratically, and their ability to accurately categorise items depended on their expertise, experience and familiarity with the goods they were recording. Indeed, the distinction between the style and the origin was probably not especially crucial in the eyes of contemporaries when it came to establishing economic value, since only a very small number of items – less than 3 per cent – specify these details.

The post-mortem inventories of artisans from Venice, Florence and Siena show that what ordinary artisans chose to wear often defies simplistic assumptions. The presence of expensive garments and accessories that were made of silk next to the cheaper ones made from plain coarse woollens, linen or waste materials simultaneously highlights the stratification of the social group and the difficulty of determining from their clothes where individuals were located in the hierarchy of wealth and status. Moreover, the differences among cities in the choice of colours, materials and styles, including foreign and exotic styles, gives a clear idea of the local diversities that characterised the Italian territory, separated by political borders, different social institutions and economic development. Consideration of the influence of different cultures via the medium of international trade on the regional territories as well as on the entirety of the European society allows for a better understanding of the forces behind consumption and dissemination of fashions across social classes well before the Industrial Revolution and of the importance global connections already possessed by the sixteenth century.[42]

Conclusion

'Everyone must dress well according to his status', prescribed Giovanni della Casa in 1558. While sumptuary laws imposed limits on the consumption of expensive textiles by the middling classes, it is clear that craftspeople and local shopkeepers could not only dress well for their status. They could go far beyond it. The picture of artisans' wardrobes is of a relative richness and variety, but its composition was influenced by several factors, including the availability of textiles, local and foreign influences, and the position within local and international commercial circuits. The distinctive qualities that set apart the society, economy and institutions of Florence, Siena and Venice, and the unique path of each city's economic development, could produce distinct consumption patterns. This concerned especially imported goods and the extent of their dissemination among the lower orders. The different textiles, colours and styles paint an extravagant and unique picture of Italian artisans' wardrobes. In commercial hubs like Florence and Venice, the taste for the fashionable was fed by foreign cultural influences as materials, ideas and trends travelled along international trade routes to the cities so that the wearing of French, Spanish and even Ottoman styles soon became a familiar part of life for even the lower orders. This also fuelled the diversification of cloth manufacturing in these cities into imitation fabrics and mixed-fibre textiles that utilised locally available materials, waste fibres and new, cheaper dyestuffs. This, in turn, stimulated new consumption patterns by making a broader range of fabrics, colours and garments accessible to even the poorest artisans. While showing local peculiarities linked to local productions and trades, the different stylistic choices give us an insight not just into what artisans wore but also into the cultural background of the popular groups, showing an already established taste for the exotic and foreign fashion influences.

The ways in which ordinary Italians mixed expensive materials with cheaper ones and combined their costly garments with inexpensive trimmings highlights the creative and personal ways in which early modern Sienese, Florentine and Venetian artisans engaged with fashion.

Notes

The research leading to these results was done as part of the 'Refashioning the Renaissance' project that has received funding from the European Research Council (ERC) under the European Union's Horizon 2020 research and innovation programme (grant agreement No. 726195).

1 Giovanni Della Casa, *Galateo*, trans. Konrad Eisenbichler and Kenneth R. Bartlett (Toronto: Centre for Reformation and Renaissance Studies, 1986), 12–13.
2 I would like to thank Fabio Andreazza, Andrea Caracausi, Giovanni Favero, Paula Hohti, Michele Nani, Andrea Rapini and Francesco Vianello for their precious advice and insights. The archival data used in the chapter are available open access at www.refashioningrenaissance.eu/database.
3 Catherine Kovesi, *Luxury and the Ethics of Greed in Early Modern Italy* (Turnhout: Brepols, 2018), xv–xix; for the European institutionalisation of fashion, see Carlo Marco Belfanti, 'Was fashion a European invention?', *Journal of Global History*, 3, 3 (2008), 419–43.
4 Margaret A. Pappano and Nicole R. Rice, 'Medieval and early modern artisan culture', *Journal of Medieval and Early Modern Studies*, 43, 3 (2013), 473–85; Denise Bezzina, *Artigiani a Genova nei secoli XII–XIII* (Florence: Florence University Press, 2015), 1–4; Anna Bellavitis, *Women's Work and Rights in Early Modern Urban Europe* (Cham: Springer International Publishing, 2018).

5 Paula Hohti, '"Conspicuous" consumption and popular consumers: Material culture and social status in sixteenth-century Siena', Renaissance Studies, 24, 5 (2010), 654–6; Pappano and Rice, 'Medieval and early modern artisan culture', 473–85.

6 Lois Green Carr and Lorena S. Walsh, 'Inventories and the analysis of wealth and consumption patterns in St. Mary's County, Maryland, 1658–1777', *Historical Methods*, 13, 2 (1980), 81–2; Bert De Munck, 'Artisans, products and gifts: Rethinking the history of material culture in early modern Europe', *Past & Present*, 224, 1 (2014), 39–74. For Siena, Paula Hohti Erichsen, *Artisans, Objects and Everyday Life in Renaissance Italy: The Material Culture of the Middling Class, Visual and Material Culture, 1300–1700* (Amsterdam: Amsterdam University Press, 2021), 31–9; for Florence, Samuel Kline Cohn, *The Laboring Classes in Renaissance Florence, Studies in Social Discontinuity* (New York: Academic Press, 1980), 65–113; for Venice, Patricia Allerston, 'Clothing and early modern Venetian society', *Continuity and Change*, 15, 3 (2000), 367–90.

7 Carlo Marco Belfanti, 'Mobilità sociale e opportunità di mercato alle origini del cambiamento', in Eugenia Paulicelli (ed.), *Moda e mode dal Medioevo al Rinascimento* (Rome: Moltemi editore, 2006), 39–40.

8 Margaret Spufford, The Great Reclothing of Rural England: Petty Chapman and Their Wares in the Seventeenth Century (London: Bloomsbury Publishing, 1984); Micheline Baulant, Anton Schuurman and Anton Servais, Inventaires après-décès et ventes de meubles: apports à une histoire de la vie économique et quotidienne (XIVe–XIXe siècle): actes du séminaire tenu dans le cadre du 9ème Congrès international d'histoire économique de Berne (1986) (Louvain: Academia, 1988); Carole Shammas, The Pre-Industrial Consumer in England and America (Oxford: Oxford University Press, 1990); Lorna Weatherill, Consumer Behaviour and Material Culture in Britain, 1660–1760 (London: Routledge,, 1996); Craig Muldrew, The Economy of Obligation: The Culture of Credit and Social Relations in Early Modern England (Basingstoke: Macmillan, 1998); Greig Parker, Probate Inventories of French Immigrants in Early Modern London (Farnham: Ashgate, 2014).

9 Stefania Montemezzo, '"Popular fashion": La ricostruzione dei consumi di moda degli artigiani nell'Italia rinascimentale', *Storiografia. Rivista Annuale Di Storia*, 27, 26 (2023), 145–67.

10 The occupational structure of artisans and small shopkeepers influences the interpretation of information on dress and appearance. At the same time, the age profile, occupations and social status of the male property owners inventoried for the Venetian Giudici di Pet*izion* are very varied, while the Florentine and Sienese Magistrati dei pupilli records, on the other hand, are dominated by individuals who were young enough to leave children under the age of majority, and who died intestate, without having appointed guardians. For a more detailed analysis of the role of institutions and the segmentation of the sample, see Montemezzo, 'Popular fashion'.

11 Not all crafts were equally represented in the inventories, as can be seen by comparing the occupational cross-section provided by the 'Refashioning the Renaissance' database with the fiscal and demographic data from the Florentine and Venetian censuses. In Florence, for instance, although food producers and retailers are more strongly represented, this is not the case in the 'Refashioning the Renaissance' database. Instead, the data show that craftsmen in the building and metallurgy trades are more represented in the inventories recorded in the children's court records. The situation was similar in Venice, with a discrepancy between the Petition Court (more textile workers) and the 1595 military census. The reasons for these differences are difficult to identify and could range from the higher mortality rates inherent in some trades (such as construction and metal) to the different support and welfare services provided by the guilds. See Montemezzo, 'Popular fashion'; and,

for the census data, Pietro Battara, 'Botteghe e pigioni nella Firenze del '500: un censimento industriale e commerciale all'epoca del Granducato mediceo', *Archivio Storico Italiano*, 95, 3 (363) (1937), 3–28. See also Richard Tilden Rapp, *Industry and Economic Decline in Seventeenth-Century Venice* (Cambridge, MA: Harvard University Press, 1976), pp. 49–106, pp. 58–63.

12 Cesare Vecellio, *De gli habiti antichi et moderni di diuerse parti del mondo libri dve*, ed. C. Chrieger (Venice: Damian Zenaro, 1590), c. 116. Translated by the author.

13 Maria Giuseppina Muzzarelli, 'Noscere ordinem et finem sui status: il valore delle vesti nella società posizionale del tardo Medioevo. Problemi di identità tra Medioevo ed età moderna : seminari e bibliografia', in Paolo Prodi and Valerio Marchetti (eds), *Problemi di identità tra Medioevo ed età moderna: seminari e bibliografia* (Bologna: CLUEB, 2001), 105–15.

14 Catherine Kovesi Killerby, 'Practical problems in the enforcement of Italian sumptuary law, 1200–1500', in Trevor Dean and Kate J. P. Lowe (eds), *Crime, Society, and the Law in Renaissance Italy* (Cambridge: Cambridge University Press, 1994), 99–120; Catherine Kovesi Killerby, *Sumptuary Law in Italy, 1200–1500* (Oxford: Clarendon Press, 2002), 1–8, 61–91; Maria Giuseppina Muzzarelli, *Gli inganni delle apparenze: disciplina di vesti e ornamenti alla fine del Medioevo* (Turin: Scriptorium, 1996), pp. 23–97; Maria Giuseppina Muzzarelli, 'Reconciling the privilege of a few with the common good: Sumptuary laws in medieval and early modern Europe', *Journal of Medieval and Early Modern Studies*, 39, 3 (2009), 597–617.

15 Paula Hohti Erichsen, 'The art of artisan fashions: Moroni's tailor and the changing culture of clothing in sixteenth-century Italy', in Rembrandt Duits (ed.), *The Art of the Poor: The Aesthetic Material Culture of the Lower Classes in Europe, 1300–1600* (London: Bloomsbury Academic, 2020), 109–16.

16 www.refashioningrenaissance.eu/database.

17 Inventory of the grinder Carlo di Girolamo Bertuzzi, 17 December 1642, State Archives of Siena (ASS), Curia del Placito, 282, fols 155r–158r.

18 Christiane Klapisch-Zuber, *Women, Family, and Ritual in Renaissance Italy* (Chicago: University of Chicago Press, 1987), 213–46; Giulia Calvi, 'Diritti e legami. Madri, figli, Stato in Toscana (XVI–XVIII secolo)', *Quaderni Storici*, 29, 86 (2) (1994), 487–510; Isabelle Chabot, 'La loi du lignage: Notes sur le système successoral Florentin (XIVe/XVe–XVIIe siècles)', *Clio*, 7 (1998), 51–72.

19 Brian Pullan, *Crisis and Change in the Venetian Economy in the Sixteenth and Seventeenth Centuries* (London: Methuen, 1968); Richard A. Goldthwaite, 'The empire of things: Consumer demand in Renaissance Italy', in William Kent, Patricia Simons and John D. Eade (eds), *Patronage, Art and Society in Renaissance Italy* (Oxford: Clarendon Press, 1987, 153–75; Richard A. Goldthwaite, 'The economy of Renaissance Italy: The preconditions for luxury consumption', *I Tatti Studies in the Italian Renaissance*, 2 (1987), 15–39.

20 For the relation between consumption habits and incomes, see Paolo Malanima and Valeria Pinchera, 'A puzzling relationship: Consumptions and incomes in early modern Europe', *Histoire & Mesure*, 27, 2 (2012), 197–222.

21 Inventory of Dal Caon, 7 February 1627, ASV, Giudici di Petizion, Inventari, 350, fol. 1r–1v.

22 Inventory Tommaso di Salvadore Mariti (date n/a) 1620, ASF, Magistrato dei Pupilli, 2717, fol. 383v-383v.

23 John H. Munro, 'The anti-red shift – to the "dark side": Colour changes in Flemish luxury woollens, 1300–1550', *Medieval Clothing and Textiles*, MPRA Paper, 3, 1 (2007), 55–98; Amedeo Quondam, *Tutti i colori del nero: moda e cultura del gentiluomo nel Rinascimento* (Costabissara (Vicenza): A. Colla, 2007), 79–118; Paula Hohti Erichsen, 'Power, black clothing, and the chromatic politics of textiles in Renaissance Europe', in Jenny Boulboullé and Sven Dupré (eds), *Burgundian Black:*

Reworking Early Modern Teechnologies (Santa Barbara: EMC Imprint, 2022), https://burgundianblack.tome.press, accessed 24 February 2023.

24 www.refashioningrenaissance.eu/database. The figures are based on only descriptions that mentioned the colour; in most cases (80 to 85 per cent) the colour of the garment is not mentioned in the inventory. These calculations include only clothes that are kept in the household for family use, while shops and fabrics stored in the home but intended for manufacturing or commercial use are excluded.

25 The total mentions for the three cities are: 63 for Florence, 118 for Siena and 169 for Venice.

26 Comparisons with Florence are harder to make because Florence suffers from a lack of precision in inventories.

27 Jacqueline Herald notes that the Sienese were noted for their ostentatious fashion in *Renaissance Dress in Italy 1400–1500* (London: Bell & Hyman, 1981), 57.

28 Inventory of Andrea q. Giacomo Padovani, 27 April 1632, ASV, Giudici di Petizion, Inventari, 353, 106, fol. 1r-1v.

29 Kovesi Killerby, *Sumptuary Law in Italy*, 148–52; Maria Giuseppina Muzzarelli, *Le regole del lusso: apparenza e vita quotidiana dal Medioevo all'età moderna* (Bologna: Società editrice Il mulino, 2020), 24–31. For examples of such complaints, or denunzie, in Siena, see Hohti Erichsen, *Artisans, Objects and Everyday Life*, 125–6.

30 As far as the materials are concerned, the composition of the garments is not specified in many cases. In 35 per cent of the cases in Venice, the inventory does not mention the material. In Tuscany, this percentage rises to 42–3 per cent.

31 See, for example, Franco Franceschi, 'Woollen luxury cloths in late medieval Italy', in Bart Lambert and Katherine Anne Wilson (eds), *Europe's Rich Fabric: The Consumption, Commercialisation, and Production of Luxury Textiles in Italy, the Low Countries and Neighbouring Territories (Fourteenth–Sixteenth Centuries)* (Farnham: Ashgate, 2016), 181–204; and Domenico Sella, 'Les mouvements longs de l'industrie lainière à Venise aux XVIe et XVIIe siècles', *Annales. Economies, Sociétés, Civilisations*, 12, 1 (1957), 29–45; Luca Molà, *The Silk Industry of Renaissance Venice* (Baltimore: Johns Hopkins University Press, 2000), 161–85; Francesco Ammannati, *Per filo e per segno: L'arte della lana a Firenze nel Cinquecento* (Florence: Florence University Press, 2020), 71–85; F. Özden Mercan, 'A diplomacy woven with textiles: Medici–Ottoman relations during the late Renaissance', *Mediterranean Historical Review*, 35, 2 (2020), 169–88.

32 Inventory of the feather dealer Joannes Antonii filii Pecini De Baratis, 15 May 1576, ASV, Cancelleria Inferiore, Miscellanea, 42, 31, fol. 2r. Other fabrics such as cotton and woollen mixes used waste materials from the production of luxury textiles to make cheaper, lower-quality materials for the less affluent, see John Styles, Chapter 1 above; and Edoardo Demo, 'Le manifatture tra medioevo ed età moderna', in Giovanni Luigi Fontana (ed.), *L'industria vicentina dal medioevo a oggi* (Padua: CLEUP, 2004), p. 34.

33 Inventory of Antonio Rossati q. Bernardini, 3 September 1555, ASV, Cancelleria Inferiore, Miscellanea, 39, 44, fol. 1r.

34 Caroline Campbell, 'Revaluing dress in history paintings for Quattrocento Florence', in Gabriele Neher and Rupert Shepherd (eds), *Revaluing Renaissance Art* (Aldershot: Ashgate, 2000), 137–43.

35 Inventory of the linen dealer Filippo di Sforzo Guerrieri linaiolo, 3 September 1627, ASF, Magistrato dei Pupilli, 2718, 2, fol. 14r.

36 Elizabeth Currie, 'Clothing and a Florentine style, 1550–1620', *Renaissance Studies*, 23, 1 (2009), 38.

37 See, for example, 'Do mude da matelo raso biancho taiado', inventory of Antonio Rossati son of Bernardini, 3 September 1555, ASV, Cancelleria inferiore, Miscellanea, 39, 44, fol. 10v.

38 Inventory of Antonio Rossati son of Bernardini, 3 September 1555, ASV, Cancelleria inferiore, Miscellanea, 39, 44, fols 6r–13r.

39 Inventory of Antonia Baldigara widow of Giovanni Marco son of Nicolò, 11 August 1551, ASV, Cancelleria inferiore, Miscellanea, 38, 57, fol. 1v; inventory of the rag dealer Antonio Rossati son of Bernardini, 3 September 1555, ASV, Cancelleria inferiore, Miscellanea, 39, 44, fol. 9r.

40 Onur Inal, 'Women's fashions in transition: Ottoman borderlands and the Anglo-Ottoman exchange of costumes', *Journal of World History*, 22, 2 (2011), 256.

41 Bronwen Wilson, 'Foggie diverse di vestire de' turchi', 97–139; Charlotte A. Jirousek, *Ottoman Dress and Design in the West: A Visual History of Cultural Exchange* (Bloomington: Indiana University Press, 2019), 79–145.

42 Beverly Lemire and Giorgio Riello, 'East & west: Textiles and fashion in early modern Europe', *Journal of Social History*, 41, 4 (2008), 887–916; Beverly Lemire, *Global Trade and the Transformation of Consumer Cultures: The Material World Remade, c. 1500–1820* (Cambridge: Cambridge University Press, 2018), 30–86.

8

Artisan attire and the politics of dress in seventeenth-century Tallinn

Astrid Wendel-Hansen

Introduction

Birgitta Falck lived and worked in Tallinn – a Swedish-ruled town on the shore of the Baltic Sea in what is now Estonia – in the latter half of the seventeenth century.[1] Over the course of three decades, she married a succession of four butchers, each more successful than his predecessor.[2] Within the confines of the corporatist system of merchant and craft guilds which regulated much of life in Tallinn, Birgitta Falck led a financially secure life and advanced steadily up the social ladder. Although butchers belonged to the lesser of the town's two craft guilds, they were some of its most influential craftspeople, and all but the first of Falck's husbands had sworn the burgher's oath.[3] Falck herself was a figure of some standing in her own right in the community. During the 1690s alone, she was named as witness or godmother at the baptisms of six children, five of whom came from artisan families, and one who was born to the town's fire marshal.[4]

Yet this is only part of the story – for Birgitta Falck was also a notable moneylender in Tallinn. Abetted by her second and fourth husbands, she offered loans to individuals across the social classes, from peasants – to whom she lent rye secured against small trinkets and silver items – and fellow artisans to merchants and members of the town's educated professions. The sums involved varied considerably; most people borrowed a few Reichsthaler (Rtl), while the secretary of the town council and royal assessor of the consistory, Conrad Akenstierna, borrowed a total of 300 Rtl on the security of some silver.[5] In 1688, Falck's second husband Hans Ostertag petitioned the Swedish governor-general of Estonia, trying to claim the 66 Rtl that a Captain Adam Johann von Burt had borrowed against some bullocks in 1682, plus interest.[6] It appears that the case was not resolved as Falck wrote another petition three years later, lamenting her situation as a poor, solitary widow who had waited patiently for the debt to be repaid.[7] While it remains unclear how Falck and her husbands acquired the capital from which they issued loans, the non-payment of one debt was by no means a threat to Falck's credit activities, which she continued until her death in 1698.

The process of drawing up a probate inventory for Falck's estate was initiated on 22 June 1698 and concluded a little more than a week later, on 1 July (Figure 8.1).[8] Among numerous personal items, the inventory lists

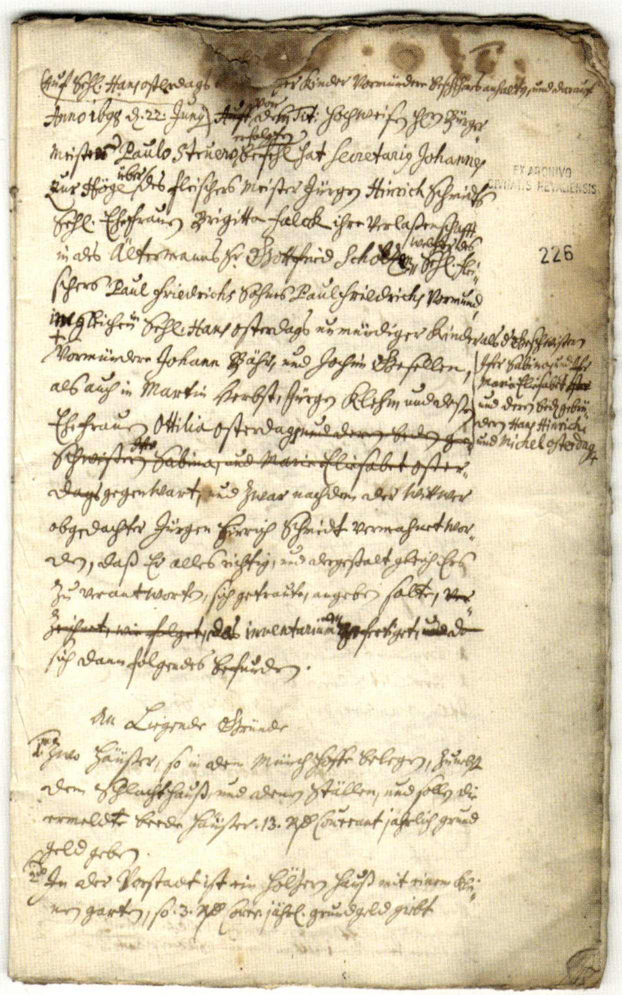

Figure 8.1 Title page of the inventory of Birgitta Falck, 22 June 1698, Tallinn City Archives, collection 230, inventory 1, No. Bt14, fol. 226r.

eighty-three items of clothing: twenty-two shirts, fourteen skirts, thirteen jackets, seven aprons, seven bonnets, six overgowns, four caps, two fur coats, one cloak, one petticoat, a set of jacket and skirt, another set listed as a jacket and bodice, one fur lining and one neckerchief. In comparing Falck's inventory with those of other artisans from a similar background, it becomes evident that she was certainly in a class far above her fellow artisan households.[9] The clothing resources available to lower-middle-class households varied widely, with some inventories including very few garments and a scattering of others containing a quantity of garments much like hers. The list attests to a level of material comfort that one might not have expected of an artisan household formally belonging to the lesser of Tallinn's two craft guilds.

Falck's probate inventory – and especially its record of her wardrobe – makes an excellent point of departure for analysing the relationship between appearances and social position in the early modern period. The traditional view is that society was rigidly hierarchical and that deep-rooted ideas about the appropriate relationship between a person's social standing and their appearance resulted in a 'coincidence of costume and social position'.[10] But Falck's inventory also shows well the kind of investment clothes were for early modern people and how they could be used as capital. Falck had a variety of garments in different stages of their life-cycle, and, as she acquired new garments, she pawned several of the old ones. At the taking of the inventory, eight garments were still pawned, indicating that she did not have a direct need for them as the clothing resources available to her were sufficient. The dynamic preserved in Falck's inventory therefore allows us to shift focus away from the role of clothes as social signifiers and to consider their full complexity, which could not easily be codified by sumptuary law.

Early modern Tallinn was a highly corporatist town, with the town council at the pinnacle of political and social influence.[11] Beneath the council were associations of merchants: the Great Guild, which consisted predominantly of merchants involved in long-distance trade, and the Brotherhood of the Blackheads (*Schwarzhäupter*), which united unmarried merchants and merchant journeymen. Artisans usually belonged to one of the two composite craft guilds. St Canute's Guild affiliated practitioners of the more esteemed crafts – those thought to require more skill – such as goldsmiths, bakers, tailors and so forth, while St Olaf's Guild represented the less prestigious trades, such as stonemasons, carpenters, butchers and furriers.[12] In the seventeenth century, approximately five per cent of Tallinn's inhabitants were burghers. Only those men not in the service of anyone else could become burghers, after swearing a 'burgher's oath' and paying a sum of money (*Bürgergeld*).[13] Being a burgher was a prerequisite for social and political advancement, and, although in theory there were no obstacles to artisans of local background swearing a burgher's oath, in reality the burgher elite was comprised mostly of merchants and artisans of German or Swedish background.

The seventeenth century witnessed increased specialisation and variation in the artisanal crafts, with hairdressers, wig-makers, comb-makers, pearl embroiderers, button-makers and ribbon-makers establishing a presence, presumably a reflection of growing popular demand for certain new luxury goods. But ideas of strict social differentiation between merchants and artisans continued to persist, and artisans' attempts to be recognised for their improved wealth and social standing ran afoul of the interests of the town council and the Great Guild.[14] As will emerge in this chapter, this new-found dynamism and the struggle for distinction in appearances was also reflected in the sumptuary laws, which went from singling out burghers and journeymen to distinguishing between several groups of merchants and artisans. The chapter therefore first examines how the appearance of Tallinn's artisans was codified and distinguished from other social groups by the series of sumptuary laws promulgated by the town council over the course of the century, and then compares this with craftspeople's wardrobes as described in household inventories. As an exhaustive analysis of the source material is beyond the scope of this chapter, focus is limited to three material categories that have previously been shown to be particularly meaningful for displaying social status in the early modern period: silk, fur and adornments.[15]

Accessing the wardrobes

Recent scholarship has shown that sumptuary legislation was not only a European but also a global phenomenon in the early modern period. Broadly speaking, these laws regulated various aspects of individual consumption and conduct, including appearance, diet and expenditure on and behaviour at weddings, funerals and baptisms.[16] In Tallinn, the texts of eight sets of ordinances issued by the town council from around 1600 until 1706 survive; two additional unpublished drafts from 1641 and 1650 have been preserved in the archive of St Canute's Guild, but were never codified into law.[17] Clothing regulations could be both sweeping and extremely detailed, and, as will emerge, there was no standard or systematic approach regarding which details of dress were regulated. For one social group, it might have been headwear; for others, cloaks or muffs; and yet others, jackets and doublets. Importantly, the regulations for each group were a combination of prohibitions and permissions, allowing certain fabrics and fashions and forbidding others. Similarly, the basis on which distinctions were drawn also softened over time: after 1690, the regulations lost most of their guild- and gender-based structure, and the latest law, issued in 1706, focused more on taxing sumptuous clothing than on forbidding it. The primary focus here is consequently on clothing regulations issued until 1665.

Examining sumptuary laws does not necessarily shed light on what people bought, how they dressed or what they thought their clothes said about them. Focusing narrowly on the prescriptions and proscriptions of the laws encourages the belief that clothing is merely a representation or manifestation of hierarchy and prioritises the perspective of those in a position of power over the intentions of those lower in the social orders.[18] Consequently, it assumes that the idea of 'order' is unproblematic – that it is absolute rather than itself socially constructed – and that people wanted primarily to dress above their social station and consume what their social superiors consumed. Scholars have rightly criticised such simplistic notions; early modern clothing practices were complex, and appearances were used by poor and rich alike to express not only social and economic aspirations but also religious, professional and familial affiliations, as well as personal identity, beliefs and values.[19] Scholarship has also demonstrated how the proliferation of 'semi-luxury' or 'populuxe' goods – from mixed silks, combs and razors to mirrors and pocket watches – among non-elite consumers was important in undermining rigid distinctions based on expensiveness or fashionableness, and in multiplying the ways in which sartorial display might be enacted and interpreted.[20]

One particularly fruitful way of placing the evidence from sumptuary laws in a wider context is to juxtapose them with the evidence that can be derived from household inventories. There is extensive literature on early modern inventories and their advantages and disadvantages as sources.[21] However, four points are of particular relevance to the sampling of inventories relied upon for this chapter.[22] First, clothing and textiles were selectively recorded in the Tallinn inventories.[23] Second, while roughly a third of the garments in the sample were described as male and a third as female, there was no indication whether the remainder would have been worn by men or women.[24] Third, unlike their western European counterparts, inventories from Tallinn rarely include financial valuations, so that we usually do not know the worth of people's wardrobes relative to the total value of their household goods. It is thus difficult to make a straightforward estimation based solely on the inventories of how far social position corresponded with expenditure on clothes. Finally, on a more general note, considering the frequent borrowing and lending of clothes between households, especially among the lower classes, we often cannot know for certain the nature of the connections between individuals and the items in their inventory – and, indeed, we can only speculate about the goods that might have been present at different times in people's lives.[25]

Silk

As silk was one of the most expensive and luxurious of dress materials, it is perhaps no surprise that in seventeenth-century Tallinn sumptuary laws sought above all to restrict the wearing of silk to the social elite and that their most fundamental distinction was between silk and woollen fabrics. The laws addressed entire outfits and individual garments made of silk, as well as silk accessories. The general principle was to regulate the cost of the fabric, but over the course of the century the wording became more detailed, and the fabrics forbidden or allowed to each group were illustrated with several examples. At the beginning of the century, silk – and especially velvet – was generally forbidden even to the social elite. While the clothing regulation issued around 1600 was not yet based on guild membership, the wearing of velvet and silk stockings by burghers and journeymen was expressly prohibited.[26] The wives and daughters of burghers were similarly barred from wearing velvet and silk skirts and capes. These proscriptions were fairly straightforward, but, as the century progressed, the regulations on the wearing of silk – and the level of detail included in them – proliferated, reflecting the intricate social order they sought to demarcate.

The 1665 regulation allowed the most senior and eminent members of the Great Guild and the Brotherhood of the Blackheads to wear suits of plain silk on Sundays and on festive occasions. Both glossy and unnapped velvet remained entirely forbidden, and plush, satin and brocade were allowed only for doublets and jackets (Figure 8.2). All other members of the above-mentioned merchant guilds could wear cloaks of cheaper silks on special occasions, but otherwise they were restricted to woollens.[27] Unlike merchants, the ordinary members of both craft guilds were forbidden silk entirely, as they had been under the regulations of 1631 and 1639. The artisans of St Canute's Guild were allowed various woollens of 'moderate value', such as broadcloth, grosgrain, floret and polymite. Only their aldermen were allowed the privilege of caffa cuffs on their coats to distinguish them from other artisans. Members of St Olaf's Guild were allowed only woollen fabrics like polymite and coarse gewand.[28]

The 1665 regulations impose a similar hierarchy for the use of silk in women's dress, albeit somewhat more leniently. Unlike their husbands and fathers, the wives and daughters of Great Guild members could wear silk also on weekdays, but certain silk fabrics like caffa, velvet, satin and plush were restricted to single garments rather than entire outfits.[29] The female family members of more esteemed artisans were limited to woollen fabrics on workdays, but for special occasions they were allowed cheaper silks such as taffeta and terzenel – a coarse, low-budget silk derived from flawed cocoons – for single garments, while silk remained wholly forbidden to their husbands and fathers.[30] The wives and daughters of artisans

Figure 8.2 Piece of silk brocade, textile sample (c. 1600–99). The sample consists of fifteen smaller pieces that have been sewn together. Width: 96.5 cm, length: 113.5 cm. Tallinn City Museum, TLM_4168 H 339.

practising less prestigious trades, meanwhile, were allowed only woollen fabrics, presumably on both workdays and festive occasions – although they were permitted a single silk ribbon on their cloaks as decoration.[31] By the time of the 1696 sumptuary law, however, the regulations had relaxed so far as to allow all women, except servants, to wear black cloaks of unspecified silks.[32]

The sartorial system envisioned by the town council was above all concerned with the type and fineness of fabric. At the same time, the evolution of the laws provides evidence of certain change occurring during the seventeenth century. Over the decades, silk undoubtedly remained an expensive commodity. However, the later laws reveal a variety of silks, including lighter and cheaper ones, some of which were also available to artisans and their household members under certain conditions, and the numbers of those who could and did wear silk increased considerably. However, the idea of silk being a clear visual marker of social distinction is complicated by the fabric's presence in household inventories. These underscore the fact that, whereas the sumptuary laws targeted specific groups perceived as especially threatening to the social order, while completely overlooking others, in reality silk goods were owned by all social classes.

The primary distinctions recorded in the inventories relate to the quantity and type of silk items owned. While it is not always possible to know with certainty whether a fabric was silk or a silk mix, an estimated 60 per cent of artisans' inventories contained at least one silk garment, while roughly 90 per cent of merchants' and professionals' inventories and all town councillors' inventories contained at least one silk garment. The most extensive collection belonged to the shoemaker Jürgen von Stahl, whose 1687 inventory contained thirteen silk garments and accessories.[33] The median numbers of silk garments in the sample of inventories were two for artisans, five or six for merchants and four for both the educated professions and town councillors. As well as having fewer silk items, artisans also generally possessed smaller ones, such as sleeves, caps and stockings. With the exception of Birgitta Falck, complete silk gowns remained the preserve of merchants, professionals and town councillors, but some wealthier artisans did own substantial outer garments, such as doublets, jackets and cloaks, already before the 1665 clothing regulation was issued. For example, in 1638 the silversmith Christof Derenthal's wife had a cloak of Florentine velvet trimmed in the front with sable, which was probably a costly item, and would have been forbidden to her by the sumptuary laws.[34] In 1658, the inventory of hat-maker Claus Reimer and his wife listed two bodices, one of damask and one of plush, a skirt made of silk tobin and silk garters.[35]

Although it is unknown when and why sumptuary laws relaxed to allow her these items, as a young woman Birgitta Falck would have been forbidden from wearing any of the silk items she owned when she died in 1698: a black floral silk skirt, a black silk jacket, a brown taffeta skirt lined with yellow linen, a gown of black taffeta skirt and jacket and a black damask cloak.[36] Her inventory may not be unique – two other artisan inventories from the last decades of the century listed more silk garments – but it illustrates the variety of silk garments appropriate for someone of influence within their community at that time.[37] An examination of artisan inventories like Birgitta Falck's reveals that individual circumstances varied greatly and challenges the notion present in Tallinn's sumptuary laws as late as 1665 that the wearing of silk was confined to the highest ranks for the urban elite. Elsewhere in Sweden, a similar ambivalence towards sumptuary laws has been shown. While the middling and lower ranks generally adhered to sumptuary laws, inventories regularly listed individual items, such as velvet caps and jackets, that the owners would have been forbidden by law to wear.[38] Although the inventories confirm that, on average, artisan households contained the least number of silk items and merchant households the greatest, they also show that even the most exclusive silks according to the sumptuary laws – velvets, damasks and brocades – were worn by artisans as well. The guild-based social hierarchy expressed through dress which appears in the sumptuary laws, and the conventional perception that people of the lower middle classes could not and did not possess silk garments, are therefore complicated and enriched by the picture that emerges from the probate inventories.

Fur

Fur was another persistently, if unsystematically, regulated element of seventeenth-century dress. Providing warmth in the cold climate of northern Europe and the uninsulated buildings of the seventeenth century, fur was an indispensable and ubiquitous part of people's wardrobes, even if few garments were made entirely from it. Furthermore, its significant financial value presented an excellent opportunity for displaying the wearer's wealth and status.[39]

Sumptuary laws of the seventeenth century repeatedly targeted the wearing of fur, restricting its use almost entirely to smaller items such as caps, muffs and various trimmings and facings. Above all sable and marten, very likely imported from Russia, remained a primary concern.[40] The men's clothing regulations from around 1600, for example, forbade all journeymen, whether merchant or artisan, from wearing sable fur caps and wide cuffs and trims of black fox fur. While sable fur caps were not forbidden to burghers, they were cautioned, in vague terms, to 'exercise restraint', and the regulations prohibited them from wearing cloaks that were lined with sable and coats with cuffs or facings of sable fur.[41] The 1665 regulation specified the maximum cost of the sable caps that members of the two merchant guilds were permitted to wear, according to their social standing. Artisans from either of the craft guilds were not specifically mentioned in that law with respect to fur, but it was declared that 'what was forbidden to merchants, was even more so forbidden to artisans'.[42] While there was no general prohibition against artisans wearing fur, they were presumably meant to stick to cheaper and lower-quality imported furs or furs of local origin.

Fur on women's dress was regulated equally vigorously. A detailed discussion of all regulated garments is unfortunately beyond the scope of this chapter, so head coverings, which were the most common fur garments listed in the inventories, will serve as an example. From 1630s onwards, the town council seems to have been particularly troubled by the tall, cylindrical headgear (*Hülle*) (Figure 8.3). Initially forbidden to all women in 1631, the council allowed it to merchants' wives in 1639.[43] However, it could not be too tall or excessively adorned and any sable used for it was not to exceed 10 Rtl in value. Wives and daughters of St Canute's Guild members were also permitted to wear these head coverings, but, if the hat was made of sable, it could not be dyed, and the total value of the cap was not to exceed 8 Rtl.[44] Concerning female family members of the artisans of St Olaf's Guild, fur headwear was not mentioned. As before, the 1665 regulation permitted sable fur caps (*Mütze*) with a maximum fur value of 20 Rtl for the wives of merchants, and their unmarried daughters were allowed only poor-quality sable fur on their caps.[45] The wives and daughters of artisans within more esteemed trades

Figure 8.3 Nicolaus Willebrandt, burgher women of Tallinn (1634). From *Stammbuch Adam Olearius*. It is possible that the drawing depicts these women wearing *Hüllen*. Landesmuseum Schleswig-Holstein, Schloss Gottorf (2006/128, p. 425).

were also permitted to use sable fur not exceeding 12 Rtl in value on their caps, and the female family members of artisans practising the less prestigious trades were allowed dyed marten or 'low-value sable' not exceeding 8 Rtl in value.[46] From 1690, fur does not appear in connection with head coverings in the sumptuary laws. Worth noting is that, in each of these cases, the laws were concerned with the cost of the sable or the entire hat, but not with whether the head covering was made entirely of fur or simply decorated with it.

An assortment of different caps and *Hüllen* made of sable, marten or fox are recorded in the inventories. The furrier Jochim Hönchen's inventory lists a cap, a 'Polish' cap and two 'boat' caps (*Bohtmütze*), all made of sable.[47] Sable fur *Hüllen* appear in the inventories of the tailor Jochim Tempelhoff and clockmaker Franciscus Zilagius.[48] Woollen or silk headgear is often described as having a fur lining or trim. The inventories of the shoemaker Jürgen von Stahl, furrier Martin Schonert, tinsmith Paul Wulff and tanner Lorentz Grawert all listed plush caps that were lined or trimmed with sable or marten fur, while the inventory of the tailor Hinrich Falck (unrelated to Birgitta Falck) mentions a striking woman's cap featuring gold lace and a sable tail.[49] While monetary values were rarely given in the Tallinn inventories, the two items of clothing that were valued in the inventory of the cordwainer Hanss Busekist were two sable fur 'boat' caps, with assessed values of 12 Rtl and 5 Rtl.[50] Birgitta Falck had four different fur caps: a cap with/made of sable tails, another with/made of dyed sable tails, an old sable cap that she had pawned and an old marten-fur cap.[51]

The cold of the northern climes made the use of fur for outer garments widespread across all social classes. Artisan inventories list no fur coats made entirely of sable or marten, but coats of squirrel do appear.[52] Birgitta Falck's inventory records two old 'peasant' fur coats, which were probably made in the neighbouring countryside from local furs, and perhaps had been pawned to her in exchange for cash.[53] Fur was also extensively used to line and trim coats, jackets, overgowns and cloaks. These could be ostentatious displays of wealth, as shown by the 'new and fine' coat of shoemaker Jürgen von Stahl, which was made of brown broadcloth, lined with exotic honey badger and decorated with golden twisted braids.[54] Detachable fur linings are also frequently recorded in the inventories. Birgitta Falck, for instance, owned not only a squirrel-lined overgown but also an old separate lining made of squirrel.[55] The popularity of such linings attests to their usefulness: they could be added for extra warmth in winter and removed for summer, and they could be transferred between different garments, extending their lifetime and saving on the expense of new clothes.

Rather than being highly exclusive and coveted, furs were available to many people. It was clearly not the case that sable and marten – the focus of many of the regulations related to fur – were restricted only to the social elite. Indeed, it is not possible to distinguish between merchants and artisans on the basis of the presence of these furs in their post-mortem inventories – some sort of sable or marten, however old or worn it might have been, appears in the inventories of all social classes except for labourers. While sable and marten were targeted in sumptuary laws because they were likely imported from Russia and consequently expensive, a wide variety of local furs were accessible to people across the entire social spectrum, confirming that the practical function of fur was as important as its ability to signify the social status of its wearer. Individuals from every walk of life in early modern Tallinn did not necessarily have to break the law in order to be warm and to display their personal tastes and preferences.

Decorative elements

Decorative elements were heavily regulated for all social groups by seventeenth-century sumptuary laws. For example, the 1631 regulation stated that members of St Canute's Guild could have one decorative ribbon on their cloaks, while members of St Olaf's Guild were allowed nothing ornamental whatsoever on their dress.[56] The same regulation forbade members of the Great Guild and the Brotherhood of the Blackheads from wearing hatbands or belts embellished with gold and silver thread or pearls, colourful garters and shoelaces, gold and silver buttons and ribbons and other excessive adornment.[57] The 1665 regulation continued in much the same vein, but was even more restrictive with respect to artisans. Ornamentation was entirely prohibited to members of the two artisan guilds, except for the aldermen of St Canute's Guild, who were allowed one silk ribbon to distinguish them from their guild brothers (Figure 8.4).[58] The 1696 regulation forbade expensive lace neckerchiefs to all men.[59]

Women were generally granted more licence in the clothing regulations than the male members of their households, but the system of distinction could be dizzying and the distinctions between what was forbidden and what was allowed rather minute, as the 1665 laws illustrate. These specified that merchants' wives and daughters could wear golden

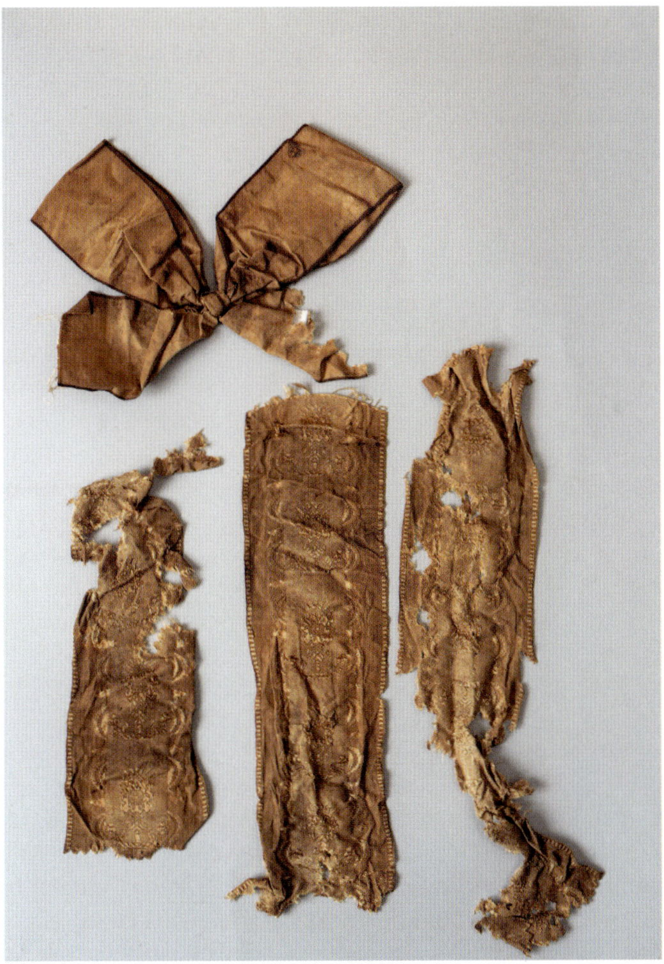

Figure 8.4 Ribbon made of brown taffeta (c. 1600–1700). The ribbon is woven in a glossy pattern and has a dark brown border. In the centre there is a plant ornament surrounded by round and triangular motifs. Tallinn City Museum, TLM_16328 H 1809.

cords, ribbons and laces on their clothes, but braids and other trimmings sewn through with gold or silver thread were forbidden.[60] Furthermore, they were allowed diamond rings and bracelets of gold and coral, but no gold ornaments on their shoes. Unmarried merchant daughters could wear pearl embroidery and pearl necklaces, but brightly coloured, gold or silver ribbons tied around the neck or on gloves were not allowed.[61] Female family members of the more esteemed artisans were entirely forbidden from wearing pearls, and their cloaks had to be without any laces, ribbons and cords, but they could enjoy silk ribbons or ribbons woven or crocheted from gold and silver thread elsewhere on their garments.[62] The wives of artisans practising less skilled crafts could wear only a single silk ribbon or cord on their cloaks, and their unmarried daughters were allowed a velvet trim with no gold or silver on their cloaks.[63] The last regulation from 1706 had lost most gender-based distinctions and taxed a wide variety of decorations, including excessively trimmed gloves, gold and silver decorations on shoes and expensive lace neckerchiefs, among others.[64]

Earlier research has distinguished between 'use value' and 'status value' of clothing and suggested that decorative elements made the garments more exclusive and bestowed more status to the wearer (Figure 8.5).[65] The organ-builder Johannes Pauls, for example, had a russet cloak of unspecified fabric decorated with a golden galloon and large gold buttons.[66] Additionally, clothing was expensive in the early modern period, so trimmings enabled people to update their wardrobes at a fraction of the cost of replacing them entirely.[67] But while Tallinn's sumptuary laws give detailed information about the size, composition and placement of adornments, the inventories are far less explicit about how clothes were decorated. In most cases, we do not even know how a jacket, doublet or waistcoat was fastened, let alone how it was decorated. In Birgitta Falck's inventory, for example, the only items of clothing with any trimmings described are two overgowns with a shared thirty-six silver buttons between the two.[68]

Indeed, buttons were the only type of decoration whose presence in the inventories appears to reflect class divisions. Gold buttons are repeatedly recorded in the inventories of merchants and of town councillors like Arendt Stippel, who owned a brown broadcloth jacket with gold buttons and gold braids, as well as a chamois leather waistcoat with gold buttons.[69] However, they never feature in the inventories of artisans or members of the educated professions or on the clothes of women from any class. Instead, gilt buttons are recorded for both professionals and artisans, while silver and tin buttons appear even more widely. For example, tin buttons were listed on the jacket of the weaver Andreas Wichman and the overgown of the shoemaker Jürgen von Stahl.[70] Although the inventories occasionally neglect to specify the types of buttons when listing them, it is unlikely that gold, silver or gilt buttons would not be mentioned. Thus, it is safe to assume that gold buttons were rather exclusive

Figure 8.5 Man dressed in mid-seventeenth-century fashion, detail from a silver tankard, workshop of goldsmith Stanislaus Schultze in Tallinn (c. 1640–60). Tallinn City Museum, TLM_4199 KA 495/V.

and rare, and the preserve of Tallinn's upper classes, while gilt and silver buttons were within the reach of artisans and other members of the town's middling sort.

Conclusion

This chapter has examined the dress of seventeenth-century artisans in Tallinn, a town in the European periphery where much of life was organised around guilds and their households. Seventeenth-century sumptuary laws sought to create distinction in appearances: primarily to limit silk fabrics as well as sable and marten furs to the upper reaches of Tallinn's social hierarchy and to regulate decorative adornments on a minute level. Household inventories, however, complicate this discursively created hierarchy. Not revealing much about adornments, or how these would have distinguished different groups of people in reality, the household inventories of middling and lower ranks show the presence of both silk garments and sable and marten furs. The observed differences in silk ownership were less in the kind and style of clothing than in the quantity of silk garments, as artisan inventories on average contained fewer silk garments than those of town councillors and merchants. While the latter had entire suits of silk, artisan inventories almost always recorded single garments, like jackets and skirts, but also smaller items, such as sleeves, caps and stockings.

The presence of silk garments and high-end furs in artisan inventories from Tallinn should prompt us to re-examine earlier assumptions that early modern artisan dress was primarily functional and durable.[71] Although individuals were certainly conditioned by prevailing notions of social hierarchy and its appropriate visualisation, we cannot ignore the role of fashion and aesthetics, personal taste and available means in people's consumption strategies. Both inventories and sumptuary laws appear to confirm John Styles's findings in Chapter 1 above that, throughout the seventeenth century, lighter and cheaper silks and silk blends increasingly complemented heavier fabrics such as velvets and brocades. At the same time, Birgitta Falck's inventory, drawn up at the close of the century, shows different silk fabrics, including silk damask, taffeta and unspecified silk appearing side-by-side, complicating the notion of linear change in fashions. For many people, especially in the middling and lower ranks, clothing remained an investment that retained its value over long periods of time. Nevertheless, despite the restrictions that Tallinn's guild-based, corporatist social structure placed on the political power and economic opportunities available to some artisans, inventories demonstrate that individuals designated lower middle class (at best) in the sumptuary laws could still enjoy substantial material security and wield significant influence within a community that stretched beyond their immediate circles.

That legal regulation and observed practice do not neatly align is not surprising. Sumptuary laws show the complexity inherent in the hierarchical order of early modern society, but the clothing owned by individuals exposes the grey areas in the system, as the boundaries it prescribed between different categories of people rarely corresponded precisely with how they actually dressed. Moreover, we are reminded that wearing and owning were two different things. While sumptuary laws were principally concerned with public appearances, they did not regulate ownership of exclusive garments. Ultimately, there were other ways in which people on all levels of society could engage with the materiality of their clothes: by wearing them in private, keeping them as a fungible resource to gift or to pawn, or even simply delighting in their materiality, while being aware that they were forbidden. Artisan inventories like Birgitta Falck's hint at the different ways in which people engaged with their clothes and also at the reasons that artisan dress should not be perceived merely as a signifier of social status. Early modern clothing culture was not a product of the mechanical workings of the social order, especially when that order itself was constantly contested and reconfigured.

Notes

1 In the seventeenth century, Tallinn was called Reval in German and Swedish.
2 St Nicolai in Reval Kirchenbuch: (Getaufte, Getraute, Verstorbene), 1652–97, Tallinn City Archives (TLA), 31, 1, No. 13, fols 101r, 110r, 147r, 153r.
3 In the seventeenth century, the aldermen of St Olaf's Guild were often elected from the ranks of its butchers and furriers, see Küllike Kaplinski, *Tallinn – meistrite linn* (Tallinn: Tallinna Kultuuriväärtuste Amet, 1995), 227–9.
4 St Nicolai in Reval Kirchenbuch: (Getaufte, Getraute, Verstorbene), 1652–97, fols 150v, 158r, 168v, 172v, 173r.
5 Falck had issued a total of two loans for a total sum of 300 Rtl. At the end of the seventeenth century, this amount would have equalled a helmsman's wages for two years, roughly nineteen men's beaver caps imported from England or sixty pairs of women's silk stockings, see Lars O. Lagerqvist, *Vad kostade det? Priser och löner från medeltid till våra dagar* (Lund: Historiska media, 2015), 121–3.
6 Petition of Hans Ostertag to the Swedish governor-general of Estonia, 10 July 1688, Estonian National Archives (RA), archive EAA, 1, 2, No. 482, fols 190r–191r.
7 Petition of Birgitta Falck to the Swedish governor-general of Estonia, 29 June 1691, RA, archive EAA, 1, 2, No. 517, fols. 340r–340v.
8 Inventory of Birgitta Falck, wife of master butcher Jürgen Hinrich Schmidt, 22 June 1698, TLA, 230, 1, No. 14, fols. 225r–237v.
9 In the sample of 105 inventories, which formed part of the source base for my doctoral dissertation, forty-four are from artisan households, see Astrid Pajur, *Dress Matters: Clothes and Social Order in Tallinn, 1600–1700* (Uppsala: Studia Historica Upsaliensia, 2020), 35–40. These include fourteen inventories drawn up for artisans in the animal-product trades: butchers, furriers, tawers, cordwainers, saddlers, chamois preparers, tanners. In this group, the median number of garments in the inventories is twenty-nine, with a minimum of six (inventory of the butcher Aloff Brande, 1620, TLA, 230, 1, Bt8/I, fols 18r–46r) and a maximum of ninety-three (inventory of the furrier Martin Schonert, 28 November 1688, TLA, 230, 1, Bt13/III, fols 229r–238v).
10 Daniel Roche, *The Culture of Clothing: Dress and Fashion in the Ancien Régime*, trans. Jean Birrell (Cambridge: Cambridge University Press, 1994), 56, 219.
11 Karsten Brüggemann and Ralph Tuchtenhagen, *Tallinn: Kleine Geschichte der Stadt* (Cologne: Böhlau, 2011), 89–93.
12 Kaplinski, *Tallinn – meistrite linn*, 249. For a full list of crafts included in each of the artisan guilds see Ernst Gierlich, *Reval 1621 bis 1645: von der Eroberung Livlands durch Gustav Adolf bis zum Frieden von Brömsebro* (Bonn: Kulturstiftung der deutschen Vertriebenen, 1991), 53–5.
13 Tiina Kala and Toomas Tamla (eds), *Tallinna ajalugu II 1561–1710* (Tallinn: Tallinna Linnaarhiiv, 2019), 67.
14 For the numerous political conflicts between artisans, merchants, and the town council in the mid-seventeenth century, see Arno Weinmann, *Reval 1646 bis 1672: Vom Frieden von Brömsebro bis zum Beginn der selbständigen Regierung Karls XI*

(Bonn: Kulturstiftung der Deutschen Vertriebenen, 1991), 86–125. For conflicts specifically around sumptuary law, see Pajur, *Dress Matters*, 109–17.

15 Mikael Alm, *Sartorial Practices and Social Order in Eighteenth-Century Sweden: Fashioning Difference* (New York and London: Routledge, 2021), 112–15; Alan Hunt, *Governance of the Consuming Passions: A History of Sumptuary Law* (Basingstoke: Macmillan, 1996), 71, 83, 122–7.

16 Giorgio Riello and Ulinka Rublack, 'Introduction', in Giorgio Riello and Ulinka Rublack (eds), *The Right to Dress: Sumptuary Laws in a Global Perspective, c. 1200–1800* (Cambridge: Cambridge University Press, 2019), 5–6.

17 The regulations can be found in the following collections: TLA, 190, 1, No. 2, Resolutione/n/ von Heerme/istern/ und Könige Lit. A, No. 14, Einband Nr. 5 (draft of 1641 clothing regulation, 19 February 1641, 15–21; draft of 1650 clothing regulation, 9 November 1650, 47–8); 191, 1, No. 19, Armen-Ordnung, Rewidierte Ordnung des allgemeinen Gottes-Kastens. Kasten-Ordnungen, Kleider-Ordnungen; 230, 1, No. 7 'I Kleider-, Hochzeits-, Kindtauf-, Begräbnis- u.s.w. Ordnungen 1497–1738'.

18 Karin Sennefelt, 'A discerning eye: Visual culture and social distinction in early modern Stockholm', *Cultural and Social History*, 12, 2 (2015), 179–95.

19 Ulinka Rublack, *Dressing Up: Cultural Identity in Renaissance Europe* (Oxford: Oxford University Press, 2010), 259–61.

20 Maxine Berg, 'New commodities, luxuries and their consumers in eighteenth-century England', in Maxine Berg and Helen Clifford (eds), *Consumers and Luxury: Consumer Culture in Europe, 1650–1850* (Manchester: Manchester University Press, 1999), 63–85; Cissie Fairchilds, 'The production and marketing of populuxe goods in eighteenth-century Paris', in John Brewer and Roy Porter (eds), *Consumption and the World of Goods* (London: Routledge, 1993), 228–48.

21 Classic works include Dean et al., *Production and Consumption in English Households, 1600–1750* (London: Routledge, 2004) and Lorna Weatherill, *Consumer Behaviour and Material Culture in Britain, 1660–1760* (London: Routledge, 1996); see also Giorgio Riello, '"Things seen and unseen": The material culture of early modern inventories and their representation of domestic interiors', in Paula Findlen (ed.), *Early Modern Things: Objects and Their Histories, 1500–1800* (New York: Routledge, 2013), 125–50.

22 For a more detailed discussion, see Pajur, *Dress Matters*, 35–40.

23 Pajur, *Dress Matters*, 38. On this point, see also Dean et al., *Production and Consumption in English Households*, 15, and Weatherill, *Consumer Behaviour and Material Culture in Britain*, 204.

24 Pajur, *Dress Matters*, 123.

25 Anne E. C. McCants, 'Porcelain for the poor: The material culture of tea and coffee consumption in eighteenth-century Amsterdam', in Findlen (ed.), *Early Modern Things*, 319.

26 1600 clothing regulation, TLA.190, 1, No. 2, fols 387–8.

27 1665 clothing regulation, TLA.191, 1, No. 19, fol. 87r.

28 1631 clothing regulation, TLA.190, 1, No 2, 396; 1639 clothing regulation, TLA, 230, 1, Bs7/I, fols 319r–320r. Polymite and gewand are referred to in the 1665 clothing regulation, see TLA, 191, 1, No. 19, fols 87v–88r. Caffa – a type of plush silk, sometimes with floral pattern; floret – light, sometimes floral woollen fabric; gewand – a type of woollen fabric, stuff; polymite – a type of camlet.

29 1665 clothing regulation, fol. 89r.

30 1665 clothing regulation, fol. 91r.

31 1665 clothing regulation, fol. 91v.

32 1696 clothing regulation, TLA, 230, 1, Bs7/I, fol. 402r.

33 Inventory of the shoemaker Jürgen von Stahl, 18 November 1687, TLA, 230, 1, Bt13/III, fols 188v–189r.

34 Inventory of the wife of silversmith Christof Derenthal, 10 February 1638, TLA, 230, 1, Bt10/III, fol. 207r.
35 Inventory of the hat-maker Claus Reimer and his wife, 25 February 1658, TLA, 230, 1, Bt11, fols 73v–74r.
36 Inventory of Birgitta Falck, fol. 231r–v.
37 Inventory of Jürgen von Stahl, fols 188v–189r; inventory of the chamois preparer Martin Friesel, 19 December 1692, TLA, 230, 1, Bt14, fols 120r–121r.
38 Eva I. Andersson, 'Foreign seductions: Sumptuary laws, consumption and national identity in early modern Sweden', in Tove Engelhardt Mathiassen et al. (eds), *Fashionable Encounters: Perspectives and Trends in Textile and Dress in the Early Modern Nordic World* (Oxford: Oxbow Books, 2014), 15–29.
39 See also Jutta Zander-Seidel, 'Nicht nur tierisch warm: Pelz in Kleidung und Mode', in Georg Ulrich Großmann (ed.), *Vom Ansehen der Tiere* (Nuremberg: Germanisches Nationalmuseum, 2009), 118–31.
40 On this, see Jarmo Kotilaine, *Russia's Foreign Trade and Economic Expansion in the Seventeenth Century: Windows on the World* (Leiden: Brill, 2005), especially chapter 4.
41 1600 clothing regulation, 387.
42 'Waß sonsten denen von der Großen-Gilde zu tragen hierinnen verbohten, dasselbe soll viellmehr von denen handtwerckern vnd andern beijden kleinen Gilden durchgehens verstanden werden', 1665 clothing regulation, fols 87v–88r. Sophie Pitman in Chapter 3 above and Experiment in focus I shows that cheaper furs imitating sable were made in Italy. This suggests how mimetic materials might have enabled individuals who were economically or legally unable to wear the finest furs to dress in early modern fashions.
43 1631 clothing regulation, 399, 401; 1639 clothing regulation, fol. 321r–v.
44 1639 clothing regulation, fol. 323v.
45 1665 clothing regulation, fol. 89v.
46 1665 clothing regulation, fols 90v, 91v.
47 Inventory of the furrier Jochim Hönchen, 25 January 1658, TLA, 230, 1, Bt11, fol. 28v.
48 Inventory of the clockmaker Franciscus Zilagius with his wife, 15 February 1658, TLA, 230, 1, Bt11, fol. 56v; inventory of the tailor Jochim Tempelhoff, 22 February 1658, TLA, 230, 1, Bt11, fol. 70v.
49 Inventory of the tanner Lorentz Grawert, c. 1600–1700, TLA, 230, 1, Bt14, fol. 247v; inventory of the tinsmith Paul Wulff, 23 August 1686, TLA, 230, 1, Bt13/II, fol. 102v; inventory of Jürgen von Stahl, fol. 189r; inventory of Martin Schonert, fol. 232r; inventory of the tailor Hinrich Falck, 14 September 1689, TLA, 230, Bt14, fol. 38v.
50 Inventory of the cordwainer Hanss Busekist, 29 April 1667, TLA, 230, 1, Bt12/I, fol. 39r.
51 Inventory of Birgitta Falck, fol. 231v.
52 Inventory of the linen weaver Engber Per(sson?)'s widow, 27 March 1601, TLA, 230, 1, Aa44, fol. 135r.
53 Inventory of Birgitta Falck, fol. 232r.
54 Inventory of Jürgen von Stahl, fol. 207r.
55 Inventory of Birgitta Falck, fol. 231v.
56 1631 clothing regulation, fol. 396.
57 1631 clothing regulation, fol. 394–5.
58 1665 clothing regulation, fols 87v–88r.
59 1696 clothing regulation, fol. 402v.
60 1665 clothing regulation, fol. 89r.
61 1665 clothing regulation, fol. 90r.
62 1665 clothing regulation, fol. 90v.
63 1665 clothing regulation, fol. 91v.

64 1706 clothing regulation, TLA, 230, 1, Bs7/I, fols 410r–411r.
65 For 'use value' and 'status value', see Gudrun Andersson, 'A mirror of oneself: Possessions and the manifestation of status among a local Swedish elite, 1650–1770', *Cultural and Social History*, 3, 1 (2006), 24.
66 Inventory of the organ-builder Johannes Pauls, 14 December 1646, TLA, 230, 1, Bt10/V, fol. 365r.
67 On this, see for example John Styles, 'Involuntary consumers? The eighteenth-century servant and her clothes', *Textile History*, 33, 1 (2002), 14.
68 Inventory of Birgitta Falck, fol. 231v.
69 Inventory of the town councillor Arendt Stippel, September–October 1696, TLA, 230, 1, Bt14, fol. 179r.
70 Inventory of the linen weaver Andreas Wichman, 30 March 1671, TLA 230, 1, Bt12/II, fol. 88r; inventory of Jürgen von Stahl, fol. 188v.
71 For criticism against this point, see Paula Hohti Erichsen, 'The art of artisan fashions: Moroni's tailor and the changing culture of clothing in sixteenth-century Italy', in Rembrandt Duits (ed.), *The Art of the Poor: The Aesthetic Material Culture of the Lower Classes in Europe 1300–1600* (London: Bloomsbury Academic, 2020), 109–16.

9

The clothing of the *contadina*: women's work, leisure and morality, 1550–1650

Elizabeth Currie and Jordan Mitchell-King

Introduction

The *contadina*, or countrywoman, became a distinctive and easily recognisable sartorial type in Italian visual culture during the second half of the sixteenth century. She can be spotted in the pages of costume books, friendship albums and images of tradespeople, often carrying baskets laden with her wares (Figure 9.1). In paintings of rural scenes, she helps with the harvest and tends to farm animals, clad in improbably white linens. The *contadina* is frequently shown as a young woman and the clothing she wears reflects her contribution to spheres of production and consumption. However, like many female subjects, her visibility is matched by her elusiveness in the archives, and very few household inventories from the period are available to relay a sense of the clothing owned or worn by these women.

This chapter analyses prevailing visual stereotypes against the backdrop of contemporary attitudes towards *contadine* as street-sellers, consumers of fashion and objects of desire. To move beyond the limitations of sources recorded by elite groups and outsiders, as part of our research we made a conjectural reconstruction of an apron, an accessory often associated with rural women, combining evidence from visual representations, archival descriptions and surviving objects. The decision processes involved in this experimental approach, discussed in the second section of the chapter, helped to shift our focus from the reception of the *contadina*'s image and moral assessments of her appearance to think more actively about how these women acquired, wore and valued their clothing.

Figure 9.1 Giuseppe Maria Mitelli, *Ortolana*, etching, from *Di Bologna l'arti per via d'Annibale Carracci* (Rome: Jacomo Rossi, 1660). Rijksmuseum, Amsterdam.

Defining rural women

The term *contadina* referred to women of varying degrees of wealth and status who lived and worked in villages and the countryside. Its broad nature is reflected in John Florio's Italian–English dictionary of 1611, which translates *contadina* simply as 'a Country lasse or wench', while offering a more derogatory view of the *contadino*: 'a country man, a swaine, a hinde, a clowne, a peasant'.[1] Given that a recent estimate suggests that 40 per cent of the Italian population lived in poverty, the majority of countrymen and countrywomen (*contadini*) would have been 'ordinary people' who, according to Joanne Ferraro, 'owned only a few garments in their entire lives, most of which were produced at home'.[2] Interestingly, an extensive analysis of the rural poor in seventeenth-century West Sussex has shown that, despite possessing very few garments, they still sometimes distinguished themselves from others using relatively cheap goods, such as coloured ribbons and stockings.[3]

While we can assume most women living in the Italian countryside wore very rudimentary dress, there is similar evidence that some owned items that were not purely functional. Paolo Malanima's analysis of archival evidence for the rural population of Tuscany in the seventeenth and eighteenth centuries has revealed that, although many individuals went without basic items like shoes, more decorative goods – such as aprons with embroidery or trimming – are occasionally listed in household records.[4] In these cases, it is possible that, as with the artisan groups studied by Sandra Cavallo, clothing formed the most 'valuable component' of their assets.[5] In 1546, Florentine sumptuary laws incorporated a separate category for the clothing of *contadine*, demonstrating that they were considered collectively as a group whose dress required regulation.[6] The subsequent laws issued by Duke Cosimo I de' Medici in 1562 noted that they applied to female members of families who were leaseholders as well as those who owned their own plots of land.[7]

Engagement with fashion

The 1562 Florentine laws banned *contadine* from wearing dresses made of silk or wool, dyed red with *grana* or crimson, but they were permitted hair nets and coifs of silk, silk ribbons and a necklace of 'silver buttons' worth up to 12 lire. They were entitled to narrow velvet belts with silver studs and buckles and silver-gilt aglets, two silver-gilt rings, decorated with 'crystal glass stones' or similar, and paternoster beads worth up to half a scudo. They could decorate the bodices and sleeves of their dresses with satin, damask or sarcenet, and wear a hat made of sarcenet or straw worth up to 1 scudo.[8] The types of clothes and accessories described here imply that some women living in the countryside and villages outside Florence incorporated fashionable elements into their dress, developing strategies to achieve this despite limited financial means.

In some respects, these were well-established patterns. Bolognese sumptuary laws in 1453, for example, divided citizens into six categories, the last comprising those who lived in the countryside and carried out 'rustic work'.[9] These women were banned from wearing dresses and ornaments in silk and cloth dyed with *grana* and crimson, but they were permitted silver buttons and ornaments weighing up to 8 oz, and embroideries or fringes valued at most 3 lire per dress.[10] Again, this suggests that rural families invested money in silver jewellery and decorative elements of clothing (compare Chapters 4 and 5 above). However, the sixteenth-century Florentine laws testify to new, fashionable trends that were prevalent further up the social scale, such as the growing diffusion of silk, in this instance restricted to sleeves, and smaller items such as ribbons and hair accessories, as well as materials that resembled more expensive ones, for example silver-gilt and glass stones. These types of ornament are very similar to the ones emphasised in Cesare Vecellio's costume book *De gli habiti antichi et moderni di diverse parti del mondo*, first printed in 1590, such as silk ribbons, coral necklaces or little silver beads sewn on to clothing or worn as jewellery.

Given that most countrywomen owned very few, plain garments, how can we explain this growing preoccupation with their appearance? Even in Italy, with its numerous urban centres, most of the population lived in rural areas, making *contadine* an important component of the overall demographic.[11] Catherine Kovesi has noted a general move to include middling and lower social groups in fifteenth century sumptuary laws across different parts of Italy. Piedmont, Ferrara and Faenza, for example, passed legislation with separate provisions for *contadini*.[12] During the sixteenth century, different components of fashionable dress, including cheaper varieties of silk, mixed textiles and the types of ornamentation described above, became accessible to a wider social range, fuelling anxieties about the erosion of sartorial distinctions. Two other factors increased the significance of the dress of rural women. First, they made valuable contributions to their household and local economies, work that sometimes gave them greater mobility and visibility. Second, the focus on their virtue or marriageability suggests that the figure of the *contadina* gradually took her place alongside other female exemplars, usually of higher social status, prominent in early modern Italian culture.

Working women

Traces of female labour are invariably bound up in representations of the bodies and clothing of *contadine*: skirts and aprons are shown tied up to allow ease of movement and to protect fabrics from dirt, clogs are worn over decorative shoes, and hats shield faces from the sun. Women pictured working in fields and kitchens, as opposed to being 'on display' as street-sellers, tend to be shown in simpler, plain outfits, with dark-coloured aprons made of more durable-looking fabrics. Typically, the older the woman depicted, the more utilitarian her dress. The food-seller from the Marca Trevigiana in the Veneto region is the only example of an older *contadina* in Vecellio.[13] Although she does wear fashionable accessories, with a belt of red or crimson silk and a bodice decorated with 'little silver-gilt studs' and tied with silk laces, the text underlines the functional nature of her dress over its desirability. We learn that she goes to Venice for the Saturday market, and that she takes off her big, wide-brimmed hat and carries it in her hand when she arrives in the city. She hitches up her skirts because she comes from the country, where it is very muddy, and wears thick leather ankle boots, also to protect against mud and water. The female vegetable-seller in *Arti di Bologna* (1646), a collection of engravings based on drawings by Annibale Carracci, is of an indeterminate age and similarly diverges from the idealised youthful femininity of many representations of *contadine*. Her clothing is plain, apart from her fringed shawl, and her silhouette is bulky and swaddled, typically a visual emblem of the poor.

Two elderly women appear in one of the very few portraits of rural women from this period, dating from 1634 by the Medici court artist Justus Suttermans. The genesis of the painting is explained in a letter from Cardinal Giovan Carlo de' Medici to his younger brother Mattias, describing an evening spent at the Medici villa at Pratolino watching Suttermans make 'portraits of those countrywomen, which he did exquisitely'.[14] Although the women's clothing appears irreproachable, being modest, clean and orderly and, so we are led to believe, the outfits they were wearing that day, Sutterman's final composition exemplifies the taste for humorous treatments of everyday people. Pietro, the African servant on the right, is shown making a sexual gesture with his hand resting on the central woman's shoulders, and further amusement is probably derived from the contrast between the eggs in her basket and her advanced years. Both women are mentioned in the Medici's accounts. Domenica delle Cascine, for example, was paid not only for her ducks and chickens but also for 'playing the fool' for the Medici.[15] As discussed below, this was not the only way that *contadine* provided entertainment for the family.

Vecellio's description of a female fruit- and vegetable-grower from Chioggia begins with praise for the people living in the villages or *borghi* around the town and their skills in cultivating produce, which they sell in 'great quantities' across Venice, Chioggia and other small neighbouring islands.[16] Many Italians, particularly those in more densely populated urban spaces such as Venice, depended on the successful harvests of *contadini* to feed their own families. It was therefore crucial that the goods and the sellers themselves were reliable and trustworthy. Concerns about the malpractices of merchants and pedlars of different social degrees were common. In *La piazza universale di tutte le professioni del mondo*, first printed in 1585, Tommaso Garzoni criticised merchants who purveyed 'counterfeit goods, with corrupt and contaminated merchandise that causes famine in the cities and countryside'.[17] Unsurprisingly, these negative attitudes also extended to female vendors, who were sometimes characterised as cunning or dishonest. The text accompanying Giuseppe Maria Mitelli's etching of a vegetable-seller highlights this, asserting that she uses her seductive powers to sell her salads at higher prices (Figure 9.1).

Despite these suspicions, women continued to operate in the marketplace.[18] Female vendors are prominent in costume books, although they tend to be a minority in panoramic street scenes and are absent from

Ambrogio Brambilla's 1582 etching of 189 street-sellers in Rome. For male artisans and merchants, clothing could play a role in attracting customers. It seems probable that the fine pinking on Giovanni Battista Moroni's tailor's doublet and his paned and lined trunk hose advertised his professional abilities.[19] Similarly, Sicillo Araldo's treatise on colours in heraldry and clothing suggests that merchants wore black to symbolise their loyalty and steadfastness.[20] Reading the appearance of countrywomen in this light, it is notable that they are often portrayed as clean, orderly, decorative, effectively as appealing as their merchandise. The emphasis on their white linen undergarments and accessories would also be seen as a sign of their hygiene and morality. As Araldo explained, 'a woman's shirt should be very white and fine, as it reflects her honesty, which should be pure and unblemished by vice'.[21] These depictions are a far cry from the more shambolic, ragged figures sometimes shown in scenes from urban life, for example by Jacques Callot. While the fortune-tellers swathed in indefinable rags in paintings by the Caravaggisti often use their clothing to conceal stolen goods, the neat silhouette of the *contadina* dispels these anxieties.

The *contadina*'s laden baskets provide evidence of her family's industry. Her clothing can also be seen to represent the fruits of her labour. Many rural women contributed to Florentine wool production, as highlighted in correspondence from Vincenzo Pitti, the overseer of the city's wool guild, to Grand Duke Ferdinando I de' Medici in 1604. Keen to emphasise how many families relied on this work, Pitti stated that in addition to those living within the city walls there were 'female wool spinners who live on the outskirts of the city, [and] female warpers and winders, whose number is difficult to calculate as they are spread out in different places'. Nevertheless, he estimated they comprised around five hundred women.[22] Countrywomen were also involved in different linen-manufacturing processes, the fabric normally used to make the shawls, scarves, coifs, shirts and aprons they are often shown wearing. Flax was grown on tenant farms and smallholdings across various regions of Italy, for the needs of individual households and to sell on. Harvesting the stems, then drying, retting, breaking, scutching and combing them to prepare the long fibres were protracted, labour-intensive activities, even before the linen thread could be spun using a spindle and distaff.

Countrywomen were frequently depicted spinning thread, sometimes in tandem with other tasks such as tending animals. A woman from the island of Ischia in Vecellio's compendium is shown with a distaff tucked into her apron. The accompanying description explains that no silk or wool is woven in the area, so the women spend their time in spinning and in cultivating the land.[23] Ann Rosalind Jones argues that Vecellio associates this particular activity with poverty and marginalisation: in this case, for example, it is a necessity as the more skilled work of weaving is not established on the island.[24] Although spinning was certainly one of the key stages of textile production assigned to rural women, it was not the only one. Fifteenth-century records of linen exporters based in Cortona and the surrounding area include women who also perhaps contributed to the manufacturing processes.[25]

Pleasure and virtue

The *contadina* is usually shown as an attractive, decoratively clad woman of marriageable age, a phase in women's lives when their behaviour was particularly scrutinised. Diane Owen Hughes has demonstrated that sumptuary legislation in some Italian cities was especially stringent towards women in the periods just before and just after they wed.[26] Vecellio's commentary often seeks to underline the modesty and virtue of countrywomen. The fruitseller from Chioggia arranges her hair in 'modest styles', while the *contadine* from the Veneto region who wear their best clothes to celebrate Ascension Day 'dress very neatly and look very lovely' (Figure 9.2).[27] These descriptions appear to follow the same precepts of

Figure 9.2 Cesare Vecellio, *Contadine di terre conconvicine a Venetia*, woodcut from Cesare Vecellio *De gli habiti antichi et moderni di diverse parti del mondo* (Venice: Giovanni Bernardo Sessa, 1598). Wellcome Collection, London.

harmony and order recommended for the clothing of women of higher ranks.[28] It certainly contrasts with narratives of slovenliness linked with the undeserving lower ranks, which became more prominent from the fifteenth century onwards.[29]

Vecellio points out the ‘beautiful needlework’ on the apron of the *contadina* from Tuscany (Figure 9.3), a detail that would have commended her to his readers, given that embroidery was an important female domestic skill. In contrast with this, Federico Luigini observed in his *Il libro della bella donna* (1554) that, while noblewomen’s embroidery brought them honour and reflected their gentility, ‘poor women’ only made practical things.[30] However, even basic sewing abilities would have been an essential part of household management. Embroidered or not, the apron *per se* was able to project female accomplishments and duties. In an account relating to a wealthier, Venetian family, we can see the same kind of simple clothing

Figure 9.3 Cesare Vecellio, *Citelle contadine di Toscana*, woodcut from Cesare Vecellio *De gli habiti antichi et moderni di diverse parti del mondo* (Venice: Giovanni Bernardo Sessa, 1598). Rijksmuseum, Amsterdam.

associated with countrywomen being held up as an ideal, in contrast to more luxurious, immoral fashions. In 1605, the wool merchant Vincenzo Zuccato defended himself against charges that he had contravened sumptuary laws by holding a lavish celebration for the birth of a child. He denied this, arguing that his mother and aunt 'went around the house that day in their usual widows' weeds of twilled wool, and wearing their white aprons, as they usually do all year round to carry out their household chores'.[31]

Beyond the personal benefits linked with owning clean, serviceable or even decorative dress, we can also trace its contribution to the public good. Just as travellers often commented on the appearances of the fashionable female elites alongside other local attractions, rural women could be a source of pride as a collective group.[32] This is especially apparent in descriptions of festive occasions. We learn from Vecellio that women from the region around Belluno often went dancing with the men they were courting, a pastime that usually ceased once they were married.[33] Additionally, the *contadine* from the Veneto who wore a distinctive apron, or *traversa*, 'take good care of themselves all the time so as to be considered beautiful at dances, which they enjoy very much, and to entertain the owners of the land on which they work so that these men will be generous to them' (Figure 9.2).[34]

The spectacle of the *contadina* was a source of entertainment for onlookers all the way up the social scale. In May 1618, the Medici family and 'all the Florentine nobility' spent successive Sundays at the Medici villas at Petraia and Castello, where they were amused by dances by 'beautiful *contadine*'.[35] The grand duke's secretary, Curzio Picchena, recorded his enjoyment with a courtly flourish: 'the countrywomen triumph and I pay humble reverence to them'.[36] A similar event is the subject of Guido Reni's *Country Dance* (1601–2), where women of different social standing all wear aprons (Figure 9.4). Like the dances themselves, the clothing of *contadine* could reflect regional traditions, upholding the ideal that one of the key functions of dress was to mark out social and geographical identities. Different styles of aprons act as regional signifiers in costume books, as well as household and trousseau inventories. For example, early seventeenth-century archival sources from Lecce, Brindisi, Taranto and the surrounding area show that women of modest social standing wore quite distinctive aprons, in bright colours such as green and yellow and decorated with netting and cords, called *vantesini*, *vantili* and *senali* in local dialect.[37]

Inevitably, female visibility sometimes brought with it dishonour. In 1581, a story reached the Medici of a monstrous birth in the Lombard countryside: 'On the night of the Feast of Saint Sebastian … a mile outside Melegnano, a married *contadina* of some beauty gave birth to a monster, looking like a pig from its head to its waist and the rest of it a human creature, however it only eats minestrone as no one was prepared to breastfeed

Figure 9.4 Guido Reni, *Country Dance*, 1601–2. Galleria Borghese, Rome.

it.'[38] In this story, which taps into wider narratives of trust and deceit in relation to female beauty, the purity associated with young *contadine* in some sources, and reinforced by their clothing, is replaced with a sense of repulsion. It is noticeable that in Florentine sumptuary legislation, the category for *contadine* was immediately followed by prostitutes (*meretrici*). The equations between *contadine* and *meretrici* as objects of desire can be traced in costume books, where rural women's bodies appear on a par with their other wares for sale and are developed further in visual tropes such as the young female egg-seller.[39] Although the apron could signal prized feminine virtues, it was also an accessory linked with sex workers. Vecellio's image of a 'public prostitute' shows her in a full-length linen apron so transparent it reveals the silhouette of her legs underneath.[40] When Pasquetta, a Venetian sex worker, was prosecuted in 1639 for contravening sumptuary legislation, her infringements included the use of black lace on not only her silk skirt, sleeves and veil but her silk apron, too.[41] The shifting use and meanings of the apron highlight the difficulty in differentiating between the clothing of 'good' and 'bad' women, a source of concern at all social levels.

The apron: touch and experience

Depictions of *contadine* conformed to a set of social conventions: they shed light on public conceptions of the appearances of rural women but tell us little about the experiences of the women themselves. To gain more understanding of their roles as consumers and producers, we chose to make an item of clothing that was essential, everyday wear for most *contadine*. We took into account the fact that most aprons surviving from the period belonged to elite consumers and considered this material evidence in conjunction with visual and written sources relating to women further down the social scale. Increasingly, scholars have shown how reconstruction as a methodology can lead to a better understanding of the material qualities of early modern garments and the skills involved in making them, as well as challenging the biases inherent in top-down historical sources.[42] Furthermore, achieving a degree of what Beverly Gordon has termed 'textile literacy', even by making a simple accessory like an apron, can help to extract forms of tacit knowledge that are rarely accessible through archival records.[43]

Images of rural women tend to show two main types of aprons: the *traversa*, which encircled the whole waist (Figure 9.2), and the *grembiule*, which covered just the front of the skirt (Figure 9.3). Both usually ended just short of the skirt hem. Some are plain, while others have embroidery, fringing and other decorative elements.[44] Several early modern aprons survive in museum collections, usually highly decorated with embroidery or needle lace; they nevertheless offer insights into a range of materials and construction techniques. All the early modern Italian aprons identified in museum collections are *grembiuli*, made of a rectangular piece of cloth gathered on to a tape tied around the waist. Several examples use multiple panels of linen to increase the width substantially, creating a fuller apron with denser gathers.

Most of the aprons held in museum collections are made of very finely woven white linen. Images of aprons also favour linen, ranging from the fine and drapey to stiff and bulky, depending not only on the quality and weave of the fabric but also presumably on laundering and starching processes and the softening that comes with wear. In contrast, inventories of the possessions of Florentine artisans and other professional groups give a much better sense of the variety and ubiquity of aprons. Here the apron's function is also sometimes described, for example for wearing outdoors or 'for making bread'; the materials used include silks such as taffeta, through to wool, cotton and leather; and the range of colours is considerable.[45] Some of these coloured aprons can be spotted in genre scenes depicting kitchens or food preparation.[46] While darker colours seem more practical for working environments as they more easily conceal stains and marks, as stated previously the clean white apron had additional meaning as an indicator of status and female virtue.[47] Some surviving aprons are decorated with vertical bands of cutwork or embroidery, suggesting they were not for practical use.[48] In contrast, visual sources often show the upper half of the apron left plain, making it possible to hitch them up or wipe hands on them without spoiling any embellishments.

We chose to make a simple *grembiule* from a rectangle of linen (Figure 9.5). Early-seventeenth century aprons in museum collections show there was a great variety in terms of dimensions. While the average length appears to have been 80–90 cm, overall they range from 66 to 110 centimetres long.[49] The dimensions used for the reconstruction were cross-referenced with the proportions seen on the apron worn by Vecellio's Tuscan *contadina* (Fig. 9.3).[50] The top width of the apron was gathered down to 20.3 cm at the waist, a similar measurement to that of a late sixteenth-century English apron in the Victoria and Albert Museum and comparable to Vecellio's depictions.[51] As modern linen widths tend to be wider than sixteenth-century ones, every side of the apron had to be cut out. Early modern aprons probably used the full selvedge width, which provided the dual benefits of conserving the fabric for possible future reuse and potentially reducing the hemming required as two edges were already finished.[52] Linen undergarments from the period reflect the variety of fabric widths current at the time, ranging from around 63 to 104 cm.[53]

On the basis of these measurements and our findings, a simple apron would require between one and one and a half braccia of linen. Most linens cost between 1 and 3 lire per braccio, although prices varied significantly according to quality. In the account books of wealthy Florentine families in the late sixteenth to early seventeenth centuries, prices range from 15 soldi per braccio for the coarsest linen, such as *canovaccio*, through to 6 lire per braccio for the most expensive imported fine linens.[54] The 1556 dowry of Claudia Mediogori, a relatively well-to-do woman from Ferrara, gives a sense of the value of finished aprons, the main distinguishing factor being the textile used.[55] Aprons were the second most common item in her trousseau, which was valued at 334 lire 2 soldi: three were made of 'good cloth' and worth a total of 3 lire 6 soldi; two were blue and valued together at 2 lire; eight were made of a linen or hemp fabric called *boracina* and together worth over 3 lire; and two were simply described as 'used' and worth 1 lira and 10 soldi.

In total, the apron took three and a half hours to make, including cutting out the fabric, hemming the edges and sewing on the tape tie.[56] An apron cut selvedge to selvedge with no hemming down the outer edge would take half as long, making this a significant time-saving method of construction. Not only was the apron relatively quick to sew, the reconstruction confirmed that it was not a resource-heavy process. Although it was carefully measured before cutting to ensure it followed the chosen dimensions, it felt far more instinctive to use the body to 'measure' proportions,

Figure 9.5 Recreation of an Italian gathered linen apron or *grembiule* by Jordan Mitchell-King.

rather than rulers and standard units. Cutting a selvedge-to-selvedge pattern piece required only the length and the width of the gathered top to be determined, which was easily done by holding the fabric up to the body. The reliance on spatial awareness and a sense of proportion is reminiscent of Claudia Kidwell's findings for eighteenth-century American garment-makers, who used notched lengths of tape rather than measurements in units and 'learned to think in spatial distances' as part of their 'individualized intuitive art'.[57]

As noted above, some images of rural women depict aprons that are highly decorated with substantial areas of embroidery and needle lace (see Experiment in focus IX).To explore the skill and labour involved in this work, we sewed sample embroideries in counted thread cross stitch, often seen on linen underwear and accessories from this period. Although large numbers of printed pattern books survive from the sixteenth century, little is known about how designs circulated further down the social scale.[58] Motifs might have been copied from other garments or possibly shared through samplers. One of the simplest motifs from Matteo Pagano's pattern book *Trionfo di Virtu. Libro Novo* (Venice, 1563) took around thirty minutes to complete (Figure 9.6). Although greater speed would come

Figure 9.6 Red silk embroidery on linen of a motif taken from Matteo Pagano, *Trionfo di Virtu. Libro Novo* (Venice, 1563).

with practice, embroidering a handful of these motifs would have been more time-consuming than making the apron itself.

Adding embroidery to an apron might require additional tools, resources and skills. An embroidery frame made the grid of the warp and weft threads clearer, as well as stabilising the fabric; while this could be hand-held, it was easiest to use placed on a trestle, as can be seen in some contemporary depictions of needleworkers.[59] Furthermore, the embroidery thread itself needs delicate handling: the silk filament used in the samples caught easily on rough skin and needed just the right amount of tension to stay smooth. The needle, thimble and silk thread had to be held with much more conscious care and a lighter touch than the linen thread of the apron. Although the counted thread technique is a simple method, it requires total concentration and good light. Fluency and spatial awareness develop during stitching, as the pattern builds up it creates points of reference, which makes counting the threads easier. Seeing the grid of warp and weft also becomes easier in the act of stitching itself, speaking to Gordon's notion of the 'hand of the maker', which works in conjunction with the eyes and begins to intuit the rhythm of the movement and the feel of the materials, reducing the mental strain of such fine needlework.[60]

The contrast between sewing the apron and embroidering it was striking. It is easy to imagine the former being made in the home: although a flat workspace is helpful, given that linen can be creased into folds and holds its shape well there is little need for pins or an iron, and all the work can be done over the lap. The embroidery, on the other hand, could call for more light, precision, concentration, equipment, time and space, making it far less compatible with the working lives of rural women. Surviving aprons hold a clue to the dichotomy between the different types of needlework, as the decorative sections on some examples are in fact separate panels stitched to the main fabric.[61] It is possible that many rural women who made their own basic aprons embellished them with bands stitched by more accomplished female relatives or purchased these from pedlars or local fairs. This offers an insight into how the production of a single accessory could combine different social and economic networks, benefiting from the skills of domestic and professional makers.

Easier and cheaper to make than a shirt, an apron was a functional item that also enabled less well-off women to participate in the fashion for linen accessories. Most aprons would have been made of local linens or linen mixes and occasionally decorated with rudimentary forms of embroidery. The desirability of these accessories is confirmed by their presence in even the very modest wardrobes detailed in the few rural household inventories and dowry records available for this period. These cursory descriptions, together with the findings from the experimental reconstruction process, suggest that visual sources, particularly costume books, reflect the experiences of a small minority of country dwellers. Vecellio's image of a Tuscan *contadina*, for example, dressed in a white shirt with a ruffled collar and an apron with multiple bands of intricate silk needlework, embodies the exemplary figure of the neatly dressed, productive young woman. Similarly, Florentine sumptuary laws allowing *contadine* to wear silver-gilt buttons and silk ribbons only affected the most affluent members of the rural population. However, as suggested by the growing scrutiny of their appearances, dress held significant material and cultural value for rural women, to the extent that even a simple rectangle of linen cloth could be transformed into a tool for social leverage.

Notes

1 John Florio, *Queen Anna's New World of Words* (London: Edward Blount and William Barret, 1611), 119.

2 Rembrandt Duits, 'Introduction: Did the poor have art?', in Rembrandt Duits (ed.), *The Art of the Poor: The Aesthetic Material Culture of the Lower Classes in Europe, 1300–1600* (London: Bloomsbury Academic, 2020), 4. Joanne M. Ferraro,

'The manufacture and movement of goods', in John Jeffries Martin (ed.), *The Renaissance World* (New York: Routledge, 2007), 96.
3 Danae Tankard, '"A pair of grass-green woollen stockings": The clothing of the rural poor in seventeenth-century Sussex', *Textile History*, 43, 1 (2012), 8–10.
4 Paolo Malanima, *Il lusso dei contadini: consumi e industrie nelle campagne toscane del Sei e Settecento* (Bologna: Il Mulino, 1990), 33–4.
5 Sandra Cavallo, 'The artisan's casa', in Marta Ajmar-Wollheim and Flora Dennis (eds), *At Home in Renaissance Italy* (London: V&A Publications, 2006), 72. Artisans and countrywomen are also linked in the depiction of *Donzelle contadine et Artigiane di Parma* in Cesare Vecellio, *De gli habiti antichi et moderni di diverse parti del mondo* (Venice: Damian Zenaro, 1590), fols 191v–192r.
6 Alberto Liva, 'Note sulla legislazione suntuaria nell'Italia centro-settentrionale', in Anna Giulia Cavagna and Grazietta Butazzi (eds), *Le trame della moda* (Rome: Bulzoni, 1995), 45.
7 'La donna contadina, che habita nel contado di Fiorenza, & lavora la terra & l'altre possessioni, o le sue in qualunque modo ...', Lorenzo Cantini (ed.), *Legislazione toscana raccolta ed illustrata dal Dottor Lorenzo Cantini*, vol. 4 (Florence: Stampa Albizziniana da Santa Maria in Campo, 1802), 406.
8 Cantini (ed.), *Legislazione toscana*, vol. 4.
9 Maria Giuseppina Muzzarelli, 'La disciplina delle apparenze', in Paolo Prodi (ed.), *Disciplina dell'anima, disciplina del corpo* (Bologna: Il Mulino, 1994), 774.
10 Maria Giuseppina Muzzarelli, *Guardaroba medievale: vesti e società dal XIII al XVI secolo* (Bologna: Il Mulino, 1999), 264.
11 Ferraro, 'The manufacture and movement of goods', 90.
12 Catherine Kovesi Killerby, *Sumptuary Law in Italy, 1200–1500* (Oxford: Clarendon Press, 2002), 82, 84, 86–7, 89.
13 Vecellio, *De gli habiti antichi et moderni*, fols 180v–181r.
14 For these details about the painting see www.uffizi.it/opere/sustermans-ritratti: 'Giusto faceva ritratti di quelle contadine, che si è portato esquisitamente.'
15 'per fare il Buffone', www.uffizi.it/opere/sustermans-ritratti.
16 Vecellio, *De gli habiti antichi et moderni*, fols 150v–151r.
17 'con le robbe falsificate, con le mercantie corrotte & appestate, che pongono carestia nelle provincie, & nelle città', Tommaso Garzoni, *La piazza universale di tutte le professioni del mondo* (Venice: Seravalle, 1605), 546.
18 Evelyn Welch, *Shopping in the Renaissance: Consumer Cultures in Italy, 1400–1600* (New Haven: Yale University Press, 2005), 34–7, and Ann Matchette, 'Women, objects, and exchange in early modern Florence', *Early Modern Women*, 3 (2008), 245–51.
19 National Gallery, London, accession no. NG697, www.nationalgallery.org.uk/paintings/giovanni-battista-moroni-the-tailor-il-tagliapanni, accessed 3 November 2022.
20 Sicillo Araldo, *Trattato dei colori nelle arme, nelle livree, e nelle divise* (Venice: Domenico Nicolino, 1565), 23.
21 'la camicia della donna poi deve essere candidissima, e sottile, che significhi l'honesta sua, che deve essere candida senz'alcuna macchia di vizio', Araldo, *Trattato dei colori*, 28.
22 'Ci sono oltre a queste le filatrici di lana che stanno nelle pendici della Città, le incannatrici e le orditrici, le quali per essere sparse in diversi luoghi non si potendo senza qualche strepito far discrivere', Maurice Carmona, 'Sull'economia Toscana del Cinquecento e del Seicento', *Archivio Storico Italiano*, 120, 1/433 (1962), 44.
23 Vecellio, *De gli habiti antichi et moderni*, fols 260v–261r.
24 Ann Rosalind Jones, 'Sleeves, purses, spindles: Fashioning women in Cesare Vecellio's costume books', in Susan Shifrin (ed.), *Re-framing Representations of Women* (Abingdon: Routledge, 2016), 113.

25 Dario Martini, 'Produzioni agricole per le manifatture medievali: a robbia e il lino di Cortona all'inizio del Quattrocento', *Rivista di Storia dell'Agricoltura*, 34, 1 (1994), 62.
26 Diane Owen Hughes, 'Sumptuary law and social relations in Renaissance Italy', in John Bossy (ed.), *Disputes and Settlements: Law and Human Relations in the West* (Cambridge: Cambridge University Press, 1983), 93–4.
27 Vecellio, *De gli habiti antichi et moderni*, fols 178v–179r.
28 Jane Bridgeman, 'Condecenti e netti ... Beauty, dress and gender in Italian Renaissance art', in Frances Ames-Lewis and Mary Rogers (eds), *Concepts of Beauty in Renaissance Art* (Aldershot: Ashgate, 1998), 44–51.
29 Samuel K. Cohn, Jr, *Paradoxes of Inequality in Renaissance Italy* (Cambridge: Cambridge University Press, 2021), 39–41.
30 Federico Luigini, *Il libro della bella donna* (Venice: Plinio Pietrasanta, 1554), 106. See Marta Ajmar-Wollheim, 'Housework', in Marta Ajmar-Wollheim and Flora Dennis (eds.), *At Home in Renaissance Italy* (London: V&A Publications, 2006), 152–63.
31 Translated in Patricia Allerston, '"Contrary to the truth and also to the semblance of reality"? Entering a Venetian "lying-in" chamber (1605)', *Renaissance Studies*, 20, 5 (2006), 636.
32 Michel de Montaigne, *Journal de voyage*, ed. Fausta Garavini (Paris: Gallimard, 1983), 206–7, 305, 306.
33 Vecellio, *De gli habiti antichi et moderni*, fols 222v–223r.
34 Vecellio, *De gli habiti antichi et moderni*, fols 178v–179r. Translation from *The Clothing of the Renaissance World: Europe, Asia, Africa, the Americas: Cesare Vecellio's Habiti Antichi et Moderni*, ed. and trans. Margaret F. Rosenthal and Ann Rosalind Jones (London: Thames and Hudson, 2008), 231.
35 State Archives of Florence (ASF), Mediceo del Principato, 6113, 100. Medici Archive Project, Doc ID 18700.
36 'le contadine trionfano, et io faccio humilissima riverenza', ASF, Mediceo del Principato, 6108, 997. Medici Archive Project, Doc ID 6764.
37 Chiara Piccolo Giannuzzi (ed.), *Fonti per il barocco leccese* (Galatina: Congedo Editore, 1995), 6, 23, 26.
38 ASF, Mediceo del Principato, 3254, 715. Medici Archive Project, Doc ID 10227.
39 On the symbolism of egg-sellers in seventeenth-century Dutch art, see Jochen Becker, 'Are these girls really so neat? On kitchen scenes and method', in David Freedberg and Jan de Vries (eds), *Art in History, History in Art: Studies in Seventeenth Century Dutch Art* (Santa Monica: Getty Center for the History of Art and the Humanities, 1991), 146–7.
40 Vecellio, *De gli habiti antichi et moderni*, fol. 146r.
41 Patricia Allerston, 'Consuming problems: Worldly goods in Renaissance Venice', in Michelle O'Malley and Evelyn Welch (eds), *The Material Renaissance* (Manchester: Manchester University Press, 2007), 23.
42 See, for example, Sarah A. Bendall, '"Take measure of your wide and flaunting garments": The farthingale, gender, and the consumption of space in Elizabethan and Jacobean England', *Renaissance Studies*, 33, 5 (2018), 712–37, and 'The case of the "French vardinggale": A methodological approach to reconstructing and understanding ephemeral garments', *Fashion Theory*, 23, 3 (2019), 363–99. For an overview of the different making methodologies used in dress history studies, see Hilary Davidson, 'The embodied turn: Making and remaking dress as an academic practice', *Fashion Theory*, 23, 3 (2019), 329–62.
43 Beverly Gordon, 'The hand of the maker: The importance of understanding textiles from the "inside out"', in *Silk Roads, Other Roads* (Proceedings of the 8th Biennial Symposium of the Textile Society of America, Northampton, MA, September 2002), 189–98.

44 See the depictions of *contadine* in Bartolomeo Grassi, *Dei veri ritratti degl'habiti di tutte le parti del mondo* (Rome: Bartolomeo Grassi, 1585).

45 References to *grembiuli* in household inventories registered with the Florentine Magistrato dei Pupilli between 1550 and 1650 can be accessed via the 'Refashioning the Renaissance' database www.refashioningrenaissance.eu/database. The database includes a total of 4,363 aprons listed in 448 inventories. It demonstrates how common they were as well as the wide range of professional purposes they served. The terms used vary according to city or region, examples being *grembiule* for Florence, *bocassino* or *traversa* for Venice and *spalagrembo* for Siena. On regional terms for aprons, see Monique Rouch, 'I vestiti dei contadini alla fine del Cinquecento e l'opera dialettale di Giulio Cesare Croce, in Jeannine Guérin Dalle Mese (ed.), *Il vestito e la sua immagine: Atti del convegno in omaggio a Cesare Vecellio nel quarto centenario della morte* (Belluno: Tipografia Piave, 2002), 119–42 and 127.

46 See, for example, Vincenzo Campi, *The Kitchen* (1590–1), Pinacoteca di Brera, Milan, accession No. 476, https://pinacotecabrera.org/en/collezione-online/opere/cucina, accessed 3 November 2022.

47 Michele Nicole Robinson, 'Dirty laundry: Caring for clothing in early modern Italy', *Costume*, 55, 1 (2021), 4.

48 See the seventeenth-century Italian apron in the Art Institute of Chicago: accession No. 1976.332, www.artic.edu/artworks/51263/apron, accessed 3 November 2022.

49 See Victoria and Albert Museum, London, accession No. 645–1896 & 348–1891, https://collections.vam.ac.uk/item/O356812 and https://collections.vam.ac.uk/item/O356803, accessed 3 November 2022.

50 Vecellio, *De gli habiti antichi et moderni*, fols 237v–238r.

51 Victoria and Albert Museum, London, accession No. T.87–1917, https://collections.vam.ac.uk/item/O113936, accessed 3 November 2022.

52 Although possibly unfinished, the selvedge is visible along the sides of the Victoria and Albert Museum apron cited above (accession No. T. 87–1917). This time-saving device would have been most likely with simple accessories, which have not survived. On selvedges and seams, see Susan North, *Sweet and Clean? Bodies and Clothes in Early Modern England* (Oxford: Oxford University Press, 2020), 191–2.

53 See examples in Janet Arnold, Jenny Tiramani and Santina M. Levey, *Patterns of Fashion 4: The Cut and Construction of Linen Shirts, Smocks, Neckwear, Headwear and Accessories for Men and Women, c. 1540–1660* (London: Macmillan, 2008), 5.

54 Linen prices between 1580 and 1610 are listed in ASF, Riccardi 55, fols 109r and 147v; loose receipts from Riccardi 224, Libri di Commercio 747, fol. 16r; and Magalotti 8, fols 96r and 290r.

55 Muzzarelli, *Guardaroba medievale*, 69–71; and see 262–8, for the function and widespread use of aprons.

56 This calculation is based on the findings of a sewist with four years of experience of historical dressmaking at the time of writing.

57 Claudia B. Kidwell, *Cutting a Fashionable Fit: Dressmakers' Drafting Systems in the United States* (Washington, DC: Smithsonian Studies in History and Technology, 1979), 4.

58 Elisa Ricci is one of the few authors to have addressed this, illustrating examples of a deer motif originally derived from a printed pattern book and interpreted for different purposes with a range of skill levels. See Elisa Ricci, *Ricami Italiani: antichi e moderni* (Florence: Felice Le Monnier, 1927), 44.

59 Alessandro Paganino, *Libro quarto de rechami* (Toscolano: Paganino, c. 1532), 2.

60 Gordon, 'The hand of the maker'.

61 See, for example, an Italian apron in the Metropolitan Museum, New York, accession No. 38.19.8, www.metmuseum.org/art/collection.

10

Practical, professional and prosperous: dressing the artisans and small shopkeepers in sixteenth- and seventeenth-century Denmark

Anne-Kristine Sindvald Larsen

Introduction

A painting by the Netherlandish artist Govert Camphuysen from the mid-1660s shows a group of men and women gathered at a town inn, probably in the region of Scania, a part of Denmark until 1658, informally dressed in layers of clothes (Figure 10.1).[1] The men wear linen shirts with collars and a combination of stockings, breeches and doublets, along with jerkins, caps or broad-brimmed hats, and slippers, boots or shoes of leather, while women are dressed in skirts, bodices, doublets, white linen partlets, caps, stockings and slippers or shoes. The clothes seem modest, practical and worn-out, dominated by coarse woollen and linen garments. They also appear bulky and loose in strong contrast to high fashion of the time. Yet the presence of small decorative details, colour and small accessories in the painting suggests that there was a desire among the lower social order to show off modest levels of prosperity or social aspirations.

Several dress historians, such as John Styles, Patricial Allerston and Paula Hohti, have shown that, even though their economic circumstances were often modest and their daily activities were dominated by work, the lower social classes were connected with contemporary fashions in major European centres such as Venice and London.[2] But it has never been explored in depth how dress fashions spread among the non-elite population in more peripheral areas of Europe, such as in early modern Denmark, and what ordinary artisans and local shopkeepers wore in their everyday lives.

This chapter explores how artisans and small shopkeepers dressed in early modern Denmark, focusing especially on clothes that they wore in their work lives.[3] Exploring the typical garments and accessories that were listed in the inventories of shoemakers, bakers, tailors, barbers and

Figure 10.1 Govert Camphuysen, *Town Inn*, 1664–65. Malmoe Art Museum.

other artisans in the port town of Elsinore in Denmark in 1550–1650, the aim is to show that appropriate garb at the lower social levels was not only designed to protect from weather but it was a key function of the clothes to help men and women of artisan rank to appear respectable. This was especially important in artisans' professional and social lives. Just as clean clothes, such as white linen aprons, held a significant material and cultural value for Italian women from artisan ranks in both cities and countryside, as Elizabeth Currie and Jordan Mitchell-King have shown in Chapter 9 above), so neat appearance and fine clothes helped urban artisans and small shopkeepers and their families in small Danish towns to appear prosperous and trustworthy – virtues that would help them to attract customers, to conduct business deals with fellow tradespeople and to express professional ambitions.

Practical and comfortable

The daily life of early modern artisans was dominated by work. Apart from Sundays and religious holidays, the majority of bakers, barbers, shoemakers, butchers and other artisans living in sixteenth- and seventeenth-century Elsinore worked from early morning until late in the evening. The clothes of artisans, therefore, had to be primarily functional and meet the requirements of manual or physical work both inside and outside of the workshop, in all weather conditions. It was vital, therefore, that their clothes were durable and constructed in a way that allowed movement of the body.[4]

These functional aspects of clothing are clearly visible in archival documentation. The majority of garments belonging to artisans in early modern Elsinore and listed in the 'Refashioning the Renaissance' database were practical, made of plain coarse materials and in modest colours. The stock of clothes of Axel Drejer, a turner from Elsinore, provides a typical example. He owned a number of garments that were well suited for his work. These included four practical linen shirts, a durable leather doublet, a grey doublet and a woollen waistcoat, as well as several pairs of leather breeches and a fur cap made of fox that would keep his head warm in the workshop.[5] An illustration from a chest from the Copenhagen Turners Guild from around 1658, depicting turners at work with one man turning wood with one bended knee and the other bent over chopping wood, shows what such practical garments might have looked like (Figure 10.2). Both men wear doublets, breeches, slippers and caps that were simply cut, most likely made of long-lasting materials suitable for physical work.

Dressing up in many layers of clothes provided a way to cope with the changing seasons and temperatures. Denmark, like the rest of Europe, was exceptionally cold in this period, due to 'the little ice age', which culminated in the sixteenth century.[6] Consequently, besides being comfortable and practical during physical work outside or indoors, it was necessary that Danish artisans' clothes were also made from materials that protected the body against changes in temperatures.[7] Garments made of a range of light and heavy woollens, mixed fabrics or leather could be used for both cold and warm climates.[8] Hats made of wool or fur, woollen stockings and gloves kept men and women warm and protected against bad weather.

Although artisans' wardrobes were often relatively modest, many artisans from Elsinore had a variety of clothes for a range of weather conditions. The strap-maker Michel Nielsen, for example, owned a mantle, a suit and a tunic of brown and grey broadcloth, several pairs of old and cheap breeches, a doublet made out of bombazine (a mix of fibres of cotton, linen, silk or hair) and a pair of linen stockings that could be worn in both warm and cold weather. In addition, he had a number of accessories for cold days, including mittens and a fine cap made of marten fur.[9]

Figure 10.2 Turners working in a workshop. Illustration on a guild chest from the turners' guild of Copenhagen, c. 1658. National Museum of Denmark, Copenhagen.

Durable, elastic and warm, leather was also a suitable material for work garments, especially outside where the body had to be protected from wind and rain.[10] The carpenter Hans Sletkrol, for example, had a yellow windproof and waterproof suit of leather, composed of doublet and breeches. He could pair this outfit with a red woollen waistcoat underneath, jerkin, grey tunic and a cap when he had to keep warm.[11] Such a leather doublet, suited for his work outdoors, might have looked something like the surviving leather doublet that used to belong to Hugo Grotius (1583–1645), a Dutch Huguenot, today preserved in the collections of Museum Rotterdam (Figure 10.3). Hugo Grotius wore this doublet when he escaped from prison in 1621, dressed as a 'bricklayer'.

Another way to cope with the changing temperatures was to insert linings of fur or wool in garments. Some of these were suited for spring and autumn weather, such as the mantles owned by the bowl-painter Oluf Nielsen, which were lined with bay (a light wool). Others, such as the tunics, overgowns and caps mentioned in the hook-maker Jacob

Figure 10.3 A seventeenth-century leather doublet worn by the Dutch Huguenot Hugo Grotius in 1621. Museum Rotterdam.

Krogemager's inventory, lined with lambskin or fox and marten fur, were used in cold weather.[12]

Most artisans wore linen shirts under their work clothes. Linen garments protected the outer garments from sweat or grease and, since linen underwear could be laundered, they supported good health and hygiene.[13] Despite the relatively high cost of linen shirts, Danish artisans in general seem to have many undershirts. The baker Kresten Hermansen's inventory, for example, mentioned ten linen shirts, some of which were described as 'clean' and others 'dirty'.[14] Visual images suggests that clean white linen was an essential feature of the artisans' appearance. In a scene from a stained-glass window from 1583 from the collection of Museum Odense (Figure 10.4), probably bequeathed to the shoemakers' guild by the master shoemaker Jesper Pedersen and installed in the shoemakers' guildhall in Odense, the central panel shows the master shoemaker Jesper Pedersen in his workshop along with a journeyman and an apprentice. The latter two wear practical white linen shirts with narrow frilled collars and rolled-up sleeves, along with protective black leather aprons.[15]

Washable linen aprons were also a common feature of work clothes. The master barber Abraham Raider, for example, owned altogether twelve aprons which he could wear when he treated his patients.[16] The majority of linen aprons in Danish artisan inventories were modest. Such plain aprons,

Figure 10.4 A shoemaker's workshop, 1583. Detail from a stained-glass window, probably donated to the Odense shoemaker's guild by the master shoemaker Jesper Pedersen. Museum Odense.

as Currie and Jordan Mitchell-King have shown, could be sewn at home relatively quickly and cheaply (Chapter 9 above). Their condition suggest that aprons, nonetheless, were used extensively until they were completely worn out. The tailor journeyman Anders Poulsen's chests, for example, contained a linen apron which was so old and worn out that it was 'not worth writing about'.[17]

Professional and respectable

Comfort and practicality were essential considerations when artisans selected clothes for the day's manual work. However, running a shop and serving clients required that artisans appeared respectable and trustworthy. Creating not only an upright professional image but also one that indicated an elevated social or financial status helped artisans appear attractive and attract customers of a greater status and wealth. Thus, selecting practical yet good quality garments was no doubt good for business.[18]

One way of conveying a respectable appearance to customers, clients and other business associates was to appear well groomed, and wear clean clothes and spotless white linen. Many artisans from Elsinore seem to be well aware of the importance of their looks. The personal belongings of the goldsmith Bastian Krammer, for example, included a range of grooming items, such as two hairbrushes and a small case to store personal grooming items.[19] The inventory of the joiner Hans Svitzer and his wife, in turn, included a brush to clean shoes as well as another brush to clean the clothes that were made of materials that could not be washed in water, such as fine silks.[20] The organ-builder Johan Lorentz, hired in 1649 by the King Christian IV to rebuild the organ in St Olaf's church in Elsinore, even kept a clothing brush in his workshop at the church in order to keep his garments clean.[21] Such examples suggest that maintaining cleanliness and a respectable and unsoiled appearance was an important matter for artisans at work. They allowed these artisans to keep a neat appearance and show their customers and business partners that they were decent businessmen as well as artisans.

A well-groomed appearance and clean, spotless linen were important especially for artisans and small shopkeepers who produced or sold food. For example, when the butcher Troels Pedersen sold meat from his stall, he could wear one of his many linen garments, including five linen shirts, eight linen collars and two linen kerchiefs. Such linen garments presented him as a capable butcher selling high-quality products, fostering his hopes that his customers would keep returning to his shop on the marketplace.[22]

It was especially important that linen was clean and fresh on the parts of the garments that were visible, such as on collars or ruffs. This provided an impression of good hygiene and a clean appearance even if the linen undergarments were modest or worn out.[23] The tailor journeyman Erich Lauridsen, for example, had four linen collars and two linen kerchiefs that were defined as more valuable and in better condition than two of his old shirts, which he wore for just hygiene and comfort under his brown woollen suit, not for show.[24]

Another way to create a professional look and emphasise reliability and success, in addition to maintaining a clean and well-groomed appearance, was to dress up according to the latest fashions. The baker Knud

Andersen, for example, had several matching suits that corresponded with the contemporary tastes, including a simple black suit made of half-woollen fabric and a yellow-brown suit. Decorated with visible linen around his neck and worn with a mantle and a costly sable cap, all listed in his inventory, garments such as these allowed him to create a look of prosperous and respectable artisan and make his social position as one of the senior bakers in town visible.[25]

A well-tailored suit, especially if it was black, was an appropriate garment for a respectable shopkeeper, since black associated the wearer with moral virtues such as trustworthiness and honesty that were regarded as essential qualities of mercantile success.[26] Visual images suggest that wearing black was an important sign of status especially for Danish master artisans and shopkeepers of a higher rank. In the stained-glass window from Museum Odense (Figure 10.4), referred to above, for example, the status of the master shoemaker Jesper Pedersen is emphasised by his fashionable ensemble, composed of a tight-fitting black suit in the Spanish style, possibly made of fine wool, and black stockings, a frilled ruff and a modishly tall felted black hat. While the black colour of his clothes associates the shoemaker with wealth, authority and good morals, his well-trimmed beard and groomed appearance enforce the messages about his masculinity. The appearance confirms his position as a skilled master artisan, an important guild member and a citizen in the local society, and a reliable and successful business owner and entrepreneur.[27]

Appearing well-dressed and in latest fashion was especially important for tailors. Well-tailored clothes made of good-quality materials provided a living advertisement of their skills and their knowledge of dress fashions. The tailor Desmer Skrædder's wardrobe is a good example of this. He owned, altogether, four linen shirts and several detachable collars, a pair of silk grosgrain breeches, a leather doublet with sleeves of *hundskot* (a light wool fabric likely produced in Hondschoote in Flanders), a doublet of white linen pinked according to the current fashion, a mantle made of black English broadcloth and a number of items of jewellery, such as a two gold rings and a gold memorial ring.[28] The finest of his garments included a mantle of black English cloth and a doublet 'in the colour of the kings' (*kollør de rois*, probably bright tawny brown), trimmed with silk braid and sleeves of *hundskot*.[29] Such garments not only held a considerable value but they enhanced his reputation as a talented tailor who had the ability to inspire his customers and create clothes with trimmings, techniques and materials that responded to the latest innovations in fashion.[30]

Ambitions, achievements and prosperity

Although the daily clothes of artisans in general did not represent 'the first flush of fashion', many local artisans in early modern Denmark were conscious of the need to maintain certain standards of decorum and fashion that were suited to their social standing.[31] Showing off the finest clothing and knowledge of fashion was important especially for journeymen and master artisans because the quality of the garments was seen as a sign of an individual's social status and success and played an important role in defining their reputation, career, marriage, wealth and public influence.[32]

While young apprentices, who ranked lowest in the artisanal hierarchy, usually wore quite modest clothes since their garments were often provided by the parents or the master, more attention was paid to the kind and quality of clothes when they became journeymen.[33] This becomes clear, for example, from an ordinance issued on 2 November 1622 at Børnehuset, an institution where orphaned and poor children were taught textile crafts in Copenhagen. It shows that, when young journeymen finished a four-year apprenticeship at the institution, they were given a new set of clothing, consisting of a suit of breeches and doublet, a plain woollen mantle, a hat, a pair of shoes, a linen collar and a shirt.[34] Receiving the new outfit gave artisans a good start when they were looking for positions in workshops outside the Børnehuset.

A close examination of inventories of some prosperous journeymen suggests that some Danish journeymen had the financial ability as well as the knowledge to dress well. One example is the barber journeyman Herbert Droff who lived and worked in the household of the master barber Jørgen Bardskærer. He appears to have possessed a strong sense of style. His inventory featured a wide variety of clothes and fashion accessories that he could use towards his professional ambitions and social aspirations. Some of his best garments included a grey woollen suit trimmed with braid, a pair of grey breeches and three doublets, made of bombazine (mixed fabric) and trimmed with braid. He also had many outer garments, including a brown mantle, a coat and a sleeveless leather buff coat that bestowed him with a masculine and military yet fashionable look.[35] He owned, furthermore, some fine accessories, such as a pair of shoes and slippers, a grey hat with two hatbands, a variety of caps, linen shirts, two fine lace collars, eleven ordinary collars, five kerchiefs, and even a pair of knitted stockings – a novelty at the time.[36]

Wearing novel fashion accessories and trimmings, such as the stockings, caps and collars worn by the barber journeyman Herbert Droff, was an effective way among artisans to make the tastes for current fashions visible. The senior shoemaker Augustinus Jørgensen, chosen as an alderman for the shoemakers' guild in 1629, made his status visible by wearing silver buttons on his brown suit and by finishing this with a ruff, a pair of knitted stockings and a black hat which was decorated with a hatband.[37]

Inventories suggest that many Danish master artisans were aware of the power of clothing to convey professional success and reflect achievements. One of them was Zacharias von Ulnitz, an armourer in Elsinore, who provided the king in 1585 with several pieces of armour.[38] His dealings with the king enabled Zacharias to possess several fine and expensive items of clothing that demonstrated his extraordinary professional achievements. His clothing included some exquisite items such as a lavish black pinked velvet doublet, a bright blue damask doublet, a pair of breeches of *trip* (a woollen velvet) and two Spanish-style caps of silk and gold.[39]

Some artisans in early modern Elsinore were aware of the social and cultural importance of fashioning an identity also by visual means. The inventory of the prosperous master barber Jurgen von Breda, for example, mentioned 'a portrait of himself and his wife'. Although the painting has not survived, it is likely that he wore his best garments in the portrait. These included, according to his inventory, a range of fineries and novelties, such as suits of *fifskaft* and caffa (a type of velvet), a black mantle of broadcloth, a pair of brown silk stockings, various types of silk and fur caps (including fine marten fur), two hats with ribbons, a hatband with small and precious pearls, two feathers for his hat, a pair of black garters in silk and two pairs of shoe roses.[40]

A surviving portrait from 1654 confirms that the most prosperous artisans were eager to use clothes in order to construct a visual identity and present a successful image of themselves and their social rank. The painting represents the art turner Jacob Jensen Nordmand and it was painted by Wolfgang Heimbach, a painter who portrayed also other members of the royal family (Figure 10.5). Jacob was a respected artist and artisan. Originally trained as a smith in Holland, he was employed as an art turner at court by Frederik III in 1649 to work on fine-art objects in materials such as ivory and to teach some of the royal family members to turn.[41] His elevated position not only allowed Jacob to have social interaction with the higher ranks but it also gave him other benefits. For example, according to his own account, when he taught the ueen to turn, he was allowed 'to eat at the royal majesty's table for some years'.[42]

His portrait conveys his professional and personal success through clothes. The turner Jacob Jensen Nordmand is depicted in a lustrous black silk doublet and a velvet jerkin, finished with a fine white linen collar and a pair of linen cuffs. By wearing garments dyed in a good-quality black, the turner Jacob Jensen could associated himself not just with moral virtues of black but also with the material and cultural prestige of the colour. Intense black was the colour of high fashion of the high-ranking elites.[43] The only features in the image that reveal his identity as an artisan are the tiny pair of golden tweezers and a *memento mori*, allegedly made by the art turner himself.[44] The portrait painting, therefore, manifests both his skills as a turner as well as his extraordinary artisanal position at court, where he climbed the social ladder and eventually became the Master of the Arsenal. He documented his journey and social uprising in his personal accounts which he completed in 1670.[45]

Figure 10.5 Wolfgang Heimbach, *Portrait of Jacob Jensen Nordmand*, 1654. The Danish Royal Collection, Rosenborg Castle, Copenhagen.

Court artisans, such as Jacob, were also often provided with fine liveries since they were regarded as an extension of the King's household.[46] In his personal account, the turner Jacob noted that during the first year of his service he was given four liveries that were 'each as good as 100 daler'. This was a significant value for one set of clothes, though evidence suggests that Jacob's own valuation of the clothes might have been slightly exaggerated.[47]

Jacob's position and social uprising did not go unnoticed. His employment and success at court allowed him to dress in garments that were otherwise considered unsuitable for individuals of unnoble birth and mingle with people way above artisanal rank. While some of his fellow artisans might have looked up to him with admiration, his privileged treatment also probably caused much envy. After noting in his accounts that he had 'eaten with counts and other great lords and other noble persons which were present in the King's Court' and 'sat at the table with clergymen, and other worldly persons', he stated that 'some were good people, but others were resentful'.[48]

Conclusion

Clothes had many functions for artisans in daily life and throughout their professional careers, in relation to their trade or craft as well their social standing. Practicality was naturally one of the primary functions of artisans' clothing. It was especially important that the clothing worn by artisans and small shopkeepers was suited for physical work and changing weather conditions, because most artisans came from modest backgrounds and their everyday lives were dominated by work. Yet visual, and written evidence suggests that consolidating a respectable image and maintaining a decent look by means of clothing and appearance were equally as important, especially when artisans and local shopkeepers performed their work in public places and dealt with their customers at the marketplace, workshops or shops. Appearing well-groomed and clean while wearing good-quality practical garments was a way to promote decency and to create an image of a reliable and trustworthy craftsman, shopkeeper or entrepreneur.

In this context, the social and cultural significance of fashion was important. Since fine clothing was associated with wealth and status, wearing both well-tailored garments or fashion accessories and novelties that corresponded to the current tastes, such as trimmings in fine lace and silk, ruffs, knitted stockings, hats or fine gloves, was an effective social signal that made their professional success and accomplishments visible in public in the local society. The most prosperous and successful artisans, such as barbers or artists working at court, could even fashion a distinctive look for themselves through visual means by commissioning portraits, providing an extraordinary status for some of the most prestigious craftsmen and artists among the artisan communities of early modern Elsinore.

Notes

The research leading to these results was done as part of the Refashioning the Renaissance project that has received funding from the European Research Council (ERC) under the European Union's Horizon 2020 research and innovation programme (grant agreement No. 726195).

1 The Danoe-Norwegian Kingdom consisted of the Kingdom of Denmark with the geographical area of Jutland, Funen and Zealand and the provinces Scania (until 1658), Blekinge and Halland (until 1645); and of the Kingdom of Norway together with the Faroe Islands, Iceland, Greenland and the duchies Schleswig and Holstein (now northern Germany). The chapter uses the term 'Denmark' or 'Danish' to refer to geographical area of the Kingdom of Denmark.

2 John Styles, *Dress of the People: Everyday Fashion in Eighteenth-Century England* (New Haven: Yale University Press, 2007); Paula Hohti, 'Dress, dissemination, and innovation: Artisan "fashions" in sixteenth- and early seventeenth-century Italy', in Evelyn Welch (ed.), *Fashioning the Early Modern: Dress, Textiles, and Innovation in Europe, 1500–1800* (Oxford: Oxford University Press, 2017), 143–65; Paula Hohti Erichsen,'The art of artisan fashions: Moroni's tailor and the changing culture of clothing in sixteenth-century Italy', in Rembrandt Duits (ed.), *The Art of the Poor: The Aesthetic Material Culture of the Lower Classes in Europe 1300–1600* (London: Bloomsbury Academic, 2020), 109–16; Paula Hohti Erichsen, 'Power, black clothing, and the chromatic politics of textiles in Renaissance Europe', in Jenny Boulboullé and Sven Dupré (eds), *Burgundian Black: Reworking Early Modern Colour Technologies* (Santa Barbara: EMC Imprint, 2022), https://burgundianblack.tome.press; Patricia Allerston, 'Clothing and early modern Venetian society', *Continuity and Change*, 15, 3 (2000), 367–90.

3 This chapter is based on my doctoral dissertation 'Clothes, culture and crafts: Dress and fashion among artisans and small shopkeepers in the Danish town of Elsinore, 1550–1650' (Aalto University, Aalto Arts Books, Espoo, 2023), conducted as part of the 'Refashioning the Renaissance' project. It is based on inventories deriving from a sample of 294 household and post-mortem inventories of artisans and their wives from Elsinore between 1573 and 1650, available online at https://refashioningrenaissance.eu/database. On how to study fashion among artisans and small shopkeepers in Denmark, Anne-Kristine Sindvald Larsen, 'Artisans and dress in Denmark 1550–1650: A preliminary exploration', in Duits (ed.), *The Art of the Poor*, 99–108. On inventories and their challenges, see Giorgio Riello, '"Things seen and unseen": The material culture of early modern inventories and their representation of domestic interior', in Paula Findlen (ed.), *Early Modern Things: Objects and Their Histories, 1500–1800* (New York: Routledge, 2013), 125–50.

4 Ole Degn and Inger Dübeck, *Håndværkets kulturhistorie: håndværket i Fremgang, perioden 1550–1700* (Copenhagen: Håndværkerrådets Forlag, 1983), 33: Maria Hayward, 'A shadow of a former self: Analysis of an early seventeenth-century doublet from Abingdon', in Tara Hamling and Catherine Richardson (eds), *Everyday Objects: Medieval and Early Modern Material Culture and Its Meanings* (Abingdon: Routledge, 2016), 110.

5 Inventory of Axel Drejer, 27 July 1616, Danish National Archives (RA), Helsingør Byfoged, Skifteprotokol, 1612–1619, fol. 238r–238v.

6 Charlotte Rimstad, 'Dragtfortællinger fra voldgraven: Klædedragten i 1600-tallets København, baseret på arkæologiske tekstiler fra Københavns Rådhusplads' (PhD dissertation, University of Copenhagen, 2018), 38; Niels Hybel, 'Klima og hungersnød i middelalderen', *Historisk Tidsskrift*, 102, 2 (2002), 267; Wolfgang Behringer, *A Cultural History of Climate* (Cambridge: Polity, 2010), 85.

7 Isabella Paresys, 'The body', in Elizabeth Currie (ed.), *A Cultural History of Dress and Fashion in the Renaissance* (London: Bloomsbury Academic, 2017), 57.

8 Records indicate that cloth in different qualities as well as mixed serge was used for winter clothes. See, for example, King Frederik II's order for some plain cloth for winter garments: Laurs Laursen (ed.). *Kancelliets Brevbøger vedrørerende Danmarks indre forhold i uddrag 1580–1583*. (Copenhagen: C. A. Reitzels Forlag, 1903), 571. Summer suits at the Swedish court were made, for example, of a light coarse silk fabric terzanelle, and a winter suit of cloth and serge, Cecilia Aneer, *Skrädderi för kungligt bruk: tillverkning av kläder vid det Svenska hovet, ca 1600–1635* (Uppsala: Uppsala Universitet, 2009), 248, 256.

9 Inventory of Michel Nielsen Remmesnider, 19 November 1642, RA, Helsingør Byfoged, Skifteprotokol, 1639–1644, fol. 366r.

10 Roy Thomson, 'The nature and properties of leather', in Marion Kite and Roy Thomson (eds), *Conservation of Leather and Related Materials* (Abingdon: Routledge, 2006), 1.

11 Inventory of Hans Sletkrol Tømmermand, 15 November 1636, RA, Helsingør Byfoged, Skifteprotokol, 1635–1639, fols 126v–127r.

12 Inventory of Oluf Nielsen Skålefarver, 9 March 1620, RA, Helsingør Byfoged, Skifteprotokol, 1621–1625, fol. 360r-360v; inventory of Jacob Krogemager, 20 August 1585, RA, Helsingør Byfoged, Skifteprotokol, 1583–1592, fol. 71r.

13 For example, Janet Arnold, Jenny Tiramani and Santina M. Levy, *Patterns of Fashion 4. The Cut and Construction of Linen Shirts, Smocks, Neckwear, Headwear and Accessories for Men and Women* (London: Macmillan, 2008), 9; Georges Vigarello, *Concepts of Cleanliness: Changing Attitudes in France Since the Middle Ages* (Cambridge: Cambridge University Press, 1988), 58–62. On laundering and wearing linen among artisanal classes see also Michele Nicole Robinson, 'Dirty laundry: Caring for clothing in early modern Italy', *Costume*, 55, 1 (2021), 3–23.

14 'Hans Skiortter som Renne var vj, Skorter Skidne iiij, skiorter', RA, Helsingør Byfoged, Skifteprotokol: 1592–1598, inventory of Kresten Hermansen Bager, 27 June 1594, fol. 372v.

15 On this window, see Jan Kock, 'Ruder med personlighed: Een mode i renæssancen', in Ole Høiris and Jens Vellev (eds), *Renæssancens verden: tænkning, kulturliv, dagligliv og efterliv* (Aarhus: Aarhus Universitetsforlag, 2006), 401–17.

16 Inventory of Mester Abraham Raider Bardskærer, 4 February 1605, RA, Helsingør Byfoged, Skifteprotokol, 1603–1610, fol. 105v.

17 'Herforuden nogen Klude och gamlt Forklede som icke er verdtt att skriffue', inventory of Anders Poulsen Skræddersvend, 14 September 1616, RA, Helsingør Byfoged, Skifteprotokol, 1612–1619, fol. 186v.

18 Patricia Allerston, 'Clothing', 380.

19 Inventory of Bastian Krammer Guldsmed, 4 July 1627, RA, Helsingør Byfoged, Skifteprotokol, 1625–1627, fol. 432r.

20 Inventory of Hans Svitzer and wife, 15 December, 1601, RA, Helsingør Byfoged, Skifteprotokol, 1599–1603, fol. 370r.

21 Inventory of Johan Lorentz Orgelbygger, 19 June 1650, RA, Helsingør Byfoged, Skifteprotokol, 1648–1650, fol. 236r–236v.

22 Inventory of Troels Pedersen Slagter, 29 February 1632, RA, Helsingør Byfoged, Skifteprotokol, 1632–1635, fol. 9r.

23 Susan North argues that this offered a way of 'cheating', see Susan North, *Sweet and Clean? Bodies and Clothes in Early Modern England* (Oxford: Oxford University Press, 2020), 138.

24 Inventory of Erich Lauridsen Skræddersvend, 18 January 1640, RA, Helsingør Byfoged, Skifteprotokol, 1639–1644, fol. 13v.

25 Inventory of Knud Andersen Bager, 9 January 1650, RA, Helsingør Byfoged, Skifteprotokol, 1648–1650, fol. 107r.

26 Hohti Erichsen, 'Power, black clothing'; Elizabeth Currie, *Fashion and Masculinity in Renaissance Florence* (London: Bloomsbury Academic, 2016), 106; John Harvey, *Men in Black* (Chicago: University of Chicago Press, 1996), 63–5. On the meaning of black dress for merchants, see also Chapter 9 above by Currie and Mitchell-King.

27 Will Fisher, 'The Renaissance beard: Masculinity in early modern England', *Renaissance Quarterly*, 54, 1 (2001), 156.

28 Inventory of Desmer Skrædder, 31 January 1596, RA, Helsingør Byfoged, Skifteprotokol, 1592–1598, fols 517r–517v, 518r. Hondschoote was a centre of say production in Flanders. See Aneer, *Skrädderi*, 217.

29 For the 'kollør de rois' (literally, the 'king's colour'), see Isis Sturtewagen, 'All together respectably dressed: Fashion and clothing in Bruges during the fifteenth and sixteenth centuries' (PhD thesis, University of Antwerp, 2016), 158.

30 Elizabeth Currie, 'Diversity and design in the Florentine tailoring trade 1550–1620', in Michelle O'Malley and Evelyn S. Welch (eds), *The Material Renaissance* (Manchester: Manchester University Press, 2007), 162.

31 Hayward, 'A shadow of', 108; Sandra Cavallo, 'The artisan's casa', in Marta Ajmar-Wollheim and Flora Dennis (eds), *At Home in Renaissance Italy* (London: V&A Publications, 2006), 73.
32 Ulinka Rublack, *Dressing Up: Cultural Identity in Renaissance Europe* (Oxford: Oxford University Press, 2010), 35–6.
33 The contract for the apprenticeship was negotiated between the master and the parents, including negotiations of clothes, Maarten Prak and Patrick Wallis, 'Introduction: Apprenticeship in early modern Europe', in Prak and Wallis (eds), *Apprenticeship in Early Modern Europe* (Cambridge: Cambridge University Press, 2019), 10. Entering adolescence made young men more fashion-conscious, see Ulinka Rublack and Maria Hayward (eds), *The First Book of Fashion: The Book of Clothes of Matthaus and Veit Konrad Schwarz of Augsburg* (London: Bloomsbury Academic, 2015), 41; Currie, *Fashion and Masculinity*, 109–27.
34 Vilhelm A. Secher (ed.), *Corpus constitutionum daniæ: forordninger. forordninger, recesser og andre kongelige breve, Danmarks lovgivning vedkommende 1558–1660*, Vol. 4 (Copenhagen: G.E. C. Gad Nielsen & Lydiche, 1897), 40. This was also specified in an earlier ordinance from 1621, see Secher (ed.), *Corpus constitutionum daniæ*, Vol. 3 (Copenhagen: G. E. C. Gad and Nielsen & Lydiche, 1891), 719.
35 Aneer, *Skrädderi*, 165; Currie, *Fashion and Masculinity*, 135; Gudrun Ekstrand, 'Some remarks on buff-coats in Sweden', in Fred Sandstedt (ed.), *17th Century War, Weaponry, and Politics: Report on the Xth Congress of the International Association of Museums of Arms and Military History* (Stockholm: The Royal Armoury and The Royal Army Museum, 1984), 190.
36 Inventory of Herbert Droff Bardskærersvend, 3 July 1630, RA, Helsingør Byfoged, Skifteprotokol, 1628–1631, fols 365v–366r.
37 Inventory of Augustinus Jørgensen Skomager, 16 May 1637, RA, Helsingør Byfoged, Skifteprotokol, 1635–1639, fols 289r –290r. He became alderman on 12 October 1629, see RA, Helsingør Skomagerlav, Lavsbog, 1623–1802, fol. 25r.
38 Laurs Laursen (ed.), *Kancelliets brevbøger vedrørerende Danmarks indre forhold i uddrag 1584–1588*. (Copenhagen: C. A. Reitzels Forlag, 1906), 396, 403. In 1599 he provided the king with one thousand pieces of armour; in 1603 and 1609, he also had large commissions from the king, Otto Blom, *Kristian den fjerdes artilleri: hans tøihuse og vaabenforraad* (Copenhagen: Thieles Bogtrykkeri, 1877), 309–10, 327.
39 He also wore a pair of white breeches, a black mantle and a sable cap, RA, Helsingør Byfoged, Skifteprotokol: 1610–1612, inventory of Zacharias von Ulnitz Plattenslager, 14 October 1612, fols 227r–228v.
40 'Sl: M: Jurgensen och Hans Sl. hustruis Contrafey 2 d', RA, Helsingør Byfoged, Skifteprotokol: 1639–1644, inventory of Mester Jurgen von Breda Bardskærer, 18 August 1641, fols 233v, 234v, 235r–236v, 238r.
41 On his life, see Henrik Carl Bering Liisberg, 'Jacob Jensen Nordmand', *Museum: Tidsskrift for Historie Og Geografi*, Første Halvbind (1893), Peter Frederik Suhm, 'En ubenævnt Dansk kunstners levnets historie', *Samlinger til den Danske Historie*, 2, 3 (1784), 134–58.
42 Suhm, 'En ubenævnt', 156–7, translated by the author.
43 Currie, *Fashion and Masculinity*, 99–101; Hohti Erichsen, 'Power, black clothing'.
44 Bering Liisberg, 'Jacob Jensen', 248–9.
45 Suhm, 'En ubenævnt', 135.
46 Degn and Dübeck, *Håndværkets kulturhistorie*, 106–7.
47 Suhm, 'En ubenævnt', 156. Records show that, as Jacob became more successful, his liveries became more expensive but never reached a high value such as this, Bering Liisberg, 'Jacob Jensen', 269–70.
48 Quoting Suhm, 'En ubenævnt', 157, translated by the author.

EXPERIMENT IN FOCUS VII: Caring for clothes

Anne-Kristine Sindvald Larsen

Introduction

Textiles were expensive in the early modern period. This meant that taking care of them was a priority in the household.[1] Clean clothes were important not only because of personal hygiene and health but also because they conveyed an image of a respectable citizen, tradesperson and upright Christian. An English conduct book from 1619, for example, insisted that ones's clothes should be kept neat and clean 'For spotted, dirty, or the like, / is lothsome to be seene'.[2]

Textiles had to be cleaned with care. Since most garments, apart from linen, could not be washed without damaging the natural dyes or the finishing, there were a range of methods for cleaning textiles without immersing them in water, including stain removal, surface cleaning and re-dyeing.[3] In addition, it was also common to use perfumes on clothes and bodies to prevent bad smells.[4] This practice seems to have been important also among artisans. In 1646, the coppersmith Gabriel, who resided in Elsinore in Denmark, had in his bedchamber a green box, with some 'bathing herbs', which could have provided a scent both pleasant and therapeutic.[5] The Italian tailor Piero di Giovanni from Florence, also wore a perfumed pendant, which made him smell clean.[6]

But how were clothes in fact refreshed, stains removed and clothing perfumed? Could this be done at home or did artisans have to take their clothes to tailors or professional cleaners in order to keep them clean? These were some of the questions explored at the workshop 'Dirty Laundry', organised by the 'Refashioning the Renaissance' project in April 2019. The aim of the workshop was to recreate some of the recipes for taking care of clothes that were circulated in cheap printed advice manuals and collections of recipes in the early modern period. As we carried out the experiments, we tested, for example, whether lemons or chanterelle mushrooms could be used to remove stains from silk. Would scented rose petals make the artisans' linens smell like those 'of a great lord'?

The workshop, held over two days, offered many new insights into the sixteenth- and seventeenth-century cleaning practices and gave a better understanding of how artisans, local shopkeepers and others from modest economic conditions could make their precious clothes last longer.

The experiment

During the sixteenth and seventeenth centuries, instructions on how to clean textiles and clothing were widely disseminated in Europe in printed collections of recipes and advice manuals. For instance, the Danish advice manual *Maangehaande artige kunstner*, published in 1578, included recipes on how to use 'water' to remove stains from gold embroidered cloth and velvet, how to clean stains from precious silk fabric, and how to remove spots of wine from woollen cloth.[7] Since these recipes, like so many other recipes in Europe, were published in cheap printed media, is it possible that such instructions circulated also among ordinary artisanal families? Could people like the Florentine butcher Della Cara, who owned a pair of breeches in taffeta defined as 'nasty', apply such instructions at home?[8]

For the experiments, we selected three sixteenth- and seventeenth-century cleaning recipes that appeared repeatedly in cheap and easy-to-obtain texts and pamphlets.[9] Such books were intended for use at home. They feature terse instructions and call for ingredients that were relatively easy for people to obtain, some of which may have even been growing in domestic gardens and pots. We followed the instructions closely to find out how well these popular recipes actually worked, using silk, linen and woollen fabrics which were stained with ink, red and white wine and oil.

First, we tried a very simple Italian recipe designed to remove stains from white wool or linen using lemon juice.[10] We found that if the stains were fresh, applied on the fabric just before treating them with lemon juice, the results on white wool were pretty good. On the set stains, on the other hand, the recipe did not seem to work. In addition, the results showed that treating the fabric with lemon juice could easily destroy the colour. The red on linen dyed with cochineal turned hot pink when rubbed with lemon juice (Figure vii.1).

Second, we tested a Danish recipe for removing stains from precious silk, using chanterelles. As the first step, the recipe advised to 'take the top of the small mushrooms, which are called *pfiffereling*' (chanterelles). Second, it was advised to soak the stained fabric with the mushrooms for two hours; the fabric should then be rinsed with clear water and left in the sun to dry.[11] Because no instructions were given on how the fabric should

Figure VII.1 A piece of stained linen treated with lemon at the 'Dirty Laundry' workshop, organised by the 'Refashioning the Renaissance' project at Aalto University in 2019.

be soaked with the chanterelles, we decided to boil the mushrooms first with water to make a paste (Figure vii.2), and then apply it on to the stains. Then it was rinsed with water. Perhaps this was not the right method, because the recipe did not seem to work. Instead, especially the ink discoloured the silk samples, making the stains even worse (Figure vii.3).

Finally, we tested an Italian recipe which promised to wash a scarlet wool and remove stains using boiled cream of tartar. As the first step, the recipe instructed to 'Boil white tartar powder in water so that it reduces by a third and then strain it and when you put it on the cloth, make sure the water is tepid and let it dry and when it is dry it will return to its colour'.[12]

This recipe was tested on samples of wool dyed with madder, kermes and cochineal and stained with white wine, oil, red wine and ink. This stain remover was probably the least clear since, it did not mention how much tartar powder or water should be used. Ultimately 50 g of tartar powder was boiled in 400 ml of water, the water was reduced by one-third and was then drained (Figure vii.4) There were lots of discussions about whether to use the water or solid portion left after boiling, and how much of the

Figure VII.2 Chanterelle mushrooms soaked in water.

Figure VII.3 Results of the sixteenth-century Danish chanterelle stain-removal recipe.

Figure VII.4 Ingredients for a sixteenth-century Italian cleaning and stain-removal recipe.

solution should be applied to the cloth. In the end, we decided to try both, but neither one of the recipes was particularly successful, yet it appear that using the water made better results, since the ink seemed to spread to other areas of the fabric and rubbing the fabric with a small linen cloth and solid portion of tartar powder damaged the cloth and made the dye come off. Neither as a stain remover was it effective; however the cloth stained with ink, became lighter, which suggest that could be used to make very dark stains lighter. Depending on where the stain was, this would have been useful to make one's finest red garment last longer.

After testing these stain-removal recipes with limited success, we decided to turn to perfume. Several early modern authors claimed not only that perfumes provided garments with a pleasing scent but that herbs could also protect clothes from vermin. A household calendar, published in Denmark in 1648, advised that 'if you want to protect your clothes from mosquitoes, cloth worms, moths and other alike, smear this [anis] oil on all sides of the chest'.[13] Some of the instructions were explicitly directed at ordinary families. Simon Paulli, a Danish doctor, botanist and anatomist, who recommended numerous remedies in his herbal book published also in 1648, was writing for the 'common man', in order to prevent moth and worms from eating and harming clothes. Adding scented herbs to chests of clothing, such as dried *origanum* (also mentioned as wild marjoram), mentioned by Paulli, protected the clothes from vermin and provided the clothes and the wearer with a 'clean' smell.[14] Being able to protect their clothes was probably a serious concern for artisanal families, for several artisan inventories mentioned garments that were 'moth-eaten'. For example, in 1592, a Danish woman was paid 22 skilling to sell two moth-eaten gowns that had previously belonged to Morten Mortensen, a son of a glazier in Elsinore.[15]

A recipe written by an anonymous sixteenth-century Italian author promises to make clothes smell like the 'scent of a great lord'. According to the instruction, the desirable scent could be obtained by mixing ingredients such as rose petals, rosewater, root of white lily, ground cloves and musk.[16] We mixed rose petals, rose water, orris root, synthetic musk, white ginger lily oil and cloves in a pot, and, after boiling and stirring the mixture, the blend was put into handsewn linen pouches (Figure vii.5).

The result was very fragrant. I stored the linen pouch in my wardrobe and got a first-hand experience of what perfumed clothes worn in the early modern period might have smelled like. At the same time, while some of us thought the smell was pleasing and others found it slightly repulsive, we were wondering how early modern Italians would have experienced such a scent in an urban environment where the air was full of strong odours from cattle, horses, smoke and strong perfumes.

Figure VII.5 A reconstructed perfumed linen bag.

Conclusion

How does recreating recipes for stain removal and scents help us to understand some of the practices that ordinary artisan families experienced when taking care of their clothes? Experiencing the past through early modern practices of cleaning, one quickly understood the extensive effort as well as the great challenges that were associated with keeping clothes clean and presentable. The experiments suggested that stains were not easy to remove, and recipes for stain removal and cleaning were often unclear and imprecise. At the same time, the experiments provided a unique insight into the kind of practical fashion knowledge that even ordinary people might have had access to through cheap print – a topic that is hard to trace in this period given the low levels of literacy, especially among artisanal women.

Finally, one of the most fruitful dimensions of the experiments was that they helped us to formulate new questions. For example, it is not clear that, even if ordinary artisan families had access to such recipes through cheap print and the ingredients required, individuals of lower social rank could read the instructions. Was such knowledge shared and transmitted between family members and neighbours in the period so that even the illiterate had access to these, or were these manuals restricted to a smaller audience, or used by the broader population at all? Such questions remain unanswered.

Notes

The Refashioning the Renaissance workshop 'Dirty laundry in early modern Italy and Denmark', organised by Michele Robinson with the support of Anne-Kristine Sindvald Larsen at Aalto University on 11–12 April 2019, was funded by the European Research Council (ERC) under the European Union's Horizon 2020 research and innovation programme (grant agreement No. 726195). See www.refashioningrenaissance.eu/experiments/dirty-laundry. We thank our advisory board members Flora Dennis and Tessa Storey and the 'Refashioning' team members for participation and discussions, and the expertise they brought into the workshop.

1 On how Italian artisans took care of their clothing see esp. Michele Nicole Robinson, 'Dirty laundry: Caring for clothing in early modern Italy', *Costume*, 55, 1 (2021), 3–23.
2 Quoting Susan North, *Sweet and Clean? Bodies and Clothes in Early Modern England* (Oxford: Oxford University Press, 2020), 34. I thank Michele Robinson for the Italian references and her assistance in writing up this experiment.
3 North, *Sweet and Clean*, 209–10.
4 Isabella Paresys, 'The body', in Elizabeth Currie (ed.), *A Cultural History of Dress and Fashion in the Renaissance* (London: Bloomsbury Academic, 2017), 71–2.
5 'En grøn Eshe med Nogle bad Vrter Vdj', Danish National Archives (RA), Helsingør Byfoged, skifteprotokoller: 1644–1648, inventory of Gabriel Riis Kobbersmed, 20 April 1646, fol. 197v.
6 'Una collana di profumi con canutiglia', State Archives of Florence (ASF), inventory of the tailor Pietro di Giovanni, 27 February 1634, Magistrato dei pupilli,

2719, fol. 119r. On the topic of perfumed accessories in the Renaissance period, see Evelyn Welch, 'Scented buttons and perfumed gloves: Smelling things in Renaissance Italy', in Bella Mirabella (ed.), *Ornamentalism. The Art of Renaissance Accessories* (Ann Arbor: University of Michigan Press, 2011), 13–39.

7 *Mange haande artige kaanster at berede gaat blick oc alle haande farffue, oc at schriffue met guld, sölff oc alle metal aff feyre: met mange andre nyttige kaanster at farffue schriffue feyre oc pergament met alle haande farffue, oc huor mand skal lade schrifft ædzis paa staal eller jern oc vaaben. Alle schriffuere, breffmalere, kniffuesmeder oc andre som slig kaanster bruge gantske nyttige at vide* (Copenhagen: Laurentz Benedicht, 1578), fol. 19r -19v.

8 '1 paio di braconi con federa di taffeta piu cattivi simili', ASF, Magistrato dei pupilli, 2709, inventory of the butcher Della Casa, 20 September 1570, fol. 18r.

9 On Renaissance cleaning recipes, see esp. Drea Leed, '"Ye shall have it cleane": Textile cleaning techniques in Renaissance Europe', in Robin Netherton and Gale T. Owen-Crocker (eds), *Medieval Clothing and Textiles 2* (Woodbridge: Boydell & Brewer, 2006).

10 Opera nova chiamata Secreti secretorum: in laquale poterai conseguire molti piaceri et utilitade: con molte cose ridiculose stampata novamente (n.p.: Stampata per Bernardino Benalio, [before 1550]), sig. Biii.

11 'Tag osen aff de smaa paddehatter, som kallis Pfifferling.' *Mange haande artige*, fol. 19v. The recipe is also found in the German recipe book *The Allerley Matkel* from 1532, the earliest recipe book on spot removing and dying. In this version the reader was instructed to use the juice of chanterelles to remove stains from silken veils, advising how to: 'Take juice of chanterelles, soak the stains therein for two hours, rinse with clear water, and allow to dry', Sidney M. Edelstein, 'The Allerley Matkel (1532): Facsimile text, translation, and critical study of the earliest printed book on spot removing and dyeing', *Technology and Culture* 5, 3 (1964), 316.

12 'A lavar uno scarlatto & tor via le machie. Recipe tartaro bianco pesto onze quatro & fallo bollire in libre doi di aqua tanto che cali il terzo & poi tu lo colarai & quando che tu lo metterai sopra lo panno fa che lacque si tevida: & lassalo sugare & quando che lo sera sutto el sara ritornato i suo color', *Opera nova chiamata Secreti secretorum: in laquale poterai conseguire molti piaceri et utilitade: con molte cose ridiculose stampata novamente* (n.p.: Stampata per Bernardino Benalio, [before 1550]), sig. Biii.

13 Jens Lauridsen Wolf, *Diarium sive calendarium ecclesiasticum, politicum et oeconomicum perpetuum. det er en evigvarende eircke, politisk oc huuszholdings caiender* (Copenhagen: Peter Hake på Boghandler Peder Andersens Bekostning, 1648), 296.

14 Simon Paulli, *Flora danica, det er: Dansk urtebog: udi huilcken efter ... Christiani IV. ... befaling ... icke alleeniste urternis historiske beskrifvelse, krafter oc virckninger med ziirligste figurer andragis, men endocsaa lægedomme til alle siugdomme gafnlige, korteligen ocklarligen antegnis, saa at den er baade en urtebog oc lægebog med største flid oc umage elaborerit aff Simone Paulli* (Copenhagen: Prentet aff Melchiore Martzan, 1648), 306 G.

15 'Er giffuid en Quinde som solldte ij kiortler som vore kommen møll vdi xxii sk', Danish National Archives (RA), Helsingør Byfoged, skifteprotokoller: 1583–1592, inventory of Morten Mortensen, 10 March 1592, fol. 277v.

16 *Opera nuova intitolata dificio de ricette* (Venice, 1529), 14.

EXPERIMENT IN FOCUS VIII: Colour

Paula Hohti

Introduction

Colour was one of the fundamental elements of fashion. While there is an assumption that visually attractive bright colours were reserved only for those who held power and authority in society, historical inventories and visual records show that ordinary Italians were receptive to colour fashions and keen to adopt the latest trends. The dress historian Ulinka Rublack has shown the speed with which colour fashions spread across social classes. For instance, when yellow became fashionable in the Swiss city of Basel at the beginning of the sixteenth century, it initially gained popularity among the wealthy but soon extended to include prostitutes, journeymen, apprentices and maidservants, as well as minor officials and artisans. By 1520, nearly everyone in the city wore yellow and the colour appeared in many innovative combinations.[1]

The 'Refashioning the Renaissance' database confirms that colours in all shades and qualities were worn by artisanal groups in the sixteenth and seventeenth centuries. As Stefania Montemezzo has shown in Chapter 7 above, Italian artisans and shopkeepers owned clothing and accessories in a wide array of colours, including intense and spectacularly dyed textiles and garments. Skirts, petticoats, aprons, shirts and sleeves were dyed in a kaleidoscope of colours, including crimson red, pink, deep black, light and medium blue, purple, light yellow and different shades of brown.[2]

However, not all colours held the same fashionable appeal. The meaning and value of colour were determined not only by the generic colour but the precise shade and brilliance defined how the colour was perceived and what it signified. The deeper or more brilliant the tone was, the more appealing it was to consumers. What kinds of colours were worn by our artisans and what did these look like?

To recover the visual look of the colours referenced in the inventories of the everyday artisans and explore how these colours were created in that time, we organised a colour workshop 'Making invisible colours visible' in 2019. Drawing on dye recipes dating back to the sixteenth and seventeenth centuries, our objective was to explore the essence of colours of different grade and quality, and what specific colouring methods and dyestuffs were available for achieving some of the most precious colours in cost-effective ways.

The experiment

Red and black were historically two of the most expensive colours. Due to the saturated tone, superb colourfastness and complex dye processes, the finest reds and blacks enjoyed extraordinary popularity and prestige throughout the Renaissance period.[3] Professional dyers made great efforts in the period to produce vibrant colours and perfect shades. This is evident in the first practical recipe book on commercial dyeing, the *Plichto*, written by Gioanventura Rosetti and published in Venice in 1548. His collection of dye recipes included thirty-five recipes for dyeing red, and twenty-one for black.[4]

Traditionally, the best scarlet and crimson reds were dyed with expensive insect dyes, 'grain', crimson dye, or cochineal.[5] However, lower artisanal classes had access to similar beautiful shades of red, which were created by using cheaper dyestuffs such as local madder or brazilwood. These alternatives were sometimes deliberately used to imitate expensive scarlet and crimson reds. For example, brazilwood, when used on plain fabric or madder-dyed fabric, served as an important source for imitating crimson silk and scarlet wool derived from grain.[6]

In his *Plichto*, Gioanventura Rosetti provided instructions on how to dye wool with madder and brazilwood to achieve a look that resembled scarlet:

> When you are about to madder, you will take 8 pounds of madder for each 12 pounds of wool, and two pails of strong [hard] water per dozen. See that the bath be tepid and put inside the madder. Stir well and quickly put in the wool. Give it a very good raking. When it has come to the boil, poke it well under and let it boil the time of eight paternosters. Then take it out and have it well sifted and washed. Then you will give it a new bath, well heated, and put inside the said wool and make it boil a little. You will take it out and have it well washed, steeping it to advantage. Then you will take 8 ounces of brazil for each dozen by weight of pounds of wool and make it boil the second. When it has boiled, set up the cauldron with a reasonable bath and when it be tepid you will put in the brazil. When it is about to raise the boil, put on the wool and have it raked and boil it until as much as you would say six paternosters. Then you take it out and you will have wool of good color.[7]

While the recipe may present challenges for modern dyers, particularly regarding details such as the exact duration denoted by 'until as much as you would say six paternosters', following the instructions using alum mordanted wool and silk revealed that such imitations could produce beautiful and intensive reds comparable to those dyed with grain or crimson, although the colours were not very lightfast (Figure viii.1).[8]

Further experiments demonstrated that vibrant and captivating colours, including pink, blue, yellow, green, as well as new shades such as the colour 'dove neck' (*colombino*) could be obtained using locally available natural dyes such as safflower, woad, weld and dyer's broom (Figures viii.2 and viii.3).[9] The pink colour achieved using safflower, for instance, was so intense and bright that it appeared almost as if it had been created using synthetic dyes (Figure viii.4).

These methods brought a wide range of affordable colours within the reach of all social groups in the market. Additionally, simple and cost-effective methods for achieving desirable colours were widely circulated in cheap printed media, enabling men and women to create colours in the comfort of their own homes.[10] One such colour was black, the colour of high fashion traditionally made by dyeing the fabric first in a blue woad solution and then dipping it in a red madder dye bath.[11] For example, a

Figure VIII.1 Dye experiments for expensive scarlet and kermes red imitations using madder and brazilwood. Recipe from the sixteenth-century dye manual *Plichto* by Gioanventura Rossetti.

Figure VIII.2 Bright yellow dyed with weld, based on a standardised historical dye recipe. Adding potash to the dye solution brightened the colour.

recipe directed at 'women so that when they have spun yarn they know how to dye it in many colours', circulated in various printed works, including the comprehensive *Plichto* and in the Dutch and French translations of *Dificio de ricette*, provided simple instructions for achieving a home-dyed 'lustrous' black using oak gall, vitriol and a small amount of gum arabic.[12]

By testing early modern recipes, we gained valuable insights not only into the appearance and production methods of colours like 'artisan black' or imitation red but also into the high level of skill required by professional dyers to achieve precise shades and specific tones. Various factors in the dyeing process influenced the outcome, resulting in different shades. For example, in the case of yellows, we noticed that the colour achieved with dyer's broom varied considerably depending on the specific crop used and the amount of potash added in the dye solution to enhance the brightness of the yellow colour.

Perhaps, then, the yellows worn in the city of Basel by wealthy individuals, artisans, maidservants and sex workers were not all equally bright or colourfast. Nevertheless, the wide range of dyeing methods

Figure VIII.3 A range of naturally dyed silk fabrics, using historical recipes, on display at the *Reconstructing Everyday Fashion* Exhibition, Aalto University, September 2021. The samples demonstrate the variety of colours available for and used by early modern artisans.

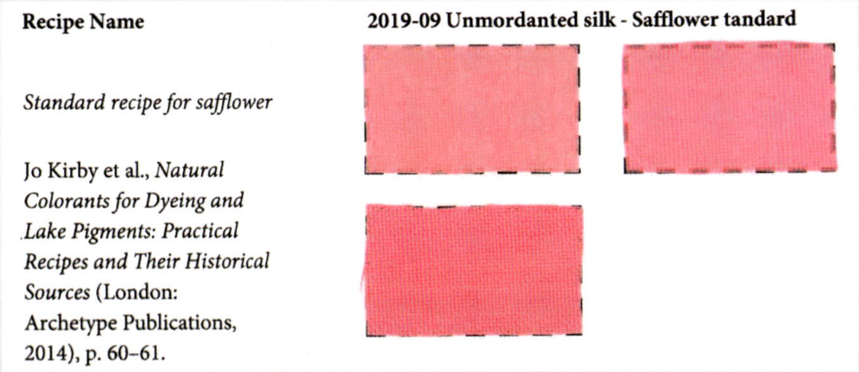

Figure VIII.4 A sheet from a sample book showing bright pink silk samples dyed with safflower, using a standardised historical recipe.

and the availability of new colourants from the sixteenth century onward allowed people from all social classes to dress up in vibrant garments and embrace evolving colour fashions which relied increasingly on producing new fashionable colours in a cheaper way for a broader range of consumers.

Notes

The 'Refashioning the Renaissance' project's workshop 'Making invisible colours visible', organised by Paula Hohti at Aalto University on 10–12 March 2020, was funded by the European Research Council (ERC) under the European Union's Horizon 2020 research and innovation programme (grant agreement No. 726195). See refashioningrenaissance.eu/experiments/making-invisible-colours-visible. We would like to thank Jo Kirby for her extensive work on leading and designing the colour experiments, and Natalia Ortega-Saez and the 'Refashioning the Renaissance' team members for participation, presentations and discussions.

1 Ulinka Rublak, 'Renaissance fashion: The birth of power dressing', *History Today*, 61, 1 (2011), www.historytoday.com/archive/renaissance-fashion-birth-power-dressing, accessed 7 June 2023,
2 See also Paula Hohti, 'Innovations and imitations of colour: Wearing and making colour in early modern Italy', in Maria Hayward, Giorgio Riello and Ulinka Rublack (eds), *A Revolution in Colour: Natural Dyes and Dress in Europe, c. 1400–1800* (London: Bloomsbury, forthcoming 2024).
3 For the superb quality and prestige of textiles dyed with kermes and carmine insects, see Dominique Cardon, *Natural Dyes: Sources, Tradition, Technology and Science* (London: Archetype Publications, 2007), 614 and 641–2.
4 Gioanventura Rosetti, *The Plichto: Instructions in the Art of the Dyers Which Teaches the Dyeing of Woolen Cloths, Linens, Cottons, and Silk by the Great Art as well as by the Common*, trans. of the first edition of 1548 by Sidney M. Edelstein and Hector C. Borghetty (Cambridge, MA: MIT Press, 1969). See also Rublack, 'Renaissance dress, cultures of making, and the period eye', *Journal of Decorative Arts, Design History, and Material Culture*, 23, 1 (2016), 6–34, at 27–31. For the materials and techniques used for dyeing silk cloths in the Renaissance period, see Luca Molà,

The Silk Industry of Renaissance Venice (Baltimore: Johns Hopkins University Press, 2000), 107–8.

5 Jo Kirby, Maartin Van Bommel and A. Verhecken (eds), *Natural Colorants for Dyeing and Lake Pigments: Practical Recipes and Their Historical Sources* (London: Archetype Publications, 2014).

6 For an example of a recipe, see Cardon, *Natural Dyes*, 284. For dye substitutions in silk trade, Molà, *The Silk Industry of Renaissance Venice*, chapter 12.

7 Recipe No. 102, Rosetti, *Plichto*, 127.

8 For historical recipes and their use in reconstructions, see Jenny Boulboullé and Sven Dupré, 'Introduction: Burgundian black', in Jenny Boulboullé and Sven Dupré (eds), *Burgundian Black: Reworking Early Modern Colour Technologies* (Santa Barbara: EMC Imprint, 2022), https://doi.org/10.55239/bb001, accessed 29 January 2024, and Pamela Smith, 'An introduction to Ms. Fr. 640 and its author-practitioner', in Pamela H. Smith et al. (eds), *Secrets of Craft and Nature in Renaissance France: A Digital Critical Edition and English Translation of BnF Ms. Fr. 640* (New York: Making and Knowing Project, 2020), available online at https://edition640.makingandknowing.org.

9 For standardised historical recipes, see Kirby et al., *Natural Colorants*; and for processes of dyeing and the material meanings of colour, Judith H. Hofenk de Graaff, *The Colourful Past: Origins, Chemistry and Identification of Natural Dyestuffs* (London: Archetype Publications, 2014); Andrea Freeser, Maureen D. Goggin and Beth F. Tobin, *The Materiality of Color: The Production, Circulation, and Application of Dyes and Pigments 1400–1800* (New York: Routledge, 2016).

10 Inexpensive printed booklets with recipes related to textiles, dyes or stain removal were included, for example, *Opera nuova intitolata dificio di ricette* (Venice, 1525). For the use and circulation of these among artisans, see Paula Hohti, 'Power, black clothing, and the chromatic politics of textiles in Renaissance Europe', in Jenny Boulboullé and Sven Dupré (eds), *Burgundian Black: Reworking Early Modern Colour Technologies* (Santa Barbara: EMC Imprint, 2022), available at https://burgundianblack.tome.press, accessed 3 June 2023; and for guild regulations covering dye practices, see Molà, *The Silk Industry in Renaissance Venice*, 107–37.

11 For the most effective methods of dying black, see Natalia Ortega-Saez, 'Black dyed wool in north-western Europe, 1680–1850: The relationship between historical recipes and the current state of preservation' (Unpublished PhD thesis, University of Antwerp, 2018), 61–2, and further recipes in Hofenk de Graaff, *The Colourful Past*, 313–23.

12 Recipe from *Recettario nuovo copiosissimo* (1550), fol. 3v, titled 'for women so that when they have spun yarn they know how to dye it in many colours', also published in *Opera nvova intitolata dificio de ricette* (Venice, 1525).

EXPERIMENT IN FOCUS IX: Lace

Elena Kanagy-Loux

Introduction

As quickly as lace-making emerged in the early sixteenth century, the openwork textile became a status symbol for the wealthy, embellishing the edges of all types of dress and home linens. The ubiquity of lace-trimmed handkerchiefs is underscored by their numerous appearances in contemporary portraits, engravings and inventories, as well as extant handkerchiefs in museum collections. Handkerchiefs were simultaneously functional and decorative objects, often given as gifts to lady's maids or as tokens of love from a suitor, and carried in all strata of society, from the aristocracy to the working poor.[1]

Although the finest Renaissance lace was an expensive luxury product, the 'Refashioning' data from Venice, Siena and Florence from 1550 to 1650, used for the analysis in many chapters of this book, show that ordinary men and women from artisanal classes owned garments and accessories trimmed with multiple types of lace. These included cuffs, handkerchiefs, ruffs, sleeves and gowns trimmed with bobbin-lace borders, edgings of reticella (an early needlepoint lace derived from cutwork), or punto in aria needle lace (which is built up from buttonhole stitches) of fine Venetian or Flemish linen thread. The probate inventories compiled in the 'Refashioning the Renaissance' database reveal the possession of numerous embellished handkerchiefs in the households of Italian artisans and merchants, including a handkerchief trimmed with gold lace owned by a Venetian innkeeper.[2] What surviving visual, material and archival evidence cannot tell us, however, is precisely what types of lace were available to the artisan class and what level of skill and sophistication was needed to make a fine lace-trimmed handkerchief.

In 2020–22 the 'Refashioning the Renaissance' project proposed to reconstruct a lace-trimmed accessory representing something a fashionable artisan, living between 1550 and 1650, would have owned based on material and visual evidence. We selected a handkerchief measuring 20 cm by 20 cm – half the scale of typical extant Renaissance handkerchiefs – to showcase four different edgings: a kind of lace sampler (Figures ix.1a and ix.1b). Building upon the archival research of the 'Refashioning the Renaissance' team, we selected each edging to represent a range of styles, techniques and fibres found in northern Italian bobbin-lace over four quarter-centuries. During this early period in lace history there were already evolutions in the design and construction of lace, as well as the development of distinct regional styles in major lace-making centres such as Venice, Milan and Genoa.

The decades on which this experiment is focused coincide with the development of bobbin and needle lace, which makes this a time period that is ripe for investigation. Bobbin-lace, which developed out of multi-strand braiding techniques related to passementerie, was made of threads wound on to bobbins to keep them organised and interlaced to create an endless variety of patterns. Simultaneously, needle lace developed from cutwork and drawnwork, an embroidery technique in which the pattern is traced on to linen fabric, threads are cut and pulled out of the design, and the voided

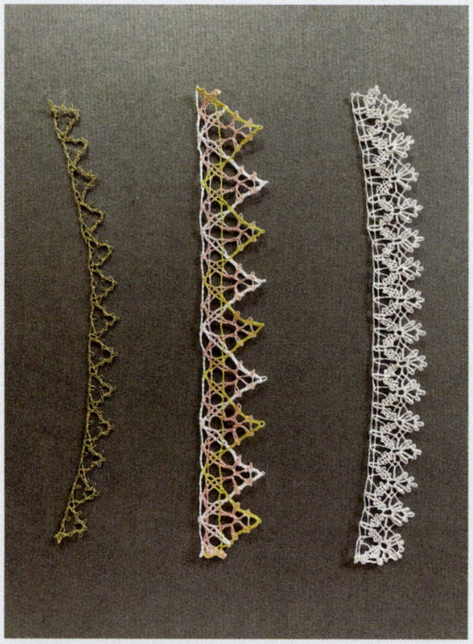

Figure IX.1a Three reconstructed bobbin-lace edgings, from left to right: metal bobbin lace, polychrome silk and linen bobbin-lace, and Parasole bobbin-lace pattern.

Figure IX.1b Handkerchief with four reconstructed bobbin-lace edgings.

areas filled in with buttonhole stitches. Eventually the base fabric was discarded entirely, and the stitches were worked directly on a parchment pattern, anchored by a couched outline thread.

The aims of this experimental reproduction project were to investigate what types of lace were available to the artisan class and identify potential differences in cost between different types of lace. In addition, I hoped to consider broader questions about the development of bobbin-lace in northern Italy, such as what drove stylistic and regional variations, and what the process of lace production entailed for the maker. During the process of making each edging, I carefully tracked the time that it took to make each repeat, how quickly each pattern could be memorised and any notable challenges that arose. Although a Renaissance lacemaker working long hours would have been much quicker than most lace-makers of today, making multiple repeats of each edging can give a general idea of production time – information that is rarely recorded in archival documents.

The experiment

In approaching an historical reproduction project, it is imperative to research the tools that would have been used by lace-makers of the period. Bobbin-lace was made on a solid pillow, which could have been handmade by the lace-maker, a relative or another craftsperson, who packed it densely with whatever materials were available – often straw, reeds, moss or horsehair.[3]

Over the centuries, bobbin-lace pillows have been made in a variety of shapes and sizes, with a taut fabric cover to create a firm surface for securing pins.[4] Given that the pillows in my collection are mostly fabric-covered ethafoam, the first step in this project was to build a cylindrical bolster pillow stuffed with barley straw. After being baked in sheets to dry, cut into segments of 5–7.5 cm and combed for bits of debris, the straw was packed into a tube of heavy muslin fabric. Lastly, the exterior was covered with a tight sheath of checked blue-and-white linen and cinched closed with a drawstring on both ends. This fabric covering was typical of the early freehand technique, wherein bobbin-lace was worked without a pattern base, following the checked or striped pattern on the pillow.[5]

As with lace pillows, bobbins varied regionally in size, shape and material, and were hand-turned on a lathe, likely by a relative or artisan in the community. Early examples depicted in paintings such as *The Lacemaker* by Nicolaes Maes from 1643 are wooden with bulbous ends, a style which persists in rural lace-making regions such as Le Puy en Velay, France.[6] I am fortunate to have inherited hand-turned nineteenth-century bobbins in a comparable shape from a friend's grandmother, as the size and weight of the bobbins can impact the lace-making process. Handmade pins, of the kind found by mudlarkers along the banks of the Seine, were not available to me, so contemporary steel straight pins had to fit the bill.

To represent the earliest quarter-century of the period, 1550–75, I opted to reproduce a metallic bobbin-lace edging from an extant handkerchief in the collection of the Victoria and Albert Museum, London (T.99–1954). Although this is catalogued as English, the metal lace was likely imported from Venice, as Venice had a thriving silk and metal thread industry in the sixteenth century, as well as both bobbin and needle lace production.[7] This style of handkerchief, with satin stitch embroidery around the border, gold or silver metal lace trimming and small tassels at each corner, is mentioned in multiple historical records, such as in a list of New Year's gifts to Queen Mary I in 1556, and in the 1599 inventory of Gabrielle d'Estrées.[8]

Gilian Dye, an expert in early bobbin-lace techniques, published a reproduction of the V&A handkerchief edging in *Sixteenth and Seventeenth Century Lace, Book 1: Gold and Silver Edgings*.[9] As it was my first time reproducing historical lace, I used this as a launching point, and I am greatly indebted to Dye's research as well as that of Rosemary Shepherd. This edging was worked in size 40/3 thread of gold-plated nickel wrapped around a rayon core from Klöppelwerkstatt, a German manufacturer. Consistent with the earliest bobbin-laces, this pattern was worked freehand, using the checked pattern on the lace pillow covering as a guide.

Initially, working with the metal thread was challenging, requiring careful tensioning or the picots (decorative loops) would collapse. The first repeat took twenty-seven minutes to complete, which was reduced to around five minutes after a few repeats, a strikingly rapid pace for lace-making. This reaffirmed that the high cost of early bobbin-lace was due to the use of precious materials rather than the production hours – an inverse of later, more elaborate, linen laces.[10] Metal laces were frequently subject to sumptuary legislation, such as a Florentine law of 1638 restricting the width of gold and silver laces to 19.4 mm.[11] Measuring 13 mm in width, this particular metal lace edging would have been safely within the reach of the artisan class.

To represent the quarter-century period from 1575 to 1600, I chose to reproduce a rare example of polychrome bobbin-lace from the collection of the Metropolitan Museum, New York (08.180.505). The Met textile conservator Giulia Chiostrini generously analysed the fibres and confirmed that the white thread is linen and the pink and yellow threads are silk. Although the original usage of this edging is unknown, based on its similarity to surviving linen objects with silk and linen bobbin-lace edgings, I postulate that it could have been attached to a linen garment or accessory such as a handkerchief. Typically, however, these bobbin-lace examples feature one colour of silk thread paired with white linen thread. Although there are a number of extant examples of polychrome silk bobbin-laces from the sixteenth century, they typically incorporate metal threads, and are often paired with corresponding embroidery.[12]

In contrast to many contemporary lace-makers who prioritise a uniform end result, production lace-makers in the Renaissance did not always work from precise stitch diagrams and may have prioritised speed over perfection. This has resulted in stitch variations between repeated motifs in historic lace edgings. Therefore, when reproducing extant historic lace, modern designers may select one repeat to focus on, or select areas of each motif to create a uniform diagram. Given that the goal of this project was to better understand lace in the context of the Renaissance rather than to design a pattern for contemporary makers to follow, I opted to work directly on a freehand drawing of the lace pattern rather than create a detailed diagram. This meant that, like historical lace-makers', my process of producing the edging involved active decision-making based on observation of the original edging.

Using materials as close to the originals as possible, I selected white Irish linen in size 40/2 and Soie Ovale, a French silk floss produced by Au Ver à Soie, in pink and yellow. Due to past experience with silk thread, I scaled the initial sample up slightly to accommodate for shrinkage when removed from the pillow. Ultimately, the tight four-strand braids that dominate the design held their shape and did not shrink to the degree anticipated, and the final design was redrawn to the precise width of the original lace (23 mm). The inclusion of sturdier linen thread in the original edging may have served a practical purpose as a kind of scaffolding to stabilise the silk floss. The first repeat of the pattern took forty minutes to complete, which gradually decreased to seventeen minutes for the final repeat, for an average of around twenty minutes each (Figure ix.2). At the resultant ten minutes per centimetre or 6 cm per hour, by my estimation, a professional lace-maker working twelve-hour shifts could likely have produced a full metre per day of this lace edging. That is arguably much quicker and thus more cost-effective than a needle-lace edging of a similar width and complexity.

The sixteenth-century bobbin-lace pattern books *Le pompe: opera nova* (1557) and *Nüw Modelbuch von allerley gattungen Däntelschnür* (1561) have previously been the subject of in-depth analysis and reproduction by Santina Levey, Milton Sonday and Laurie Waters, among others.[13] For this project I was interested in examining the bobbin-lace patterns of Isabella Catanea Parasole's 1610 pattern book *Fior d'ogni virtu per le nobili et honeste matrone* to represent the quarter-century period from 1600 to 1625. As one of few published female lace pattern designers of her day Parasole is known for her elaborate needle-lace, cutwork and embroidery designs, but her patterns for bobbin-lace are lesser known. Described by Santina Levey as 'spidery' and 'spiky', Parasole's bobbin-lace designs

Figure IX.2 Polychrome bobbin-lace reconstruction in progress on linen-covered bolster pillow.

were likely inspired by fine needle-lace scallops and intended to be paired with straight cutwork bands.[14]

Renaissance pattern-book designers such as Giovani Antonio Tagliente (active 1522–45) and Alessandro Paganino (active 1511–38) described several methods for transferring the designs, many of which involved removing pages, rendering intact early pattern books a rarity.[15] The most straightforward option was to work directly on to a facsimile of the pattern, but after twelve hours of work on a preliminary sample in size 100/2 linen thread it became apparent that the scale was smaller than comparable extant lace edgings of the period. Additionally, the first pattern I selected on the bottom left of page 34 had numerous ambiguities in the design that made it difficult to execute without reinterpretation. Rather than struggle to decipher the design, I opted to select another pattern on the far-left side of page 35 and scaled it up 20 per cent to be worked in size 60/2 linen thread (Figure ix.3).

However, this pattern had incongruities as well, enough to indicate that Parasole lacked a thorough understanding of bobbin-lace construction. Bobbin-lace requires management of numerous threads that travel across the design, and the pattern should clearly indicate the continuous paths linking motifs. As a rule, early bobbin-lace patterns did not

Figure IX.3 *Fior d'ogni virtu per le nobili et honeste matrone* by Isabella Catanea Parasole, page 35, 1610. The Metropolitan Museum of Art, New York, 37.1 (1–40).

provide instructional information, and lace-makers would have had to be experienced in order to find creative solutions to these obstacles. Once the pattern was resolved, I referenced parts of a diagram on page 79 of Rosemary Shepherd's *An Early Lace Workbook* for the first three repeats until the pattern was committed to memory. The final edging varied from seventy-seven minutes to thirty-one minutes for each repeat, for a total of ten hours and fifty-five minutes to create 20 cm of lace.

The final edging representing 1625–50 was by far the most challenging, highlighting the technical development of bobbin-lace over the previous century. Based on an extant seventeenth-century handkerchief from the collection of the Metropolitan Museum of Art (39.123.1), the central fabric is linen with an embroidered silk border and a scalloped bobbin-lace trim that is gathered at the corners (Figure ix.4). As is often the case with early linen laces, centuries of use have caused the fibres to blend together, making it difficult to decipher the precise stitches under magnification. After careful analysis, I measured out the 50 mm width and drew the pattern freehand. Again, rather than draw up a precise stitch diagram, I opted to interpret the pattern as I worked, referencing photos of the original. Mounting twenty-two pairs of bobbins on the bolster pillow proved to be cumbersome, as a cylindrical base is more suited to laces requiring fewer bobbins, so I opted to switch to a large, flat, contemporary pillow.

Figure IX.4 Handkerchief. Linen fabric with silk embroidery, linen bobbin-lace. Italian, seventeenth century; 45.1 cm by 45.1 cm. The Metropolitan Museum of Art, New York, 39.123.1.

Each scallop is worked by travelling along the top half of the motif and leaving pairs behind to pick up when working the bottom half. Sewings, a technique in which a fine hook is used to link new threads into old stitches, are much less common in early bobbin-lace, although analysis of the original edging revealed one between each scallop. The challenge with Genoese-style edgings is in gracefully hiding the many pairs required to form the wide scallops in the narrow space between each motif; in this case, fourteen pairs of 100/2 thread in a narrow strip of cloth stitch. The first three repeats took between six and seven hours each to complete, but the subsequent three were closer to five hours each, for a total of thirty-four hours and thirty-eight minutes (Figure ix.5). Although a professional seventeenth-century lace-maker would have been able to work this pattern more quickly, this is a sharp increase from the pace of the previous edgings.

Arguably, the Genoese scallop represents the first quarter-century period in which, broadly speaking, the value of the labour to create bobbin-lace not only far surpasses that of the materials but is also comparable to its more time-consuming competitor, needle lace. A complementary experiment of four needle-lace edgings in a similar scale from 1550 to 1650 could further underscore the differences between these two techniques during their development. Additionally, a more in-depth examination of Parasole's bobbin-lace patterns is warranted, particularly in contrast to *Le pompe* and *Nüw Modelbuch*.

Conclusion

What can a humble handkerchief teach us about the value of labour during the Italian Renaissance? As with the dress of the artisan class, the experiences and perspectives of lace-makers are rarely accessible through research that draws solely on written texts, and practice-led research provides an avenue to explore the choices and challenges that they would have faced in their daily lives. Rather than reconstructing a single object and illuminating just one type of lace from one moment in time, four edgings were selected to illustrate the breadth of possibilities for lace as it developed during the late Renaissance, thus creating the possibility of a comparison between periods. The resulting lace represents a spectrum of quality and design aimed at all different levels of society, from costly needle lace visible among the nobility to simpler bobbin-lace edgings, which were more affordable even when made in precious metals due to their quicker production. Overall, this experiment highlights the relative affordability of simpler bobbin-lace-trimmed handkerchiefs for the artisan class during the Renaissance. As is so often the case, the discoveries in this project have led to more questions, which I hope to explore in future research.

The technical understanding that is revealed through hands-on reconstruction also sheds light on to lacemakers' innovations that could not have

Figure IX.5 Reconstruction of Genoese-style linen bobbin-lace handkerchief edging in progress.

been gleaned from traditional research methodologies. Whether labouring over a length of dimensional needle lace for many months or hastening through yardage of simple bobbin-lace trim, lace-makers had a degree of creative liberty in the execution of their work. They were not mere fabricators but had to have a significant amount of knowledge and skill to interpret design ambiguities. Physically embodying their gestures can help to articulate their decision-making process – for example, which threads or stitches are more suitable or more difficult within specific techniques – and shed light on to how stylistic changes evolved. Although the names of multiple lace designers survive, namely those of pattern books, and it is easy to attribute innovations to them, through reproductions of lace, we can return some agency to the lace-makers themselves.

Notes

The ‘Refashioning the Renaissance’ lace reconstruction experiment, carried out in 2020–22, was funded by the European Research Council (ERC) under the European Union’s Horizon 2020 research and innovation programme (grant agreement No. 726195). See www.refashioningrenaissance.eu/early-modern-lace.

1 Bella Mirabella, ‘Embellishing herself with a cloth: The contradictory life of a handkerchief’, in Bella Mirabella (ed.), *Ornamentalism: The Art of Renaissance Accessories* (Ann Arbor: University of Michigan Press, 2011) 59.
2 Paula Hohti, ‘“Monstrous” ruffs and elegant trimmings: Lace and lacemaking in early modern Italy’, in Emma Cormack and Michele Majer (eds), *Threads of Power: Lace from the Textilmuseum St. Gallen* (New York: Yale University Press, Bard Graduate Center, 2023), 86. Further examples can be found in the ‘Refashioning the Renaissance’ database at refashioningrenaissance.eu/database/.
3 Lena Dahrén, *Med kant av guld och silver: en studie av knypplade bårder uddar av metall 1550–1640* (Uppsala: Uppsala Universitet, 2010), 58.
4 Rosemary Shepherd, *An Early Lace Workbook: Bobbin Lace Techniques Before the Baroque* (Sydney: Lace Daisy Press, 2009), 17–18.
5 Shepherd, An Early Lace Workbook, 8.
6 See also the earliest surviving bobbins from the Batavia shipwreck of 1629, now in the Western Australian Museum (BAT6363).
7 Santina M. Levey, *Lace: A History* (London: Victoria and Albert Museum, 1983), 9.
8 See, respectively, Levey, *Lace*, 15; and Pat Earnshaw, *Lace in Fashion* (Dublin: Gorse Publications, 1991), 18.
9 Gilian Dye, *Sixteenth and Seventeenth Century Lace, Book 1: Gold and Silver Edgings* (London: Cleveden Press, 2012), 24–5.
10 Anne Rosalind Jones and Peter Stallybrass (eds), *Renaissance Clothing and the Materials of Memory* (Cambridge: Cambridge University Press, 2000), 25.
11 Hohti, ‘“Monstrous Ruffs”’, 86.
12 For an example, see the edging on a pair of sixteenth-century Italian trousers in The Met Costume Institute collection (10.124.3).
13 Santina M. Levey, *Le pompe* (1559): *Patterns for Venetian Bobbin Lace* (Bedfordshire: Ruth Bean Publishers, 1983); the Milton Sonday Archive at the Antonio Ratti Textile Center, The Metropolitan Museum of Art; and Laurie Waters, ‘A new interpretation of certain bobbin lace patterns in Le Pompe, 1559’, *Textiles and Politics: Textile Society of America 13th Biennial Symposium Proceedings, Washington, DC, September 18 – September 22, 2012*. https://digitalcommons.unl.edu/tsaconf/754/.
14 Levey, *Lace*, 7.
15 Femke Speelberg, ‘Fashion and virtue: Textile patterns and the print revolution, 1520–1620’, *The Metropolitan Museum of Art Bulletin*, 73, 2 (Fall 2015), 23.

Bibliography

Archives and museums

Italy

State Archives of Florence (Archivio di Stato di Firenze, ASF)
Magistrato dei Pupilli, 2649–2720, 1550–1652.
Pratica Segreta, 176, 1638–1642.
Mediceo del Principato, 229; 3254; 5080; 6113.
Otto di Guardia e Balia del Principato, 2228–2247.

State Archives of Siena (Archivio di Stato di Siena, ASS)
Curia del Placito, 262–286, 1586–1651.
Balia, 830, *Libro dei Bandi*, 1576.
Regolatori, 767, *Bandi*, 1594.

State Archives of Turin (Archivio di Stato di Torino, AST)
Consolato di Commercio, *Atti del Consolato in materia di arti e mestieri*, 37, 'Regia opera delle povere figlie raminghe', 24 March 1783, fol. 87r.
Materie economiche, *Arti e Manifatture*, 9, 22, 23 December 1755.

State Archives of Venice (Archivio di Stato di Venezia, ASV)
Cancelleria inferiore, *Miscellanea*, 38–45, 1550–1630.
Giudici di Petizion, *Inventari*, 337–363, 1581–1650.
Giudice dell Esaminador, *Inventari* 1–2, 1601–1649.

United Kingdom

Cambridgeshire Archives
P68/12/32–7: St Mary Ely, parish workhouse, spinning accounts, 1736–39.

Castle Museum, York
Handbill for Nicholas Forster, weaver, Alnwick, Northumberland, 1800.

Cumbria Archive Service (Kendal)
WD/TE/11/16: Browne of Troutbeck Mss., Account book of an Ambleside shopkeeper, 1701–9.

Denbighshire Record Office
PD/39/1/81: Pattern book of William Jones and William Jones junior, jobbing weavers, Holt, 1775–82.

Essex Record Office
Calendars of Essex Assize Files in the Public Record Office (typescript), vol. 4, 1660–1685, and vol. 5, 1684–1714.
Calendars of Essex Sessions Rolls (typescript), vols 1–18, 1536–1610, and vols 22–5, 1654–99.

Kent Archives
L/M/4/1: Returns of men liable for service in the militia, made by each parish or 'borough', Wingham subdivision, 1764.

Lambeth Palace Archives
Letter from John Knyveton to the Earl of Shrewsbury, 6 July 1574, Shrewsbury Papers MS.697.

Lancashire Archives
DP 385: Account book of Richard Latham of Scarisbrick, 1723–1767.
WRW 1687: Archdeaconry of Richmond, Probate Records, inventory of Christopher Hewit/Hewet, Forton, chapman, 1687.

London Metropolitan archives
Acts of the Court of Common Council, COL/CC/01/01/29/01.

North Yorkshire County Record Office
Z.371: The weaver's guide, linen designs of Ralph Watson of Aiskew, late eighteenth century.

Oxfordshire History Centre
QSR: Oxfordshire Quarter Sessions rolls, 1700–79.

The National Archives
ASSI 2: Assizes, Oxford Circuit, Crown Minute Books, 1714–79.
ASSI 5: Assizes, Oxford Circuit, Indictment Files, 1714–79.

University of Leeds, Brotherton Library Special Collections
Yorkshire Archaeological Society, MS 2022: John Brearley's Memorandum Book for 1772–73.

West Yorkshire Archive Service Bradford
33D80/7: Shop book of Stephen Hudson of Fewston parish, Yorkshire, 1751–59.

Wiltshire and Swindon History Centre
927/1: John Jeffries and John Usher, Cloth Making Book, 1721–26.
1720/744: Personal and household account book of Hester Soame, 1753–62.

Worcestershire Archive and Archaeology Service
110 BA1: Worcestershire Quarter Sessions Rolls, 1700–79.

United States

Cooper Hewitt, National Design Museum, New York
1958–30–1: Weaver's thesis book, England, 17th–18th century.

Huntington Library, San Marino, CA
HM 31192: William Smedley's executors, account book of farm receipts and disbursments, Derbyshire, 1741–52.

Lewis Walpole Library, Farmington, CT
Mss Vol. 212: Account book of Reverend Miles Tarn, Rector of Dean, Cumberland, 1719–97.

France

National Archives, Paris
F12/677A: Lettre de l'inspecteur des manufactures à Rouen sur le mérite comparitif de la filature de la laine au rouet ou à la quenouille, 1762.
F12/677A: Mémoire sur le commerce et les manufactures de la Provence, 1781.
F12/1341: Projet d'Etablissement de filature de Lins, Chanvres et Cottons dans les Villages qui avoisinent la Ville de Troyes, 1770.

Denmark

National Archive in Copenhagen (Rigsarkivet, RA)
Helsingør Byfoged, Skifteprotol: 1583–1592, 1592–1598, 1599–1603, 1610–1612, 1612–1619, 1621–1625, 1625–1627, 1635–1639, 1639–1644, 1648–1650.
Helsingør Skomagerlav, Lavsbog: 1623–1802.

Estonia

National Archives in Tallinn
Archive EAA
Inventory 1 *Schwedischer Generalgouverneur von Estland.*
No. 482 *Eingelaufene Schreiben in Verwaltungs- und Gerichtsangelegenheiten (Suppliken)*, 1688.
No. 517 *Eingelaufene Schreiben in Verwaltungs- und Gerichtsangelegenheiten (Suppliken)*, 1691.

Tallinn City Archives (TLA)
Collection 190 *Revaler St. Kanuti Gilde*
Inventory 1

No. 2 *Resolutione/n/ von Heerme/istern/ und Könige Lit. A*, 1471–1767.
No. 14 *Einband Nr. 5*, 1631–99.
Collection 191 *Revaler Kaufmanns- oder Grosse Gilde*
Inventory 1
No. 19 *Armen-Ordnung, Rewidierte Ordnung des allgemeinen Gottes-Kastens. Kasten-Ordnungen, Kleider-Ordnungen*, 1609–90.

Collection 230 *Revaler Magistrat*
Inventory 1
No. Aa *Protokoll des Niedergerichts/Niedergerichts-Protokoll*, 1600–03.
No. Bs7/I *Kleider-, Hochzeits-, Kindtauf-, Begräbnis u.s.w. Ordnungen*, 1497–1738.
No. Bt *Inventarien*, 1500–1700.

Collection 31 *Die Gemeinde der Revaler St. Nikolai-Kirche*
Inventory 1
No. 13 *St. Nicolai in Reval Kirchenbuch: (Getaufte, Getraute, Verstorbene)*, 1652–97.

Journals and newspapers

Bonhams Magazine
Ipswich Journal
Cambridge Chronicle
Derby Mercury
Journals of the House of Commons
Norfolk Chronicle

Primary and secondary sources

Ackerman, Phyllis, 'A note on suf or camlet', *Bulletin of the American Institute for Iranian Art and Archaeology*, 5, 3 (1938), 254–56.

Adamson, Glenn, Giorgio Riello and Sarah Teasley (eds), *Global Design History* (London: Routledge, 2011).

Advice to the Maidens of LONDON: To Forsake Their Fantastical TOP-KNOTS; Since they are become so Common with Billings-gate Women, and the Wenches that cryes Kitchin-Stuff: Together with the Wanton Misses of the Town (London. 1685–88), Magdalene College, Pepys Ballads 4.365, EBBA 22029.

Ajmar, Marta, 'The Renaissance in material culture: Material mimesis as force and evidence of globalization', in Tamar Hodos (ed.), *The Routledge Handbook of Archaeology and Globalization* (London: Routledge, 2017), 669–86.

Ajmar, Marta and Dora Thornton, 'When is a portrait not a portrait? Belle donne on maiolica and the Renaissance praise of local beauties', in Nicholas Mannand and Luke Syson (eds), *The Image of the Individual: Portraits in the Renaissance* (London: British Museum, 1998), 138–53.

Ajmar-Wollheim, Marta, 'Housework', in Marta Ajmar-Wollheim and Flora Dennis (eds), *At Home in Renaissance Italy* (London: V&A Publications, 2006), 152–63.

Allen, Robert C., and Jacob Weisdorf, 'Was there an "industrious revolution" before the industrial revolution? An empirical exercise for England, c. 1300–1830', *Economic History Review*, 64 (2011), 715–29.

Allerley Mackel (Mainz: Peter Jordanim, 1532). Transl. Sidney M. Edelstein in 'The Allerley Matkel (1532): Facsimile Text, Translation, and Critical Study of the Earliest Printed Book on Spot Removing and Dyeing', *Technology and Culture*, 5, 3 (1964), 317.

Allerston, Patricia, 'Clothing and early modern Venetian society', *Continuity and Change*, 15, 3 (2000), 367–90.

Allerston, Patricia, 'Consuming problems: Worldly goods in Renaissance Venice', in Michelle O'Malley and Evelyn Welch (eds), *The Material Renaissance* (Manchester: Manchester University Press, 2007), 11–46.

Allerston, Patricia, '"Contrary to the truth and also the semblance of reality"? Entering a Venetian "lying-in" chamber (1605)', *Renaissance Studies*, 20, 5 (2006), 629–39.

Allerston, Patricia, 'L'abito come articolo di scambio nella società dell'età moderna: alcune implicazioni', in Anna Giulia Cavagna and Grazietta Butazzi (eds), *Le trame della moda* (Rome: Bulzoni Editore, 1994), 109–24.

Allerston, Patricia, 'Reconstructing the second-hand clothes trade in sixteenth- and seventeenth-century Venice', *Costume*, 33, 1 (1999), 46–56.

Allerston, Patricia, 'The market in second-hand clothes and furnishings in Venice, c. 1500–c. 1650', 1996, https://cadmus.eui.eu/handle/1814/5818.

Alm, Mikael, *Sartorial Practices and Social Order in Eighteenth-Century Sweden: Fashioning Difference* (New York and London: Routledge, 2021).

Ammannati, Francesco, 'Craft guild legislation and woollen production: The Florentine Arte della Lana in the fifteenth and sixteenth centuries', in Karel Davids and Bert De Munck (eds), *Innovation and Creativity in Late Medieval and Early Modern European Cities* (Farnham: Ashgate, 2014).

Ammannati, Francesco, *Per filo e per segno: l'arte della lana a Firenze nel Cinquecento* (Florence: Florence University Press, 2020).

Andersen, Vivi Lena, 'Between cobbles, bunions, shoe last and fashion: An appropriation and adaption analysis of footwear from the Middle Ages, Renaissance and Absolutism from archaeological excavations in Copenhagen' (PhD dissertation submitted for University of Copenhagen, 2017).

Andersson, Eva I., 'Foreign seductions: Sumptuary laws, consumption and national identity in early modern Sweden', in Tove Engelhardt Mathiassen et al. (eds), *Fashionable Encounters: Perspectives and Trends in Textile and Dress in the Early Modern Nordic World* (Oxford: Oxbow Books, 2014), 15–29.

Andersson, Gudrun, 'A mirror of oneself: Possessions and the manifestation of status among a local Swedish elite, 1650–1770', *Cultural and Social History*, 3, 1 (2006), 21–44.

Aneer, Cecilia, *Skrädderi för kungligt bruk: tillverkning av kläder vid det Svenska hovet, ca 1600 1635* (Uppsala: Uppsala Universitet, 2009).

Anton, Dr., 'Spinning in Lusatia', *Annals of Agriculture*, 10 (1788), 312–15.

Araldo, Sicillo, *Trattato dei colori nelle arme, nelle livree, e nelle divise* (Venice: Domenico Nicolino, 1565).

Arnobio, Cleandro, *Il tesoro delle gioie: trattato marauiglioso in torno alle vertuti e proprieta più rare di tutte le gioie, perle, gemme* (Venice: Appresso Gio. Battista Ciotti, 1602).

Arnold, Janet, Jenny Tiramani and Santina M. Levy, *Patterns of Fashion 4. The Cut and Construction of Linen Shirts, Smocks, Neckwear, Headwear and Accessories for Men and Women* (London: Macmillan, 2008).

Ashley, William J., *An Introduction to English Economic History and Theory, Part 2* (London: Longmans, 1893).

At a meeting of the master weavers, living in Hexham, and parts adjacent, it was unanimously agreed, that the prices for weaving of plain cloth, sheeting, ticking, huckaback, etc. be as follows (Hexham, 1792).

Awais-Dean, Natasha, 'Bejewelled: The male body and adornment in early modern England' (PhD thesis, Queen Mary, University of London, 2012).

Awais-Dean, Natasha, *Bejewelled: Men and Jewellery in Tudor and Jacobean England* (London: British Museum Press, 2017).

Awais-Dean, Natasha, 'Redressing the balance: Dress accessories of the non-elites in early modern England', in Toby F. Martin and Rosie Weetch (eds), *Dress and Society: Contributions from Archaeology* (Oxford: Oxbow Books, 2017), 151–69.

Baart, Jan, *Opgravingen in Amsterdam: 20 Jaar Stadskernonderzoek* (Haarlem: Fibula-Van Dishoeck, 1977).

Bacci, Andrea, *Le XII pietre pretiose …* (Rome: Giovanni Martinelli, 1587).

Bailey, Mark, *Medieval Suffolk: An Economic and Social History, 1200–1500* (Woodbridge: Boydell, 2007).

Barker, Joseph, Peter M. K. Kitson, Leigh Shaw-Taylor and E. A. Wrigley, *1762 Northumberland Militia Ballot List Database* (Cambridge: Cambridge Group for the History of Population and Social Structure, 2013).

Bartels, Victoria and Katy Bond, 'Dress & Material Culture,' in *A Companion to Late Medieval and Early Modern Augsburg*, edited by B. Ann Tlusty and Mark Haberlein (Leiden: Brill, 2020).

Bartels, Victoria, 'Masculinity, Arms and Armour, and the Culture of Warfare in Sixteenth-Century Florence' (PhD Thesis, University of Cambridge, 2020).

Barthes, Roland, *The Fashion System*, translated by Matthew Ward and Richard Howard (Berkeley: University of California Press, 1990).

Battara, Pietro, 'Botteghe e pigioni nella Firenze del '500: Un censimento industriale e commerciale all'epoca del granducato mediceo', *Archivio Storico Italiano*, 95, 3/363 (1937), 3–28.

Baulant, Micheline, Anton Schuurman and Anton Servais, *Inventaires après-décès et ventes de meubles: apports à une histoire de la vie économique et quotidienne (XIVe–XIXe siècle): actes du séminaire tenu dans le cadre du 9ème Congrès international d'histoire économique de Berne (1986)* (Louvain: Academia, 1988).

Becker, Jochen, 'Are these girls really so neat? On kitchen scenes and method', in David Freedberg and Jan de Vries (eds), *Art in History, History in Art: Studies in Seventeenth Century Dutch Art* (Santa Monica: Getty Center for the History of Art and the Humanities, 1991), 139–74.

Behringer, Wolfgang, *A Cultural History of Climate* (Cambridge: Polity Press, 2010).

Belfanti, Carlo Marco, 'Branding before the brand: Marks, imitations and counterfeits in pre-modern Europe', *Business History*, 60, 8 (2018), 1127–46.

Belfanti, Carlo Marco, *Calze e maglie*: *moda e innovazione nell'industria della maglieria del Rinascimento a oggi* (Mantua: Tre Lune Edizioni, 2005).

Belfanti, Carlo Marco, *Civiltà della moda* (Bologna: Il mulino, 2008).

Belfanti, Carlo Marco, 'Clothing and social inequality in early modern Europe: Introductory remarks', *Continuity and Change*, 15, 3 (2000), 259–65.

Belfanti, Carlo Marco, 'Fashion and innovation: The origins of the Italian hosiery industry in the sixteenth and seventeenth centuries', *Textile History*, 27, 2 (1996), 132–47.

Belfanti, Carlo Marco, 'Mobilità sociale e opportunità di mercato alle origini del cambiamento', in Eugenia Paulicelli (ed.), *Moda e mode dal Medioevo al Rinascimento* (Rome: Moltemi editore, 2006), 39–40.

Belfanti, Carlo Marco, 'The civilization of fashion: At the origins of a western social institution', *Journal of Social History*, 43, 2 (2009), 261–83.

Belfanti, Carlo Marco, 'Was fashion a European invention?', *Journal of Global History*, 3, 3 (2008), 419–43.

Bell, Rudolph, *How to Do It: Guides to Good Living for Renaissance Italians* (Chicago: University of Chicago Press, 1999).

Bellavitis, Anna, *Women's Work and Rights in Early Modern Urban Europe* (Cham: Springer International Publishing, 2018).

Bendall, Sarah A., *Shaping Femininity: Foundation Garments, the Body and Women in Early Modern England* (London: Bloomsbury, 2021).

Bendall, Sarah A., '"Take measure of your wide and flaunting garments": The farthingale, gender and the consumption of space in Elizabethan and Jacobean England', *Renaissance Studies*, 33, 5 (2018), 712–37.

Bendall, Sarah A., 'The case of the "French Vardinggale": A methodological approach to reconstructing and understanding ephemeral garments', *Fashion Theory*, 23, 3 (2019), 363–99.

Benhamou, Reed, 'Imitation in the decorative arts of the eighteenth century', *Journal of Design History*, 4, 1 (1991), 1–13.

Berg, Maxine, 'From imitation to invention: Creating commodities in eighteenth-century Britain', *Economic History Review*, 55, 1 (2002), 1–30.

Berg, Maxine, *Luxury and Pleasure in Eighteenth-Century Britain* (Oxford: Oxford University Press, 2007).

Berg, Maxine, 'New commodities, luxuries and their consumers in eighteenth-century England', in Maxine Berg and Helen Clifford (eds), *Consumers and Luxury: Consumer Culture in Europe, 1650–1850* (Manchester: Manchester University Press, 1999).

Berg, Maxine and Elizabeth Eger, *Luxury in the Eighteenth Century: Debates, Desires and Delectable Goods* (Basingstoke: Palgrave Macmillan, 2002).

Bering, Lüsberg, and Henrik Carl, 'Jacob Jensen Nordmand', *Museum. Tidsskrift for Historie og Geografi*, Første Halvbind (1893), 245–84.

Bettoni, Barbara, 'Fashion, tradition, and innovation in button manufacturing in early modern Italy', *Technology and Culture*, 55, 3 (2014), 675–710.

Bettoni, Barbara, *Perle di vetro e gioie false: produzioni e cultura del gioiello non prezioso nell'Italia moderna* (Venice: Marsilio, 2017).

Bettoni, Barbara, 'Usefulness, ornamental function and novelty: Debates on quality in button and buckle manufacturing in northern Italy (eighteenth to nineteenth centuries)', in Bert De Munck and Dries Lyna (eds), *Concepts of Value in European Material Culture, 1500–1900* (London: Routledge, 2015).

Bezzina, Denise, *Artigiani a Genova nei secoli XII–XIII* (Florence: Florence University Press, 2015).

Biringuccio, Vannoccio, *The Pirotechnia of Vannoccio Biringuccio: The Classic Sixteenth-Century Treatise on Metals and Metallurgy*, ed. and trans. Cyril Stanley Smith and Martha Teach Gnudi (New York: Dover Publications, 1990).

Boccia, Lionello G., *Armi difensive dal Medioevo all'Eta Moderna* (Florence: Centro Di, 1980).

Boccia, Lionello G., and Eduardo T. Coelho, *Armi bianche italiane* (Milan: Bramante Editrice, 1975).

Bol, Marjolijn, 'Coloring topaz, crystal and moonstone: Factitious gems and the imitation of art and nature, 300–1500', in Marco Beretta and Maria Conforti (eds), *Fakes!? Hoaxes, Counterfeits and Deception in Early Modern Science* (Sagamore Beach, MA: Science History Publications, 2014), 108–29.

Bol, Marjolijn and Emma C. Spary, *The Matter of Mimesis: Studies of Mimesis and Materials in Nature, Art and Science* (Leiden: Brill, 2023).

Boulboullé, Jenny and Sven Dupré (eds), *Burgundian Black: Reworking Early Modern Colour Technologies* (Santa Barbara: EMC Imprint, 2022), https://burgundianblack.tome.press, accessed 3 June 2023.

Brackett, John K., *Criminal Justice and Crime in Late Renaissance Florence, 1537–1609* (Cambridge: Cambridge University Press, 1992).

Brandenburgh, Chrystel, 'Een paar zijden kousen', in Birgit van den Hoven and Iris Toussaint (eds), *Wereldvondsten uit een Hollands schip: Basisrapportage BZN17/Palmhoutwrak* (Leiden: Provincie Noord-Holland, 2019), 258–69.

Braun, Melanie, Luca Costigliolo, Susan North, Claire Thornton and Jenny Tiramani, *17th-Century Men's Dress Patterns 1600–1630* (London: Thames and Hudson, 2016).

Braun, Rudolf, *Industrialisation and Everyday Life*, trans. S. Hanbury Tenison (Cambridge: Cambridge University Press, 1979).

Breeze, Camille, M. 'Digitally printed textiles: Their potential use in costume collections and living-history museums', *Waac Newsletter*, 22, 2 (May 2002).

Breward, Christopher, 'Cultures, identities, histories: Fashioning a cultural approach to dress', *Fashion Theory*, 2, 4 (1998), 301–13.

Bridgeman, Jane, '*Condecenti e netti* ... Beauty, dress and gender in Italian Renaissance art', in Francis Ames Lewis and Mary Rogers (eds), *Concepts of Beauty in Renaissance Art* (Aldershot: Ashgate, 1998), 44–51.

Brüggemann, Karsten and Ralph Tuchtenhagen, *Tallinn: Kleine Geschichte der Stadt* (Cologne: Böhlau, 2011).

Buck, Anne, 'Variations in English women's dress in the eighteenth century', *Folk Life*, 9 (1971), 5–28.

Burr Litchfield, Robert, *Emergence of a Bureaucracy: The Florentine Patricians, 1530–1790* (Princeton: Princeton University Press, 2014).

By the King: A Proclamation Prohibiting the Wearing, Buying, or Selling of Counterfeit Iewels (London: John Bill, 1636), STC 9062.

Caferro, William P., 'Warfare and economy in Renaissance Italy, 1350–1450', *Journal of Interdisciplinary History*, 39, 2 (2008), 167–209.

Calendar of State Papers, Domestic Series, of the Reign of Charles I: Addenda – March 1625 to January 1649, ed. William Douglas Hamilton and Sophia Crawford Lomas (London: Her Majesty's Stationery Office, 1897) at British History Online, www.british-history.ac.uk/cal-state-papers/domestic/chas1/addenda/1625-49, accessed 7 January, 2024.

Calvert, Sophie., et al., 'Digital 3D analysis of historical textile fragment', *Proceedings for Huddersfield University Conference: Transition- Re-thinking Textiles and Surfaces*, Yorkshire, 2 November 2014.

Calvi, Giulia, 'Abito, genere, cittadinanza nella Toscana Moderna (Secoli XVI–XVII)', *Quaderni Storici*, 37, 110 (2002), 477–503.

Calvi, Giulia, 'Diritti e legami. Madri, figli, Stato in Toscana (XVI–XVIII secolo)', *Quaderni Storici*, 29, 86 (2) (1994), 487–510.

Campbell, Caroline, 'Revaluing dress in history paintings for Quattrocento Florence', in Gabriel Neher and Rupert Shepherd (eds), *Revaluing Renaissance Art* (Aldershot: Ashgate, 2000), 137–43.

Campbell, Erin J., *Old Women and Art in the Early Modern Italian Domestic Interior* (Abingdon: Routledge, 2016).

Cantini, Lorenzo (ed.), *Legislazione toscana raccolta ed illustrata dal Dottor Lorenzo Cantini*, vols 4–5 (Florence: Stampa Albizziniana da Santa Maria in Campo, 1800–3).

Capwell, Tobias and Sydney Anglo, *The Noble Art of the Sword: Fashion and Fencing in Renaissance Europe, 1520–1630* (London: The Wallace Collection, 2012).

Caracausi, Andrea, 'Beaten children and women's work in early modern Italy', *Past & Present*, 222, 1 (2014), 95–128.

Caracausi, Andrea, *Dentro la bottega: culture del lavoro in una città d'età moderna* (Venice: Marsilio, 2008).

Caracausi, Andrea, 'Fashion, capitalism and ribbon-making in early modern Europe', in Thomas Max Safley (ed.), *Labor before the Industrial Revolution: Work, Technology and Their Ecologies in an Age of Early Capitalism* (London: Routledge, 2018), 48–69.

Caracausi, Andrea, 'Fra sistema a domicilio e manifattura accentrata : l'istituto degli orfani Nazzareni di Padova nella prima metà del Seicento', *Mediterranea Ricerche Storiche*, 48 (2020), 123–42.

Caracausi, Andrea, *Nastri, nastrini, cordelle: l'industria serica nel padovano, secc. XVII.–XIX* (Padua: CLEUP, 2004).

Caracausi, Andrea, 'Textiles manufacturing, product innovations and transfers of technology in Padua and Venice between the sixteenth and eighteenth centuries', in Karel Davids and Bert de Munck (eds), *Innovation and Creativity in Late Medieval and Early Modern European Cities* (Aldershot: Ashgate, 2014), 131–60.

Caracausi, Andrea, 'Una compagnia di cordelle di seta nella seconda metà del Seicento', *OS. Opificio della Storia*, 2, 2 (2021), 8–21.

Cardano, Girolamo, *The De Subtilitate of Girolamo Cardano*, ed. and trans. J. M. Forrester, 2 vols (Tempe,: Arizona Center for Medieval and Renaissance Studies, 2013).

Cardon, Dominique, *Natural Dyes: Sources, Tradition, Technology and Science* (London: Archetype Publications, 2007).

Carmona, Maurice, 'Sull'economia Toscana del Cinquecento e del Seicento', *Archivio Storico Italiano*, 120, 1/433 (1962), 32–46.

Cavallo, Sandra, 'The artisan's casa', in Marta Ajmar-Wollheim and Flora Dennis (eds), *At Home in Renaissance Italy* (London: V&A Publications, 2006), 66–75.

Cavallo, Sandra, *Artisans of the Body in Early Modern Italy: Identities, Families and Masculinities* (Manchester: Manchester University Press, 2007).

Cecchini, Isabella, 'A world of small objects: Probate inventories, pawns, and domestic life in early modern Venice', *Renaissance and Reformation*, 35, 3 (2012), 39–61.

Cecchini, Isabella, *Quadri e commercio a Venezia durante il Seicento: uno studio sul mercato dell'arte / Isabella Cecchini* (Venice: Marsilio, 2000).

Cerasuolo, Angela, *Literature and Artistic Practice in Sixteenth-Century Italy* (Leiden: Brill, 2017).

Cerri, Monica, 'Sarti toscani nel Seicento: attività e clientela', in Anna G. Cavagna and Grazietta Butazzi (eds), *Le trame della moda* (Roma: Bulzoni Editore, 1995), 421–35.

Chabot, Isabelle, 'La loi du lignage. Notes sur le système successoral florentin (XIVe/XVe–XVIIe siècles)', *Clio*, 7 (1998), 51–72.

Cherry, John, *Goldsmiths* (London: British Museum Press, 1992).

Chevis, Hugh, 'Why early modern English clothiers started using Spanish wool', *Textile History*, 52 (2021), 122–43.

Chorley, Patrick, 'The "Draperies légères" of Lille, Arras, Tournai, Valenciennes: new materials for new markets?', in Marc Boone and Walter Prevenier (eds), *La Draperie Ancienne des Pays-Bas: débouchés et stratégies de survie (14e–16e siècles)* (Leuven: Garant, 1993), 151–66.

Clapham, John H., *An Economic History of Modern Britain; The Early Railway Age, 1820–1850* (Cambridge: Cambridge University Press, 1930).

Clark, Christopher, *The Roots of Rural Capitalism: Western Massachusetts, 1780–1860* (Ithaca: Cornell University Press, 1992).

Clark, Gregory, 'The consumer revolution: Turning point in human history, or statistical artifact?', working paper, Munich Personal RePEc Archive, Paper No. 25467 (September 2010).

Clark, Hazel, 'Uncut: The materiality of textiles and the politics of sustainment in fashionable clothing', in *Textile Society of America Symposium Proceedings*, 9, 1 (2012), Textile Society of America.

Clifford, Helen, 'Concepts of invention, identity and imitation in the London and provincial metal-working trades, 1750–1800', *Journal of Design History*, 12, 3 (1999), 241–55.

Cockburn, James (ed.), *Calendar of Assize Records: Essex Indictments, Elizabeth I* (London: HMSO, 1978–79).

Cockburn, James (ed.), *Calendar of Assize Records: Kent Indictments, Charles II* (London: HMSO, 1995; Woodbridge and London: Boydell, 1997).

Cohn, Samuel K., Jr, *Paradoxes of Inequality in Renaissance Italy* (Cambridge: Cambridge University Press, 2021).

Cohn, Samuel K. *The Laboring Classes in Renaissance Florence* (New York: Academic Press, 1980).

Coleman, Donald C., 'An innovation and its diffusion: "The new draperies"', *Economic History Review*, 22 (1969), 417–29.

Coleman, Donald C., 'Proto-industrialization: A concept too many', *Economic History Review*, 36 (1983), 435–48.

Colenbrander, Sjoukje. 'Woollen velvets and their terminology: Panne, peluche, moquette, tripp and velours d'Utrecht', in Anna Jolly (ed.), *Furnishing Textiles: Studies on Seventeenth- and Eighteenth-Century Interior Decoration* (Riggisberg: Abegg-Stiftung, 2009), 211–22.

Cooper, Tarnya, *Citizen Portrait: Portrait Painting and the Urban Elite of Tudor and Jacobean England and Wales* (New Haven: Yale University Press for the Paul Mellon Centre for Studies in British Art, 2012).

Corner, David, 'The tyranny of fashion: The case of the felt-hatting trade in the late seventeenth and eighteenth centuries', *Textile History*, 22, 2 (1991), 153–78.

Corporation of London book of fines, 1517–1628, London Metropolitan Archives, COL/CHD/CM/10/001.

Cortese, Isabella, *I secreti* (Venice: Giacomo Cornetti, 1584).

Coryat, Thomas, *Coryats Crudities* ... (London: William Stansby, 1611).

Cox, Nancy and Karin Dannehl, *Dictionary of Traded Goods and Commodities, 1550–1820* (Wolverhampton: University of Wolverhampton, 2007), at British History Online, www.british-history.ac.uk.

Cox, Nancy, and Karin Dannehl, *Perceptions of Retailing in Early Modern England* (Aldershot: Ashgate, 2007).

Cunningham, William, *The Growth of English Industry and Commerce in Modern Times: The Mercantile System* (Cambridge: Cambridge University Press, 1903).

Curatolo, Paola, *Struttura, crisi e trasformazione di un sistema produttivo urbano: le corporazioni auroseriche milanesi (1570–1720)* (Milan: Cooperativa universitaria editrice, 1996).

Currie, Elizabeth (ed.), *A Cultural History of Dress and Fashion in the Renaissance* (London: Bloomsbury, 2017).

Currie, Elizabeth, *Fashion and Masculinity in Renaissance Florence* (London: Bloomsbury Academic, 2016).

Currie, Elizabeth, 'Fashion networks: Consumer demand and the clothing trade in Florence from the mid-sixteenth to early seventeenth centuries', *Journal of Medieval and Early Modern Studies*, 39, 3 (2009), 483–509.

Currie, Elizabeth, 'Clothing and a Florentine style, 1550–1620', *Renaissance Studies*, 23, 1 (2009), 33–52.

Currie, Elizabeth, 'Diversity and design in the Florentine tailoring trade, 1560–1620', in Michelle O'Malley and Evelyn Welch (eds), *The Material Renaissance* (Manchester: Manchester University Press, 2007), 154–73.

Currie, Elizabeth, 'Health hazards: Clothing's impact on the body in Italy and England, 1550–1650', *Bulletin of the John Rylands Library*, 95, 2 (2019), 115–33.

Currie, Elizabeth, 'Ink, print, repeat: A hands-on approach to early printed textiles', Victoria and Albert Museum blog (October 2019), https://www.vam.ac.uk/blog/museum-life/ink-print-repeat-a-hands-on-approach-to-early-printed-textiles.

Cybulska, Maria, 'To see the unseen: Computer graphics in visualisation and reconstruction of archaeological and historical textiles', in Nobuhiko Mukai (ed.), *Computer Graphics* (London: InTech, Open, 2012), 213–29.

Dahrén, Lena, *Med kant av guld och silver: en studie av knypplade bårder uddar av metall 1550–1640* (Uppsala: Uppsala Universitet, 2010).

Darlington, Ida (ed.), *London Consistory Court Wills, 1492–1547* (London: London Record Society, 1967).

Davanzo Poli, Doretta, *Abiti antichi e moderni dei Veneziani* (Vicenza: Neri Pozza, 2001).

Davidson, Hilary, 'The embodied turn: Making and remaking dress as an academic practice', *Fashion Theory*, 23, 3 (2019), 329–62.

Davis, Karel and Bert de Munck (eds), *Innovation and Creativity in Late Medieval and Early Modern European Cities* (Aldershot: Ashgate, 2014), 131–60.

Davis, Natalie Zemon, *Fiction in the Archives: Pardon Tales and Their Tellers in Sixteenth-Century France* (Cambridge: Polity, 1987).

Davis, Robert C., 'The Renaissance goes up in smoke', in John Jeffries Martin (ed.), *The Renaissance World* (New York: Routledge, 2007).

De Boodt, Anselm, *Lapidary, or, The History of Pretious Stones ...*, trans. Thomas Nicols (Cambridge: Thomas Buck, 1652).

De Munck, Bert, 'Artisans, products and gifts: Rethinking the history of material culture in early modern Europe', *Past & Present*, 224, 1 (2014), 39–74.

De Vries, Jan, 'Between purchasing power and the world of goods: Understanding the household economy in early modern Europe', in John Brewer and Roy Porter (eds), *Consumption and the World of Goods in the Seventeenth and Eighteenth Centuries* (London: Routledge, 1993), 85–132.

De Vries, Jan, 'Rethinking protoindustry: Human capital and the rise of modern industry', in Kristine Bruland, Anne Gerritsen, Pat Hudson and Giorgio Riello (eds), *Reinventing the Economic History of Industrialisation* (Montreal and Kingston: McGill-Queen's University Press, 2020), 107–26.

De Vries, Jan, 'The Industrial Revolution and the Industrious Revolution', *Journal of Economic History*, 54 (1994), 249–70.

De Vries, Jan, *The Industrious Revolution: Consumer Behaviour and the Household Economy, 1650 to the Present* (Cambridge: Cambridge University Press, 2008).

Dean, Darron, Andrew Hann, Mark Overton and Jane Whittle, *Production and Consumption in English Households, 1600–1750* (London: Routledge, 2004).

Defoe, Daniel, *A Brief Deduction of the Original, Progress, and Immense Greatness of the British Woollen Manufacture* (London, 1727).

Defoe, Daniel, *A Plan of the English Commerce* (London: 1728).

Degn, Ole, and Inger Dübeck, *Håndværkets kulturhistorie: håndværket i fremgang, perioden 1550–1700* (Copenhagen: Håndværkerrådets Forlag, 1983).

Dekker, Thomas, *The Guls Horne-booke* (London: R.S., 1609).

Della Casa, Giovanni, *Galateo*, trans. Konrad Eisenbichler and Kenneth R. Bartlett (Toronto: Centre for Reformation and Renaissance Studies, 1986).

Deloney, Thomas, *A Most Excellent and Vertuous Ballad of the Patient Grissell* (London, ?1624), Magdalene College, Pepys Ballads 1.34–35, EBBA 20160.

Demo, Edoardo, 'Le manifatture tra Medioevo ed Età Moderna', in Giovanni L. Fontana (ed.), *L'industria vicentina dal Medioevo a oggi* (Padua: CLEUP, 2004), II, 21–126.

Denton, Michael J. and Paul N. Daniels, *Textile Terms and Definitions* (Manchester: Textile Institute, 2002).

Diderot, Denis and Jean le Rond d'Alembert, *L'Encyclopédie: Recueil de planches sur les sciences, les arts libéraux et les arts méchaniques, avec leur explication – Arts de l'habillement* (Paris, 1757).

Diplomentarium Helsingoranum : en samling af aktstykker til byen Helsingørs historie (Copenhagen: Trykt i Bianco Lunos bogtrykkeri ved F.S. Muhle, 1861).

Dolan, Alice, 'The fabric of life: Time and textiles in an eighteenth-century plebeian home', *Home Cultures*, 11 (2014), 353–74.

Dolce, Lodovico, *Libri tre ne i quali si tratta delle diverse sorti delle gemme che produce la Natura ...* (Venice: Gio. Battista, Marchiot Sessa, e Fratelli, 1565).

Donzel, Catherine, and Sabine Marchal, *L'art de la passementerie et sa contribution à l'histoire de la mode et de la décoration* (Paris: Chêne, 1992).

Duits, Rembrandt, 'Introduction: Did the poor have art?', in Rembrandt Duits (ed.), *The Art of the Poor: The Aesthetic Material Culture of the Lower Classes in Europe, 1300–1600* (London: Bloomsbury Academic, 2020), 1–22.

DuPlessis, Robert S., *Lille and the Dutch Revolt: Urban Stability in an Era of Revolution, 1500–1582* (Cambridge: Cambridge University Press, 1991).

Dupré, Sven, 'The art of glassmaking and the nature of stones: The role of imitation in Anselm De Boodt's classification of stones', in Isabella Augart, Maurice Saß and Iris Wenderholm (eds), *Steinformen: Materialität, Qualität, Imitation* (Berlin: De Gruyter, 2019), 207–20.

Dupré, Sven, Anna Harris and Julia Kursell (eds), *Reconstruction, Replication and Re-enactment in the Humanities and Social Sciences* (Amsterdam: Amsterdam University Press, 2020).

Dye, Gilian, *Sixteenth and Seventeenth Century Lace, Book 1: Gold and Silver Edgings* (London: Cleveden Press, 2012).

Eamon, William, *Science and the Secrets of Nature: Books of Secrets in Medieval and Early Modern Culture* (Princeton: Princeton University Press, 1994).

Earnshaw, Pat, *Lace in Fashion* (Dublin: Gorse Publications, 1991).

Edelstein, Sidney M., 'The Allerley Matkel (1532): Facsimile text, translation, and critical study of the earliest printed book on spot removing and dyeing', *Technology and Culture*, 5, 3 (1964), 297–321.

Edgcumbe, Richard, 'O, those jewels! The pride and glory of this Kingdom!', in Olga Dmitrieva and Tessa Murdoch (eds), *Treasures of the Royal Court: Tudors, Stuarts and the Russian Tsars* (London: V&A Publishing, 2013), 144–57.

Elias, Norbert, *The Civilizing Process* (Cambridge, MA: Blackwell, 1994, originally printed 1939).

Endrei, Walter, *L'Évolution des techniques du filage et du tissage du moyen âge à la révolution industrielle* (Paris: Mouton, 1968).

Epstein, Stephen R., 'The late medieval crisis as an "integration crisis"', in Maarten Prak (ed.), *Early Modern Capitalism: Economic and Social Change in Europe 1400–1800* (London: Routledge, 2001), 25–50.

Ertl, Thomas and Barbara Karl (eds), *Inventories of Textiles – Textiles in Inventories: Studies on Late Medieval and Early Modern Material Culture* (Göttingen: V&R Unipress, 2017).

Evans, Dorinda, 'Leslie, Charles Robert (1794–1859), literary genre painter and author', *Oxford Dictionary of National Biography* (Oxford: Oxford University Press, 2004).

Ezell, Margaret J. M., *The Patriarch's Wife: Literary Evidence and the History of the Family* (Chapel Hill: University of North Carolina Press, 1987).

Faiers, Jonathan, *Fur: A Sensitive History* (New Haven: Yale University Press, 2020).

Fairchilds, Cissie, 'The production and marketing of populuxe goods in eighteenth-century Paris', in John Brewer and Roy Porter (eds), *Consumption and the World of Goods* (London: Routledge, 1993), 228–48.

Fasano Guarini, Elena, 'Center and Periphery', *Journal of Modern History*, 67, supplement, *The Origins of the State in Italy, 1300–1600* (1995), S74–S96.

Fasano Guarini, Elena, 'Considerazioni su giustizia stato e società nel Ducato di Toscana del Cinquecento', in Christine Smith and Salvatore I. Camporeale (eds), *Florence and Venice: Comparisons and Relations. Acts of Two Conferences at Villa I Tatti in 1976–1977*, vol. 2 (Florence: Villa I Tatti, 1980), 135–68.

Ferraro, Joanne M., 'The manufacture and movement of goods', in John Jeffries Martin (ed.), *The Renaissance World* (New York: Routledge, 2007), 87–100.

Ferretti, Emanuela. *Acquedotti e fontane del Rinascimento in Toscana. Acqua, architettura e città al tempo di Cosimo I dei Medici* (Florence: Olschki, 2016).

Fioravanti, Leonardo, *Del compendio de i secreti rationali: libri 5* (Venice: Vincenzo Valgrisi, 1564).

Firmin, Thomas, *Some Proposals for the Imployment of the Poor* (London: 1681).

Fisher, Will, 'The Renaissance beard: Masculinity in early modern England', *Renaissance Quarterly*, 54, 1 (2001), 155–87.

Florio, John, *Queen Anna's New World of Words, Or, Dictionarie of the Italian and English Tongues* (London: Edward Blount and William Barret, 1611).

Forsyth, Hazel, *London's Lost Jewels: The Cheapside Hoard* (London: Philip Wilson Publishers, 2013).

Foster, Charles F., *Seven Households: Life in Cheshire and Lancashire, 1582–1774* (Northwich: Arley Hall Press, 2002).

Fowler, Joseph T. (ed.), 'The Account-Book of William Wray', *The Antiquary*, 32 (1896).

Franceschi, Franco, 'Woollen luxury cloths in late medieval Italy', in Bart Lambert and Katherine A. Wilson (eds), *Europe's Rich Fabric: The Consumption, Commercialisation, and Production of Luxury Textiles in Italy, the Low Countries and Neighbouring Territories (Fourteenth–Sixteenth Centuries)* (Farnham: Ashgate, 2016), 181–204.

Franco, Giacomo, *Habiti delle donne venetiane intagliate in rame nuovamente* (Venice, c. 1610).

Freeser, Andrea, Maureen D. Goggin, and Beth F. Tobin, *The Materiality of Color: The Production, Circulation, and Application of Dyes and Pigments 1400–1800* (New York: Routledge, 2016).

French, David, 'A sixteenth century English merchant in Ankara?', *Anatolian Studies*, 22 (1972), 241–7.

Frick, Carole C., *Dressing Renaissance Florence: Families, Fortunes, and Fine Clothing* (Baltimore: Johns Hopkins University Press, 2002).

Frick, Carole C., 'The Florentine "rigattieri": Second hand clothing dealers and the circulation of goods in the Renaissance', in Alexandra Palmer and Hazel Clark (eds), *Old Clothes, New Looks: Second Hand Fashion* (Oxford: Berg Publishers, 2005), pp. 13–28.

Fulke, William, *Two Treatises Written Against the Papists* (London: Thomas Vautrollier, 1577).

Gallucci, Margaret A., *Benvenuto Cellini: Sexuality, Masculinity, and Artistic Identity in Renaissance Italy* (New York: Palgrave Macmillan, 2003).

Gary, Kathryn E. and Mats Olsson, 'Men at work. Wages and industriousness in southern Sweden, 1500–1850', *Scandinavian Economic History Review*, 68 (2020), 112–28.

Garzoni, Tommaso, *La piazza universale di tutte le professioni del mondo* (Venice: Seravalle, 1605).

Gentilcore, David, '"Cool and tasty waters": Managing Naples's water supply, c. 1500–c. 1750', *Water History*, 11 (2019), 125–51.

George, Edwin and Stella George (eds), *Bristol Probate Inventories, Part 1: 1542–1650* (Bristol: Bristol Record Society, 2002).

Giannuzzi, Chiara Piccolo (ed.), *Fonti per il barocco leccese* (Galatina: Congedo Editore, 1995).

Gierlich, Ernst, *Reval 1621 bis 1645: von der Eroberung Livlands durch Gustav Adolf bis zum Frieden von Brömsebro* (Bonn: Kulturstiftung der deutschen Vertriebenen, 1991).

Glennie, Paul, *'Distinguishing men's trades': Occupational Sources and Debates for Pre-Census England* (Cheltenham: Historical Geography Research Group, 1990).

Glennie, Paul, 'The social shape of the market for domestic linens in early modern England', working paper for the conference 'Clothing and Consumption in England and America, 1600–1800', Victoria and Albert Museum, 1992.

Goldthwaite, Richard A., *The Building of Renaissance Florence* (Baltimore: Johns Hopkins University Press, 1982).

Goldthwaite, Richard A., *The Economy of Renaissance Florence* (Baltimore: Johns Hopkins University Press, 2009).

Goldthwaite, Richard A., 'The Economy of Renaissance Italy: The preconditions for luxury consumption', *I Tatti Studies in the Italian Renaissance*, 2 (1987), 15–39.

Goldthwaite, Richard A., 'The empire of things: Consumer demand in Renaissance Italy', in William Kent, Patricia Simons and John D. Eade (eds), *Patronage, Art and Society in Renaissance Italy* (Oxford: Clarendon Press, 1987).

Gordon, Beverly, 'The hand of the maker: The importance of understanding textiles from the "inside out"', in *Silk Roads, Other Roads* (Proceedings of the 8th Biennial Symposium of the Textile Society of America, Northampton, MA, September 2002), 189–98.

Grancsay, Stephen V., *Arms and Armor: Essays from the Metropolitan Museum of Art Bulletin, 1920–1964* (New York: The Metropolitan Museum of Art, 1986).

Grandt-Nielsen, Finn, 'Laugshusets Pryd: Malede Vinduer fra Odense, 1563–1641', *Fynske Minder* (1998), 77–96.

Grassi, Bartolomeo, *Dei veri ritratti degl'habiti di tutte le parti del mondo* (Rome: Bartolomeo Grassi, 1585).

Grasskamp, Anna, 'Unpacking foreign ingenuity: The German conquest of artful objects with "Indian" provenance', in Richard Oosterhoff, José Ramón Marcaida and Alexander Marr (eds), *Ingenuity in the Making: Matter and Technique in Early Modern Europe* (Pittsburgh: University of Pittsburgh Press, 2021), 203–18.

Gray, Todd, *Devon Household Accounts, 1627–59, Part 2* (Exeter: Devon and Cornwall Record Society, 1996).

Green Carr, Lois and Lorena S. Walsh, 'Inventories and the analysis of wealth and consumption patterns in St. Mary's County, Maryland, 1658–1777', *Historical Methods*, 13, 2 (1980), 81–104.

Gregory, Sharon and Sally Anne Hickson (eds), *Inganno – the Art of Deception: Imitation, Reception, and Deceit in Early Modern Art* (Farnham: Ashgate, 2012).

Grendler, Paul F., *Schooling in Renaissance Italy: Literacy and Learning 1300–1600* (Baltimore: Johns Hopkins University Press, 1989).

Gritt, Andrew J., 'The farming and domestic economy of a Lancashire smallholder: Richard Latham and the agricultural revolution, 1724–67', in Richard W. Hoyle (ed.), *The Farmer in England, 1650–1980* (Farnham: Ashgate, 2013), 101–34.

Guerzoni, Guido, 'Social emulation, product imitation and technical innovation', Paper presented at the Center for Science and Innovation Studies, University of California, Davis, 2 April 2014, www.academia.edu/7833070.

Hackenbroch, Yvonne, *Enseignes: Renaissance Hat Jewels* (Florence: Studio per edizioni scelte, 1996).

Hamilton, Polly, 'Haberdashery for use in dress, 1550–1800' (PhD thesis, University of Wolverhampton, 2007).

Hardy, Evelyn, *Survivors of the Armada* (London: Constable, 1966).

Harrison, William, *The Description of England: The Classic Contemporary Account of Tudor Social Life, ed. Georges Edelen* (Washington, DC, and New York: Folger Shakespeare Library and Dover, 1994).

Harte, Negley B., 'The rise of protection and the English linen trade, 1690–1790', in Negley B. Harte and Kenneth G. Ponting (eds), *Textile History and Economic History: Essays in Honour of Julia de Lacy Mann* (Manchester: Manchester University Press, 1973), 74–112.

Harvey, John, *Men in Black* (Chicago: University of Chicago Press, 1996).

Hatley, Victor A. and Brian G. Statham, 'Nassaburgh militia lists, 1762', in Edmund King (ed.), *A Northamptonshire Miscellany* (Northampton: Northamptonshire Record Society, 1983), 109–46.

Havinden, Michael A., *Household and Farm Inventories in Oxfordshire, 1550–1590* (London: HMSO, 1966).

Hayward, Maria, 'A shadow of a former self: Analysis of an early seventeenth-century doublet from Abingdon', in Tara Hamling and Catherine Richardson (eds), *Everyday Objects: Medieval and Early Modern Material Culture and Its Meanings* (Abingdon: Routledge, 2016), 107–18.

Hayward, Maria, *Rich Apparel: Clothing and the Law in Henry VIII's England* (Abingdon: Routledge, 2016).

Herald, Jacqueline, *Renaissance Dress in Italy 1400–1500* (London: Bell & Hyman, 1981).

'Hertfordshire Militia Ballot Lists', PDF files on compact disk (Watford: Hertfordshire Family History Society, 2008).

Heywood, John, *The Play Called the Four PP*, ed. John S. Farmer (London: The Tudor Facsimile Texts, 1908).

Hibbert, Christopher, *Florence: The Biography of a City* (London: Penguin, 2004).

Hill, George F., 'The Whitcombe green plaquettes', *The Burlington Magazine for Connosieurs*, 30, 168 (1917), 103–10.

Hindle, Steve, *The Social Topography of a Rural Community: Scenes of Labouring Life in Seventeenth Century England* (Oxford: Oxford University Press, 2023).

Hinton, Jack, 'By sale, by gift: Aspects of the resale and bequest of goods in late-sixteenth-century Venice', *Journal of Design History*, 15, 4 (2002), 245–62.

Hodgen, Margaret T., 'Fairs of Elizabethan England', *Economic Geography*, 18, 4 (1942), 389–400.

Hofenk de Graaff, Judith H., *The Colourful Past: Origins, Chemistry and Identification of Natural Dyestuffs* (London: Archetype Publications, 2014).

Hohti, Paula E., *Artisans, Objects and Everyday Life in Renaissance Italy: The Material Culture of the Middling Class* (Amsterdam: Amsterdam University Press, 2020).

Hohti, Paula, 'Artisans, pawn-broking and the circulation of material goods in sixteenth-century Siena', in M. Ascheri, G. Mazzini, and F. Nevola (eds), *Siena nel Rinascimento: l'ultimo secolo della Repubblica, II: Arte, architettura, cultura*, Acts of the Conference held in Siena, 28–30 September 2003 and 16–18 September 2004 (Siena: Accademia Senese degli Intronati, 2008), 271–81.

Hohti, Paula, 'Cheap magnificence? Imitation and low-cost luxuries in Renaissance Italy', in Catherine Kovesi (ed.), *Luxury and the Ethics of Greed in Early Modern Italy* (Turnhout: Brepols, 2018), 277–97.

Hohti, Paula, '"Conspicuous" consumption and popular consumers: Material culture and social status in sixteenth-century Siena', *Renaissance Studies*, 24, 5 (2010), 654–70.

Hohti, Paula, 'Dress, dissemination, and innovation: Artisan "fashions" in sixteenth- and early seventeenth-century Italy', in Evelyn Welch (ed.), *Fashioning the Early Modern: Dress, Textiles, and Innovation in Europe, 1500–1800* (Oxford: Oxford University Press, 2017), 143–65.

Hohti, Paula, 'Innovations and imitations of colour: Wearing and making colour in early modern Italy', in Maria Hayward, Giorgio Riello and Ulinka Rublack (eds), *A Revolution in Colour: Natural Dyes and Dress in Europe, c. 1400–1800* (London: Bloomsbury, forthcoming 2024).

Hohti, Paula 'Knitting history through reconstruction: The making and meaning of early modern stockings', in Sophie Pitman and Paula Hohti (eds), *Remaking Dress and Textile History: Applying Reconstruction Methods to Early Modern Textiles and Clothing*, Special Issue, *Textile History* (forthcoming 2024).

Hohti, Paula, 'Material culture, shopkeepers and artisans in sixteenth-century Siena' (PhD thesis, University of Sussex, 2006).

Hohti, Paula, '"Monstrous" ruffs and elegant trimmings: Lace and lacemaking in early modern Italy', in Emma Cormack and Michele Majer (eds), *Threads of Power: Lace from the Textilmuseum St. Gallen* (New York: Yale University Press, Bard Graduate Center, 2023), 69–87.

Hohti, Paula E., 'Power, black clothing, and the chromatic politics of textiles in Renaissance Europe', in Jenny Boulboullé and Sven Dupré (eds), *Burgundian Black: Reworking Early Modern Colour Technologies* (Santa Barbara: EMC Imprint, 2022), https://doi.org/10.55239/bb001.

Hohti, Paula E., 'The art of artisan fashions: Moroni's tailor and the changing culture of clothing in sixteenth-century Italy', in Rembrandt Duits (ed.), *The Art of the Poor: The Aesthetic Material Culture of the Lower Classes in Europe 1300–1600* (London: Bloomsbury Academic, 2020), 109–16.

Hohti, Paula, 'The innkeeper's goods: The use and acquisition of household property in sixteenth century Siena', in M. O'Malley and E. Welch (eds), *The Material Renaissance, 1450–1600* (Manchester: Manchester University Press, 2007), 242–59.

Huang, Angela L., 'Hanseatic textile production in 15th century long distance trade', in Carsten Jahnke and Angela L. Huang (eds), *Textiles and the Medieval Economy: Production, Trade and Consumption of Textiles, 8th to 16th Centuries* (Oxford: Oxbow Books, 2015), 204–15.

Hughes, Diane Owen, 'Sumptuary law and social relations in Renaissance Italy', in John Bossy (ed.), *Disputes and Settlements: Law and Human Relations in the West* (Cambridge: Cambridge University Press, 1983), 69–99.

Hunt, Alan, *Governance of the Consuming Passions: A History of Sumptuary Law* (Basingstoke: Macmillan, 1996).

Husbands, Christopher, 'Standards of living in north Warwickshire in the seventeenth century', *Warwickshire History*, 4, 6 (1980–1), 203–15.

Hybel, Niels, 'Klima og hungersnød i middelalderen', *Historisk Tidsskrift*, 102, 2 (2002), 265–80.

Inal, Onur, 'Women's fashions in transition: Ottoman borderlands and the Anglo-Ottoman exchange of costumes', *Journal of World History*, 22, 2 (2011), 243–72.

Innis, Harold Adams, *The Fur Trade in Canada: An Introduction to Canadian Economic History* (Toronto: University of Toronto Press, 1999).

Jeggle, Christof, 'Copier et contrefaire des textiles au temps de la Renaissance', in Pascale Mounier and Colette Nativel (eds), *Copier et contrefaire à la Renaissance: Faux et usage de faux* (Paris: Honoré Champion, 2014), 293–306.

Jeggle, Christof, 'Labelling with numbers? Weavers, merchants and the valuation of linen in seventeenth-century Münster', in Bert De Munck and Dries Lyna (eds), *Concepts of Value in European Material Culture, 1500–1900* (London: Routledge, 2015), 171–206.

Jenks, Stuart, 'The missing link: the distribution revolution of the 15th century', in Carsten Jahnke and Angela L. Huang (eds), *Textiles and the Medieval Economy: Production, Trade and Consumption of Textiles, 8th to 16th Centuries* (Oxford: Oxbow, 2015), 230–52.

Jeremy, David J., 'British and American yarn count systems: An historical analysis', *Business History Review*, 45 (1971), 336–68.

Jirousek, Charlotte A., *Ottoman Dress and Design in the West: A Visual History of Cultural Exchange* (Bloomington: Indiana University Press, 2019).

Johansen, Katia, *Ti kongers tøj: tre århundreders kongedragter på rosenborg slot, 1596–1863* (Aarhus: Aarhus University Press, 2020).

Jones, Ann Rosalind, 'Sleeves, purses, spindles: Fashioning women in Cesare Vecellio's costume books', in Susan Shifrin (ed.), *Re-framing Representations of Women: Figuring, Fashioning, Portraiting and Telling in the 'Picturing' Women Project* (Abingdon: Routledge, 2016), 99–114.

Jones, Ann Rosalind and Peter Stallybrass (eds), *Renaissance Clothing and the Materials of Memory* (Cambridge: Cambridge University Press, 2000).

Jonson, Ben, *Poetaster; or The Arraignment* (London: M. Lownes, 1602) in *The Cambridge Edition of the Works of Ben Jonson* Online, https://universitypublishingonline.org/cambridge/benjonson.

Kala, Tiina and Toomas Tamla (eds), *Tallinna ajalugu II 1561–1710* (Tallinn: Tallinna Linnaarhiiv, 2019).

Kalm, Pehr, Kalm's Account of His Visit to England on His Way to America in 1748, trans. Joseph Lucas (London: Macmillan, 1892).

Kaplinski, Küllike, *Tallinn – meistrite linn* (Tallinn: Tallinna Kultuuriväärtuste Amet, 1995).

Kerridge, Eric, *Textile Manufactures in Early Modern England* (Manchester: Manchester University Press, 1985).

Kidwell, Claudia B., *Cutting a Fashionable Fit: Dressmakers' Drafting Systems in the United States* (Washington, DC: Smithsonian Studies in History and Technology, 1979).

King, Rachel, '"The beads with which we pray are made from it": Devotional ambers in early modern Italy', in Wietse de Boer and Christine Göttler (eds), *Religion and the Senses in Early Modern Europe* (Leiden: Brill, 2013), 153–76.

Kirby, Jo, Maartin Van Bommel and A. Verhecken, (eds), *Natural Colorants for Dyeing and Lake Pigments: Practical Recipes and Their Historical Sources* (London: Archetype Publications, 2014).

Kitson, Peter, 'The male occupational structure of Northumberland, 1762–1871: A preliminary report', working paper, Cambridge Group for the History of Population and Social Structure, 2005.

Klapisch-Zuber, Christiane, *Women, Family, and Ritual in Renaissance Italy* (Chicago: University of Chicago Press, 1985).

Kock, Jan, 'Ruder med personlighed: Een mode i Renæssancen', in Ole Høiris and Jens Vellev (eds), *Renæssancens verden: tænkning, kulturliv, dagligliv og efterliv* (Aarhus: Aarhus University Press, 2006), 401–17.

Kotilaine, Jarmo, *Russia's Foreign Trade and Economic Expansion in the Seventeenth Century: Windows on the World* (Leiden: Brill, 2005).

Kovesi, Catherine (ed.), *Luxury and the Ethics of Greed in Early Modern Italy* (Turnhout: Brepols, 2018).

Kovesi Killerby, Catherine, 'Practical problems in the enforcement of Italian sumptuary law, 1200–1500', in Trevor Dean and Kate J. P. Lowe (eds), *Crime, Society, and the Law in Renaissance Italy* (Cambridge: Cambridge University Press, 1994), 99–120

Kovesi Killerby, Catherine, *Sumptuary Law in Italy, 1200–1500* (Oxford: Clarendon Press, 2002).

Krohn, Deborah L., 'Rites of passage: Art objects to celebrate betrothal, marriage, and the family', in Andrea Bayer (ed.), *Art and Love in Renaissance Italy* (New York: Metropolitan Museum of Art, 2008), 60–7.

Kulikoff, Allan, *Agrarian Origins of American Capitalism* (Charlotte: Virginia University Press, 1992).

Kunz, George Frederick and Charles Hugh Stevenson, *The Book of the Pearl: The History, Art, Science, and Industry of the Queen of Gems* (New York: The Century Co., 1908).

Kussmaul, Anne, *A General View of the Rural Economy of England, 1538–1840* (Cambridge: Cambridge University Press, 1990).

Kuzmichev, Viktor et al., 'Computer reconstruction of 19th century trousers', *International Journal of Clothing Science and Technology*, 29, 4 (2017), 594–606.

Lagerqvist, Lars O., *Vad kostade det? Priser och löner från medeltid till våra dagar* (Lund: Historiska media, 2015).

Landriani, Marsilio, *Relazioni di Marsilio Landriani sui progressi delle manifatture in Europa alla fine del Settecento*, ed. M. Pessina (Milan: Il Polifilo, 1981).

Lauridsen Wolf, Jens, *Diarium sive calendarium ecclesiasticum, politicum et oeconomicum perpetuum. Det er en evigvarende eircke, politisk oc huuszholdings caiender* (Copenhagen: Peter Hake på Boghandler Peder Andersens Bekostning, 1648).

Laursen L. (ed.), *Kancelliets brevbøger vedrørende Danmarks indre forhold, 1571–1575, 1580–1583, 1584–1588* (Copenhagen: C. A. Reitzels Forlag, 1898; 1903; 1906).

Leed, Drea, '"Ye shall have it cleane": Textile cleaning techniques in Renaissance Europe', in Robin Netherton and Gale T. Owen-Crocker (eds), *Medieval Clothing and Textiles 2* (Woodbridge: Boydell & Brewer, 2006), 101–20.

Legon, Edward, 'Bound up with meaning: The politics and memory of ribbon wearing in Restoration England and Scotland', *Journal of British Studies*, 56, 1 (2017), 27–50.

Leino, Marika A., 'Italian Renaissance plaquettes in context' (PhD thesis, University of London, 2003).

Lemire, Beverly, *Dress, Culture and Commerce: The English Clothing Trade Before the Factory, 1660–1800* (Basingstoke: Macmillan, 1997).

Lemire, Beverly, *Global Trade and the Transformation of Consumer Cultures: The Material World Remade, c. 1500–1820* (Cambridge: Cambridge University Press, 2018).

Lemire, Beverly and Giorgio Riello, 'East & west: Textiles and fashion in early modern Europe', *Journal of Social History*, 41, 4 (2008), 887–916.

Levey, Santina M., *Lace: A History* (London: Victoria and Albert Museum, 1983).

Levey, Santina M., 'Lace in the early modern period, c. 1500–1780', in David Jenkins (ed.), *The Cambridge History of Western Textiles* (Cambridge: Cambridge University Press, 2003), vol. 1, 585–96.

Levey, Santina M., *Le Pompe, 1559: Patterns for Venetian Bobbin Lace* (Bedfordshire: Ruth Bean Publishers, 1983).

Lipson, Ephraim, *The History of the Woollen and Worsted Industries* (London: A. and C. Black, 1921).

Liva, Alberto, 'Note sulla legislazione suntuaria nell'Italia centro-settentrionale', in Anna Giulia Cavagna and Grazietta Butazzi (eds), *Le trame della moda* (Rome: Bulzoni, 1995), 31–52.

Lomazzo, Giovanni Paolo, *Trattato dell'arte de la pittura di Gio. Paolo Lomazzo … diviso in sette libri*, 3 vols (Rome: Presso S. Del-Monte, 1844).

Lord Mayor's Ordinance, 28 October 1576 (London: John Day). Short Title Catalogue 16707.

Luigini, Federico, *Il libro della bella donna* (Venice: Plinio Pietrasanta, 1554).

Maitte, Corine, 'Donner du travail aux pauvres: les logiques laborieuses dans les institutions charitables florentines aux XVIIe et XVIIIe siècle', *Mediterranea Ricerche Storiche*, 48 (2020), 97–122.

Maitte, Corine, 'Imitation, copie, contrefaçon, faux: Définitions et pratiques sous l'Ancien Régime', *Entreprises et Histoire*, 78, 1 (2015), 13–26.

Malanima, Paolo and Valeria Pinchera, 'A puzzling relationship: Consumptions and incomes in early modern Europe', *Histoire & Mesure*, 27, 2 (2012), 197–222.

Malanima, Paolo, *Il lusso dei contadini : consumi e industrie nelle campagne toscane del Sei e Settecento* (Bologna: Il Mulino, 1990).

Malcolm-Davis, Jane, 'An early modern mystery: A pilot study of knitting, napping and capping', *Archaeological Textile Review*, 58 (2016), 57–74.

Mann, Julia de Lacy, *The Cloth Industry in the West of England from 1640 to 1880* (Oxford: Oxford University Press, 1971).

Markham, Gervase, *Country Contentments, or The English Huswife* (London, 1623).

Martin, Janet B. (ed.), *The Account Book of Clement Taylor of Finsthwaite, 1712–1753* (Liverpool: Record Society of Lancashire and Cheshire, 1997).

Martin, Luc N. D., 'Textile manufactures in Norwich and Norfolk, 1550–1622' (PhD thesis, University of Cambridge, 1991).

Martin, Luc, 'The rise of the new draperies in Norwich, 1550–1622'. in Negley B. Harte (ed.), *The New Draperies in the Low Countries and England, 1300–1800* (Oxford: Oxford University Press, 1997), 245–74.

Martin, Richard, 'Review of Yeohlee: Energetics: Clothes and Enclosures, Ades Gallery, Berlin, Germany, 1998', *Fashion Theory*, 2, 3 (1998), 287–93.

Martini, Dario, 'Produzioni agricole per le manifatture medievali : la robbia e il lino di Cortona all'inizio del Quattrocento', *Rivista di Storia dell'Agricoltura*, 34, 1 (1994), 62.

Masi, Gianluca, 'I rapporti tra il Granducato di Toscana e il Principato di Transilvania (1540–1699)' (PhD thesis, Ca' Foscari Venezia, 2013).

Matchette, Ann, 'Credit and credibility: Used goods and social relations in sixteenth-century Florence', in Michelle O'Malley and Evelyn Welch (eds), *The Material Renaissance* (Manchester: Manchester University Press, 2007), 225–41.

Matchette, Ann, 'To have and have not: The disposal of household furnishing in Florence', *Renaissance Studies*, 20, 5 (2006), 701–16.

Matchette, Ann, 'Women, objects, and exchange in early modern Florence', *Early Modern Women*, 3 (2008), 245–51.

Mathewson, R. Duncan, III, *Treasure of the Atocha: Sixteen Dramatic Years in Search of the Historic Wreck* (London: Sidgwick & Jackson, 1986).

McCall, Timothy, *Brilliant Bodies: Fashioning Courtly Men in Early Renaissance Italy* (University Park: Penn State University Press, 2022).

McCall, Timothy, 'Brilliant bodies: Material culture and the adornment of men in north Italy's Quattrocento Courts', *I Tatti Studies in the Italian Renaissance*, 16, 1–2 (2013), 445–90.

McCall, Timothy, *Making the Renaissance Man: Masculinity in the Courts of Renaissance Italy* (Chicago: University of Chicago Press, 2023).

McCall, Timothy, 'Material fictions of luxury in Sforza Milan', in Catherine Kovesi (ed.), *Luxury and the Ethics of Greed in Early Modern Italy* (Turnhout: Brepols, 2018), 239–76.

McCants, Anne E. C., 'Porcelain for the poor: The material culture of tea and coffee consumption in eighteenth-century Amsterdam', in Paula Findlen (ed.), *Early Modern Things: Objects and Their Histories* (New York: Routledge, 2013), 316–42.

McKendrick, Neil, John Brewer and J. Plumb, *The Birth of a Consumer Society: The Commercialization of Eighteenth-Century England* (Bloomington: Indiana University Press, 1982).

McNeil, Peter and Melissa Bellanta, 'Fashion, embodiment and the "making turn"', *Fashion Theory: The Journal of Dress, Body & Culture*, 23, 3 (May 2019), 325–8.

McShane, Angela and Claire Backhouse, 'Top-knots and lower sorts: Popular print and promiscuous consumption in late seventeenth-century England', in Michael Hunter (ed.), *Printed Images in Early Modern Britain: Essays in Interpretation* (Farnham: Ashgate, 2010), 337–57.

Medlycott, Mervyn (ed.), *Index to Dorset Militia Ballot Lists, 1757 to 1799*, vol. 2 (Weymouth: Somerset and Dorset Family History Society, 1999).

Mendels, Franklin, 'Proto-industrialization: The first phase of the industrialization process', *Journal of Economic History*, 32 (1972), 241–61.

Meneghin, Alessia, *The Social Fabric of Fifteenth-Century Florence: Identities and Change in the World of Second-Hand Dealers* (New York: Routledge, 2019).

Mercan, F. Özden, 'A diplomacy woven with textiles: Medici–Ottoman relations during the late Renaissance', *Mediterranean Historical Review*, 35, 2 (2020), 169–88.

Merrill, Michael, 'Cash is good to eat: Self-sufficiency and exchange in the rural economy of the United States', *Radical History Review*, 3 (1977), 42–71.

Mirabella, Bella, 'Embellishing herself with a cloth: The contradictory life of a handkerchief', in Bella Mirabella (ed.), *Ornamentalism: The Art of Renaissance Accessories* (Ann Arbor: University of Michigan Press, 2011), 59–82.

Mirabella, Bella (ed.), *Ornamentalism: The Art of Renaissance Accessories* (Ann Arbor: University of Michigan Press, 2011).

Mitchell, David, '"What d'ye lack ladies? Hoods, ribbands, very fine silk stockings": The silk trade in restoration London', in Dagmar Schafer, Giorgio Riello and Luca Molà (eds), *Threads of Global Desire: Silk in the Pre-Modern World* (Woodbridge: Boydell Press and Pasold Research Fund, 2018), 187–224.

Molà, Luca, 'A luxury industry: The production of Italian silks, 1400–1600', in Bart Lambert and Kathrine A. Wilson (eds), *Europe's Rich Fabric: The Consumption, Commercialisation, and Production of Luxury Textiles in Italy, the Low Countries and Neighbouring Territories (Fourteenth–Sixteenth Centuries)* (Farnham: Ashgate, 2016), 205–34.

Molà, Luca, *The Silk Industry of Renaissance Venice* (Baltimore: Johns Hopkins University Press, 2000).

Monnas, Lisa, 'A study in contrasts: Silk consumption in Italy and England during the fifteenth and sixteenth centuries', in Dagmar Schäfer, Giorgio Riello and Luca Molà (eds), *Threads of Global Desire: Silk in the Pre-Modern World* (Woodbridge: Boydell Press, 2018), 149–86.

Monnas, Lisa, 'Loom widths and selvedges prescribed by Italian silk weaving statutes, 1265–1512: A preliminary investigation', *Bulletin de Liason du Centre International d'Etude des Textiles Anciens*, 66 (1988), 35–44.

Montaigne, Michel de, *Journal de voyage*, ed. Fausta Garavini (Paris: Gallimard, 1983).

Monte, Piero, *The Arms, Armour and Fighting Techniques of a Fifteenth-Century Soldier*, trans. Jeffrey L. Forgeng (Woodbridge: Boydell & Brewer, 2018).

Montemezzo, Stefania, '"Popular fashion": la ricostruzione dei consumi di moda degli artigiani nell'Italia rinascimentale', *Storiografia. Rivista Annuale di Storia*, 27, 26 (2023), 145–67.

Montgomery, Florence, *Textiles in America, 1650–1870* (New York: W. W. Norton, 1984).

Moore, Sarah E. H., *Ribbon Culture: Charity, Compassion, and Public Awareness* (Basingstoke: Palgrave Macmillan, 2008).

Moorhouse, Stephen and John Hurst, 'An imported stoneware spindlewhorl, with some preliminary comments on stoneware spindlewhorls in England, their dating and origin', *Transactions of the London and Middlesex Archaeological Society*, 32 (1981), 124–27.

Moskvin, Aleksei et al., 'Digital replicas of historical skirts', *The Journal of the Textile Institute*, 110, 12 (2019), 1810–26.

Mottu-Weber, Liliane, *Economie et refuge à Genève au siècle de la réforme: la draperie et la soierie (1540–1630)* (Geneva: Champion, 1987).

Mottu-Weber, Liliane, 'Production et innovation en Suisse et dans les états allemands', in Simoneetta Cavaciocchi (ed.), *La seta in Europa, sec. XII–XX: Atti Della 'Ventiquattrimesa Settimana Di Studi', 4–9 Maggio 1992*, 24 (Florence: Le Monnier, 1993), 141–61.

Muldrew, Craig, *Food, Energy and the Creation of Industriousness: Work and Material Culture in Agrarian England, 1550–1780* (Cambridge: Cambridge University Press, 2011).

Muldrew, Craig, '"Th'ancient Distaff" and "Whirling Spindle": Measuring the contribution of spinning to household earnings and the national economy of England, 1550–1770', *Economic History Review*, 65 (2012), 498–526.
Muldrew, Craig, *The Economy of Obligation: The Culture of Credit and Social Relation in Early Modern England* (Basingstoke: Macmillan, 1998).
Muller, Priscilla E., *Jewels in Spain, 1500–1800* (Madrid: Ediciones El Viso, 2012).
Munro, John, 'Medieval woollens: Textiles, textile technology and industrial organisation, c. 800– 1500', in David Jenkins (ed.), *The Cambridge History of Western Textiles* (Cambridge: Cambridge University Press, 2003), vol. 2, 181–227.
Munro, John, 'Medieval woollens: The western European woollen industries and their struggles for international markets, c. 1000–1500', in David Jenkins (ed.), *The Cambridge History of Western Textiles* (Cambridge: Cambridge University Press, 2003), vol. 2, 228–324.
Munro, John H., 'The anti-red shift – to the "dark side": Colour changes in Flemish luxury woollens, 1300–1550', *Medieval Clothing and Textiles*, MPRA Paper, 3, 1 (2007), 55–98.
Munro, John, 'The origins of the English "New Draperies": The resurrection of an old Flemish industry, 1270–1570', in Negley Harte (ed.), *The New Draperies in the Low Countries and England, 1300–1800* (Oxford: Oxford University Press, 1997), 35–127.
Murphy, Hannah, 'Calligraphy and metamorphosis: Invention and imitation in a sixteenth-century craft', in Richard Oosterhoff, José Ramón Marcaida and Alexander Marr (eds), *Ingenuity in the Making: Matter and Technique in Early Modern Europe* (Pittsburgh: University of Pittsburgh Press, 2021), 79–93.
Muzzarelli, Maria Giuseppina, 'From the closet to the wallet: Pawning clothes in Renaissance Italy', *Renaissance and Reformation*, 35, 3 (2012), 23–38.
Muzzarelli, Maria Giuseppina, *Gli inganni delle apparenze: disciplina di vesti e ornamenti alla fine del Medioevo* (Turin: Scriptorium, 1996).
Muzzarelli, Maria Giuseppina, *Guardaroba medievale: vesti e società dal XIII al XVI secolo* (Bologna: Il Mulino, 1999).
Muzzarelli, Maria Giuseppina, 'La disciplina delle apparenze', in Paolo Prodi (ed.), *Disciplina dell'anima, disciplina del corpo* (Bologna: Il Mulino, 1994), 757–86.
Muzzarelli, Maria Giuseppina, *Le regole del lusso: apparenza e vita quotidiana dal Medioevo all'età moderna* (Bologna: Società editrice Il mulino, 2020).
Muzzarelli, Maria Giuseppina, 'Noscere ordinem et finem sui status: il valore delle vesti nella società posizionale del tardo Medioevo. Problemi di identità tra Medioevo ed età moderna: seminari e bibliografia', in Paolo Prodi and Valerio Marchetti (eds), *Problemi di identità tra Medioevo ed età moderna: seminari e bibliografia* (Bologna: CLUEB, 2001), 105–15.
Muzzarelli, Maria Giuseppina, 'Reconciling the privilege of a few with the common good: Sumptuary laws in medieval and early modern Europe', *Journal of Medieval and Early Modern Studies*, 39, 3 (2009), 597–617.
Muzzarelli, Maria Giuseppina, 'Una seconda chance per le persone e per le cose: i pegni consegnati ai Monti di Pietà alla fine del Medioevo: casi', *Anuario de Estudios Medievales*, 52, 1 (2022), 235–51.
Navagero, Andrea, *Opera omnia*, ed. Gaetano Volpi and Giovanni Antonio Volpi (Venetiis: Ex Typographia Remondiniana, 1754).

North, Susan, '"Galloon, inkle and points": Fashionable dress and accessories in rural England, 1552–1665', in Richard Jones and Christopher Dyer (eds), *Farmers, Consumers, Innovators* (Hatfield: University of Hertfordshire Press, 2016), 104–23.

North, Susan, *Sweet and Clean? Bodies and Clothes in Early Modern England* (Oxford: Oxford University Press, 2020).

O'Connell Edwards, Lesley, 'Knitted wool stockings in the Museum of London: A study of 16th century construction', *Archaeological Textiles Review*, 60 (2018), 42–50.

O'Connell Edwards, Lesley, 'The stocking knitting industry of later sixteenth-century Norwich', *Textile History*, 52 (2021), 144–64.

O'Malley, Michelle and Evelyn Welch (eds), *The Material Renaissance* (Manchester: Manchester University Press, 2007).

Odstrčilová, Sylvie, 'Early modern stockings in museums in the Czech Republic', *Archaeological Textiles Review*, 60 (2018), 51–63.

Ogilvie, Sheilagh, 'Consumption, social capital, and the "Industrious Revolution" in early modern Germany', *Journal of Economic History*, 70 (2010), 287–325.

Ogilvie, Sheilagh C. and Markus Cerman, 'The theories of proto-industrialization', in Sheilagh C. Ogilvie and Markus Cerman (eds), *European Proto-Industrialization: An Introductory Handbook* (Cambridge: Cambridge University Press, 1996), 1–11.

Oldland, John, '"Fyne worsted whech is almost like silke": Norwich's double worsted', *Textile History*, 42 (2011), 181–99.

Oldland, John, 'The economic impact of clothmaking on rural society, 1300–1550', in Martin Allen and Matthew Davies (eds), *Medieval Merchants and Money* (London: Institute of Historical Research, 2016), 229–52.

Oldland, John, *The English Woollen Industry, c. 1200–c. 1560* (Abingdon: Routledge, 2019).

Opera nova chiamata Secreti secretorum: in laquale poterai conseguire molti piaceri et utilitade: con molte cose ridiculose stampata novamente (Venice: B. Benalius, 1520).

Opera nuova intitolata dificio di ricette (Venice, 1525).

Ormrod, David, *The Rise of Commercial Empires: England and the Netherlands in the Age of Mercantilism, 1650–1770* (Cambridge: Cambridge University Press, 2003).

Orsi Landini, Roberta N., *Moda a Firenze: Lo stile di Eleonora di Toledo e la sua influenza* (Florence: Edizioni Polistampa, 2007).

Ortega-Saez, Natalia, 'Black dyed wool in north-western Europe, 1680–1850: The relationship between historical recipes and the current state of preservation' (Unpublished PhD thesis, University of Antwerp, 2018).

Paganino, Alessandro, *Libro quarto de rechami* (Toscolano: Paganino, *c.* 1532).

Pagi, Hembo, 'When data becomes information: Visualizing archaeological textiles', in B. D. Frischer, J. W. Crawford and D. Koller (eds), *Making History Interactive: Computer Methods and Quantitative Methods in Archaeology*, Proceedings of the 37th International Conference of the CAA, Williamsburg, Virginia, United States, March 22–26, 2009 (Oxford: Archaeopress, 2010).

Pajur, Astrid, 'Costume and control: Sumptuary laws and social order in seventeenth-century Tallinn' (master's thesis, Uppsala University, 2014).

Pajur, Astrid, *Dress Matters: Clothes and Social Order in Tallinn, 1600–1700* (Uppsala: Studia Historica Upsaliensia, 2020).

Palumbo Fossati, Isabella, *Dentro le case: abitare a Venezia nel Cinquecento* (Venice: Gambier&Keller, 2013).

Pappano, Margaret A. and Nicole R. Rice, 'Medieval and early modern artisan culture', *Journal of Medieval and Early Modern Studies*, 43, 3 (2013), 473–85.

Paresys, Isabella, 'The body', in Elizabeth Currie (ed.), *A Cultural History of Dress and Fashion in the Renaissance* (London: Bloomsbury Academic, 2017), 57–73.

Parker, Greig, *Probate Inventories of French Immigrants in Early Modern London* (Farnham: Ashgate, 2014).

Patterson, Angus, *Fashion and Armour in Renaissance Europe: Proud Lookes and Brave Attire* (London: V&A Publishing, 2009).

Paulicelli, Eugenia, *Writing Fashion in Early Modern Italy: From Sprezzatura to Satire* (Farnham: Ashgate, 2014).

Paulli, Simon, *Flora danica, det er: dansk urtebog: udi huilcken efter … Christiani IV. … befaling … icke alleeniste urternis historiske beskrifvelse, krafter oc virckninger med ziirligste figurer andragis, men endocsaa lægedomme til alle siugdomme gafnlige, korteligen ocklarligen antegnis, saa at den er baade en urtebog oc lægebog med største flid oc umage elaborerit aff Simone Paulli* (Copenhagen: Prentet aff Melchiore Martzan, 1648).

Pedrocco, Filippo, 'Iconografia delle cortigiane di Venezia', in Doretta Davanzo Poli et al., *Il gioco dell'amore: Le cortigiane di Venezia dal trecento al settecento* (Milan: Berenice Art Books, 1990), 81–94.

Pennant, Thomas, *A Tour in Scotland; 1769* (London, 1776).

Pezzolo, Luciano, 'The Venetian economy', in Eric Dursteler (ed.), *A Companion to Venetian History, 1400–1797* (Leiden: Brill, 2013), vol. IV, 255–89.

Pfister, Ulrich, 'Craft guilds and technological change: The engine loom in the European silk ribbon industry in the seventeenth and eighteenth centuries', in Stephen Epstein and Maarten Prak (eds), *Guilds, Innovation, and the European Economy, 1400–1800* (Cambridge: Cambridge University Press, 2008), 172–98.

Philiatros, *Natura Exenterata: Or Nature Unbowelled by the Most Exquisite Anatomies of Her* (London: H. Twiford, 1655).

Pitman, Sophie. 'Black color for dyeing, and the place of textiles in Ms. Fr. 640', in Pamela H. Smith et al. (eds), *Secrets of Craft and Nature in Renaissance France: A Digital Critical Edition and English Translation of BnF Ms. Fr. 640*, Making and Knowing Project, 2020, https://edition640.makingandknowing.org, accessed 7 January 2024.

Pitman, Sophie, 'Codes: Redressing London – sumptuary laws and the control of clothing in the early modern city', in Peter Goodrich (ed.), *A Cultural History of Law in the Early Modern Age* (London: Bloomsbury, 2019), 65–86.

Pitman, Sophie, 'Prodigal years?: Negotiating luxury and fashioning identity in a seventeenth-century account book', *Luxury: History, Culture, Consumption*, 3, 1–2 (November 2016), 7–31.

Pitman, Sophie, 'Reconstructing fashion: The mock-velvet doublet of a seventeenth-century Florentine waterseller', in Sophie Pitman and Paula Hohti (eds), *Remaking Dress and Textile History: Applying Reconstruction Methods to Early Modern Textiles and Clothing*, Special Issue, *Textile History*, forthcoming 2024.

Pitman, Sophie, 'The making of clothing and the making of London, 1560–1660' (PhD thesis, University of Cambridge, 2017).

Pitman, Sophie and Paula Hohti (eds), *Remaking Dress History: Applying Reconstruction Methods to Early Modern Textiles and Clothing*, Special Issue, *Textile History*, forthcoming 2024.

Pitt, William, *General View of the Agriculture of the County of Stafford* (London, 1794).

Pliny the Elder, *Natural History: A Selection*, trans. John F. Healy (London: Penguin, 1991).

Plon, Eugène, *Benvenuto Cellini, orfèvre, medailleur, sculpteur: recherches sur savie, sur son oeuvre et sur les pièces qui lui sont attribuées* (Paris: Plon, 1883).

Plummer, Alfred, *The London Weavers' Company, 1600–1970* (London: Routledge, 2006).

Poni, Carlo, 'Archéologie de la fabrique: La diffusion des moulins à soie "alla bolognese" dans les États vénitiens, du XVe au XVIIIe siècle', *Annales: Economies, Sociétés, Civilisations*, 27, 6 (1972), 1475–96.

Poni, Carlo, 'Fashion as flexible production: The strategies of the Lyons silk merchants in the eighteenth century', in Charles Sabel and Jonathan Zeitlin (eds), *World of Possibilities: Flexibility and Mass Production in Western Industrialization* (Cambridge: Cambridge University Press, 1997), 37–74.

Poni, Carlo, 'Standards, trust and civil discourse: Measuring the thickness and quality of silk thread', *History of Technology*, 23 (2001), 1–16.

Poppy, Pat, 'The clothing accessory choices of Rachel, Countess of Bath, and other mid-seventeenth-century women', *Costume*, 54 (2020), 3–29.

Porter, Roy and John Brewer, *Consumption and the World of Goods* (London: Routledge, 1993).

Prak, Maarten and Patrick Wallis (eds), *Apprenticeship in Early Modern Europe* (Cambridge: Cambridge University Press, 2019).

Prammatica sopra le perle, gioie, drappi, ricami, et altro, per la citta di Pisa e Territorio dello Stato Pisano (Pisa: Per il Tanagli, & il Dote, 1638).

Priestley, Ursula, 'Norwich stuffs, 1600–1700', in Negley Harte, *The New Draperies in the Low Countries and England, 1300–1800* (Oxford: Oxford University Press, 1997), 275–88.

Pullan, Brian, *Crisis and Change in the Venetian Economy in the Sixteenth and Seventeenth Centuries* (London: Methuen, 1968).

Pullan, Brian, 'Wage-earners and the Venetian economy, 1550–1630', *Economic History Review*, 16, 3 (1964), 407–26.

Pylkkänen, Riitta, *1600-luvun kuolinpukuja Turun tuomiokirkkomuseossa* (Turku: Turun kaupungin Historiallinen Museo, 1955).

Pylkkänen, Riitta, *Barokin pukumuoti Suomessa 1620–1720* (Helsinki: Suomen Muinaismuistoyhdistys, 1970).

Pylkkänen, Riitta, *Säätyläispuku Suomessa Vanhalla Vaasa-ajalla 1550–1620*, with a short summary in English, 'The costume of the nobility, clergy and burghers in the earlier Vasa period, 1550–1620', 505–22 (Helsinki: Suomen Muinaismuistoyhdistys, 1956).

Quondam, Amedeo, *Tutti i colori del nero: moda e cultura del gentiluomo nel Rinascimento* (Costabissara Vicenza: A. Colla, 2007).

Raber, Karen, 'Chains of pearls: Gender, property, identity', in Bella Mirabella (ed.), *Ornamentalism: The Art of Renaissance Accessories* (Ann Arbor: University of Michigan Press, 2011), 159–81.

Raffaella Sarti, *Europe at Home: Family and Material Culture 1500–1800* (New Haven: Yale University Press, 2002).

Ragosta, Rosalba, *Napoli, città della seta: produzione e mercato in età moderna* (Rome: Donzelli Editore, 2009).

Ragosta, Rosalba, 'Specializzazione produttiva a Napoli nei secoli XVI e XVII', in Simonetta Cavaciocchi (ed.), *La seta in Europa, Sec. XII–XX: Atti Della 'Ventiquattrimesa Settimana Di Studi', 4–9 Maggio 1992*, 24 (Florence: Le Monnier, 1993), 339–49.

Ramsay, George D., *The English Woollen Industry, 1500–1750* (London: Macmillan, 1982).

Rangström, Lena, *Modeleijon: manligt ode 1500-tal 1600-tal 1700-tal* (Stockholm: Livrustkammaren/Atlantis, 2002).

Rapp, Richard T., *Industry and Economic Decline in Seventeenth-Century Venice* (Cambridge, MA: Harvard University Press, 1976).

Raybould, Wendy, 'Open for business: Textile manufacture in Northamptonshire, c.1685–1800' (PhD thesis, University of Leicester, 2005).

Reddy, William, *The Rise of Market Culture: The Textile Trade and French Society, 1750–1900* (Cambridge: Cambridge University Press, 1984).

Reed, Michael (ed.), *The Ipswich Probate Inventories, 1583–1631* (Woodbridge: The Boydell Press for the Suffolk Records Society, 1981).

Ribeiro, Aileen, *The Gallery of Fashion* (London: National Portrait Gallery, 2000).

Ricci, Elisa, *Ricami Italiani: antichi e moderni* (Florence: Felice Le Monnier, 1927).

Riello, Giorgio, '"Things seen and unseen": The material culture of early modern inventories and their representation of domestic interiors', in Paula Findlen (ed.), *Early Modern Things: Objects and Their Histories, 1500–1800* (New York: Routledge, 2013), 125–50.

Riello, Giorgio, 'The world of textiles in three spheres: European woollens, Indian cottons and Chinese silks, 1300–1700', in Marie-Louise Nosch, Zhao Feng and Lotika Varadarajanpp (eds), *Global Textile Encounters* (Oxford: Oxbow Books, 2014), 93–106.

Riello, Giorgio and Prasannan Parthasarathi, *The Spinning World: A Global History of Cotton Textiles, 1200–1850* (Oxford: Oxford University Press / Pasold, 2009).

Riello, Giorgio and Peter McNeil (eds), *The Fashion History Reader: Global Perspectives* (London: Routledge, 2010).

Riello, Giorgio and Ulinka Rublack (eds), *The Right to Dress: Sumptuary Laws in a Global Perspective, c. 1200–1800* (Cambridge: Cambridge University Press, 2019).

Riforma, e Prammatica sopra l'uso delle perle, gioie, vestire, et altro per la Città & Contado di Firenze (Florence: Massi e Landi, 1638).

Rijks, Marlise, 'Gems and counterfeited gems in early modern Antwerp: From workshops to collections', in Michael Bycroft and Sven Dupré (eds), *Gems in the Early Modern World* (Cham: Springer International Publishing, 2019), 309–42.

Rimstad, Charlotte, 'Dragtfortællinger fra voldgraven: Klædedragten i 1600-tallets København, baseret på arkæologiske tekstiler fra Københavns Rådhusplads' (PhD thesis, University of Copenhagen, 2018).

Ringgaard, Maj, 'Hosekoner og sålede strixstrømper', *Dragtjournalen*, 8, 11 (2014), 4–12.

Rinne, Katherine Wentworth, *The Waters of Rome: Aqueducts, Fountains, and the Birth of the Baroque City* (New Haven: Yale University Press, 2010).
Robinson, Michele Nicole. 'Dirty laundry: Caring for clothing in early modern Italy', *Costume*, 55, 1 (2021), 3–23.
Roche, Daniel, *The Culture of Clothing: Dress and Fashion in the 'Ancien Régime'*, trans. Jean Birrell (Cambridge: Cambridge University Press, 1996).
Rodini, Elizabeth, 'Baroque pearls', *Art Institute of Chicago Museum Studies*, 25, 2 (2000), 68–106.
Rogers, James E. Thorold, *A History of Agriculture and Prices in England: 1583–1702* (Oxford: Clarendon Press, 1887).
Romanin, Samuele, *Storia documentata di Venezia*, 3rd edition (Venice: Filippi, 1972).
Rosenthal, Margaret F., 'Clothing, fashion, dress, and costume in Venice (c. 1450–1650)', in Eric R. Dursteler (ed.), *A Companion to Venetian History, 1400–1797* (Leiden: Brill, 2013), 889–928.
Rosenthal, Margaret F, 'Cultures of clothing in later medieval and early modern Europe', *Journal of Medieval and Early Modern Studies*, 39, 3 (2009), 459–81.
Rosetti, Gioanventura, *The Plichto: Instructions in the Art of the Dyers Which Teaches the Dyeing of Woolen Cloths, Linens, Cottons, and Silk by the Great Art as Well as by the Common*, trans. of the first edition of 1548 by Sidney M. Edelstein and Hector C. Borghetty (Cambridge, MA: MIT Press, 1969).
Rothstein, Natalie, 'Fashion, silk and the worshipful company of weavers', in Simonetta Cavaciocchi (ed.), *La seta in Europa, Sec. XIII–XX: Atti Della 'Ventiquattrimesa Settimana Di Studi', 4–9 Maggio 1992*, 24 (Florence: Le Monnier, 1993), 471–2.
Rothstein, Natalie, 'Silk in the early modern period, c. 1500–1780', in David Jenkins (ed.), *The Cambridge History of Western Textiles* (Cambridge: Cambridge University Press, 2003), vol. 2, 528–61.
Rouch, Monique, 'I vestiti dei contadini alla fine del Cinquecento e l'opera dialettale di Giulio Cesare Croce', in Jeannine Guérin Dalle Mese (ed.), *Il vestito e la sua immagine: Atti del convegno in omaggio a Cesare Vecellio nel quarto centenario della morte* (Belluno: Tipografia Piave, 2002).
Rublack, Ulinka, *Dressing Up: Cultural Identity in Renaissance Europe* (Oxford: Oxford University Press, 2010).
Rublack, Ulinka, 'Matter in the material Renaissance', *Past & Present*, 219 (2013), 41–85.
Rublack, Ulinka 'Renaissance fashion: The birth of power dressing', *History Today*, 61, 1 (2011), www.historytoday.com/archive/renaissance-fashion-birth-power-dressing.
Rublack, Ulinka, 'Renaissance dress, cultures of making, and the period eye', *Journal of Decorative Arts, Design History, and Material Culture*, 23, 1 (2016), 6–34.
Rublack, Ulinka, Maria Hayward and Jenny Tiramani (eds), *The First Book of Fashion: The Book of Clothes of Matthaus and Veit Konrad Schwarz of Augsburg* (London: Bloomsbury Academic, 2015).
Ruscelli, Girolamo, *De' secreti del reuerendo donno Alessio Piemontese* (Venice: Antonio de gli Antonii, 1563).
Ruscelli, Girolamo, *The Secrets of the Reuerend Maister Alexis of Piemont, Containing Excellent Remedies against Diuerse Diseases, Wounds, and Other Accidents, trans. William Ward* (London: Thomas Wright, 1595).

Ryder, Michael L., 'Fleece grading and wool sorting: The historical perspective', *Textile History*, 26 (1995), 3–22.

Savary des Brûlons, Jacques, *Dictionnaire universel du commerce* (Copenhagen, 1759).

Scuro, Rachele, 'Shaping identity through glass in Renaissance Venice', in Susanna Burghartz et al. (eds), *Materialized Identities in Early Modern Culture, 1450–1750* (Amsterdam: Amsterdam University Press, 2021), 99–136.

Sear, Joanne and Ken Sneath, *The Origins of the Consumer Revolution in England: From Brass Pots to Clocks* (Abingdon: Routledge, 2020).

Secher, Vilhelm Adolf (ed.), *Corpus constitutionum daniæ: forordninger: forordninger, recesser og andre kongelige breve, Danmarks lovgivning vedkommende 1558–1660*, vol. 4 (Copenhagen: G. E. C. Gad Nielsen & Lydiche, 1897).

Sella, Domenico, 'Les mouvements longs de l'industrie lainière à Venise aux XVIe et XVIIe siècles', *Annales. Economies, Sociétés, Civilisations*, 12, 1 (1957), 29–45.

Sennefelt, Karin, 'A discerning eye: Visual culture and social distinction in early modern Stockholm', *Cultural and Social History*, 12, 2 (2015), 179–95.

Sennett, Richard, *Il declino dell'uomo pubblico* (Milan: Mondador, 2006).

Sframeli, Maria (ed.), *I gioielli dei Medici: dal vero e in ritratto* (Livorno: Sillabe, 2003).

Shakespeare, William, *The Complete Works of William Shakespeare*, Shakespeare Head Press Edition (Ware: Wordsworth Editions, 1999).

Shammas, Carole, 'The decline of textile prices in England and British America prior to industrialization', *Economic History Review*, 47 (1994), 483–507.

Shammas, Carole, *The Pre-Industrial Consumer in England and America* (Oxford: Oxford University Press, 1990).

Shaw, James E., 'Writing to the prince: Supplications, equity and absolutism in sixteenth-century Tuscany', *Past & Present*, 215 (2012), 51–83.

Shepherd, Rosemary, *An Early Lace Workbook: Bobbin Lace Techniques Before the Baroque* (Sydney: Lace Daisy Press, 2009).

Sindvald Larsen, Anne-Kristine, 'Artisans and dress in Denmark 1550–1650: A preliminary exploration', in Rembrandt Duits (ed.), *The Art of the Poor: The Aesthetic Material Culture of the Lower Classes in Europe 1300–1600* (London: Bloomsbury Academic, 2020), 99–108.

Sindvald Larsen, Anne-Kristine, 'Clothes, culture and crafts: Dress and fashion among artisans and small shopkeepers in the Danish town of Elsinore, 1550–1650' (PhD dissertation, Aalto University, Aalto ARTS Books, 2023).

Smith, Pamela, *The Body of the Artisan* (Chicago: University of Chicago Press, 2004).

Smith, Pamela, 'Historians in the laboratory: Reconstruction of Renaissance art and technology in the Making and Knowing project', *Art History*, 39.2 (2016), 210–33.

Smith, Pamela et al. (eds), *Secrets of Craft and Nature in Renaissance France: A Digital Critical Edition and English Translation of BnF Ms. Fr. 640* (New York: Making and Knowing Project, 2020), available online at https://edition640.makingandknowing.org.

Smith, Pamela, Amy R. W. Mayers and Harold J. Cook (eds), *Ways of Making and Knowing: The Material Culture of Empirical Knowledge* (Chicago: University of Chicago Press / Bard Graduate Centre, 2017).

Speelberg, Femke (ed.), 'Fashion and virtue: Textile patterns and the print revolution, 1520–1620', special issue *The Metropolitan Museum of Art Bulletin*, 73, 2 (Fall 2015).

Spenceley, Geoffrey F. R., 'The origins of the English pillow lace industry', *Agricultural History Review*, 21 (1973), 81–93.

Sperling, Jutta, 'Dowry or inheritance? Kinship, property, and women's agency in Lisbon, Venice, and Florence (1572)', *Journal of Early Modern History*, 11, 3 (2007), 197–238.

Spilman, John and William Herrick, Bill for repairing the Crown Jewels for the Coronation of King James the First of England, June 4 1604, Edinburgh University EU Laing II 525, as reproduced in Historical Manuscripts Commission, *Report on the Laing Manuscripts Preserved in the University of Edinburgh*, vol. 1 (London: HMSO, 1914).

Springer, Carolyn, *Armour and Masculinity in the Italian Renaissance* (Toronto: University of Toronto Press, 2010).

Spufford, Margaret, 'Fabric for seventeenth-century children and adolescents' clothes', *Textile History*, 34 (2003), 47–63.

Spufford, Margaret, *The Great Reclothing of Rural England: Petty Chapmen and Their Wares in the Seventeenth Century* (London: Hambledon, 1984).

Standley, Eleanor R., 'Spinning yarns: The archaeological evidence for hand spinning and its social implications, c. AD 1200–1500', *Medieval Archaeology*, 60 (2016), 266–99.

Stapleford, Richard, *Lorenzo de Medici at Home: The Inventory of the Palazzo Medici in 1492* (University Park: Penn State University Press, 2013).

Staples, Kate K. 'The significance of the secondhand trade in Europe, 1200–1600', *History Compass*, 13, 6 (2015), 297–309.

Statuti et ordini della magnifica città di Pistoia sopra il vesire delle donne pubblicati il di diciotto settembre MDLVIII (n.p., 1563).

Steer, Francis W. (ed.), *The Memoirs of James Spershott* (Chichester: Chichester City Council, 1962).

Stow, John, *A Survey of London, Reprinted from the Text of 1603*, ed. Charles Lethbridge Kingsford, 2 vols (Oxford: Clarendon Press, 1971).

Strickland, Henry E., *A General View of the Agriculture of the East Riding of Yorkshire* (York, 1812).

Sturtewagen, Isis, 'All together respectably dressed: Fashion and clothing in Bruges during the fifteenth and sixteenth centuries' (PhD thesis, University of Antwerp, 2016).

Styles, John, *Dress of the People: Everyday Fashion in Eighteenth-Century England* (New Haven: Yale University Press, 2007).

Styles, John, 'Fashion and innovation in early-modern Europe', in Evelyn Welch (ed.), *Fashioning the Early Modern: Dress, Textiles, and Innovation in Europe, 1500–1800* (Oxford: Oxford University Press, 2016), 33–55.

Styles, John, 'Involuntary consumers? The eighteenth-century servant and her clothes', *Textile History*, 33, 1 (2002), 9–21.

Styles, John, 'Product innovation in early modern London', *Past & Present*, 168 (2000), 124–69.

Styles, John, 'Spinners and the law: Regulating yarn standards in the English worsted industries, 1550–1800', *Textile History*, 44, 2 (2013), 145–70.

Styles, John, 'What were cottons for in the Industrial Revolution?', in Giorgio Riello and Prasanan Parthasarathi (eds), *The Spinning World: A Global History of Cotton Textiles* (Oxford: Oxford University Press / Pasold Research Fund, 2011), 307–26.

Suhm, Peter Frederik, 'En ubenævnt Dansk kunstners levnets historie', *Samlinger til den Danske Historie*, 2, 3 (1784), 134–58.

Tankard, Danae, '"A pair of grass-green woollen stockings": The clothing of the rural poor in seventeenth-century Sussex', *Textile History*, 43, 1 (2012), 5–22.

Taylor, Lou, *The Study of Dress History* (Manchester: Manchester University Press, 2002).

Teisseyre-Sallmann, Line, *L'industrie de la soie en Bas-Languedoc: XVIIe–XVIIIe siècles* (Paris: Ecole des Chartes, 1995).

The Pedlars Case Stated, Or, Some Remarks upon the Pedlars, and All Their Carrying of Goods Abroad to Proffer to Sale in All Cities, Towns, and Places throughout England and Wales, &c., in Order to the Prevention Thereof (London, 1691).

The Weavers' True Case (London, 1719).

Thijs, Alfons K. L., 'Les textiles au marché anversois au XVIe siècle', in Erik Aerts and John Munro (eds), *Textiles of the Low Countries in European Economic History* (Proceedings of the Tenth International Economic History Congress, Leuven, August 1990), 76–86.

Thirsk, Joan, *Economic Policy and Projects: The Development of a Consumer Society in Early Modern England* (Oxford: Oxford University Press, 1978).

Thirsk, Joan, 'Knitting and knitwear, c. 1500–1780', in David Jenkins (ed.), *The Cambridge History of Western Textiles* (Cambridge: Cambridge University Press, 2003), 562–84.

Thirsk, Joan, 'The fantastical folly of fashion: The English stocking knitting industry, 1500–1700', in Negley B. Harte and Kenneth B. Ponting (eds), *Textile History and Economic History: Essays in Honour of Miss Julia de Lacy Mann* (Manchester: Manchester University Press, 1973), 50–73.

Thomson, Roy, 'The nature and properties of leather', in Marion Kite and Roy Thomson (eds), *Conservation of Leather and Related Materials* (Abingdon: Routledge, 2006), 1–3.

Thornton, Claire, Jenny Tiramani, Melanie Braun, Dave Budd, Jordan Colls and Ruth Gilbert, *The Mockado Doublet Project Report* (unpublished document for the 'Refashioning' project, 2021).

Tiramani, Jenny, 'Reconstructing a Schwartz outfit', in Ulinka Rublack, Maria Hayward and Jenny Tiramani (eds), *The First Book of Fashion: The Book of Clothes of Matthaus and Veit Konrad Schwarz of Augsburg* (London: Bloomsbury Academic, 2015), 373–96.

Tlusty, B. Ann, *The Martial Ethic in Early Modern Germany: Civic Duty and the Right of Arms* (Basingstoke: Palgrave Macmillan, 2011).

Tosi Brandi, Elisa, *L'arte del sarto nel Medioevo: quando la moda diventa un mestiere* (Bologna: Società editrice Il mulino, 2017).

Turnau, Irene, 'La bonneterie en Europe du XVIe au XVIIIe siècle', *Annales. Economies, Sociétés, Civilisations*, 26, 5 (1971), 1118–32.

Turnau, Irene, 'The diffusion of knitting in medieval Europe', in Negley B. Harte and Kenneth G. Pointing (eds), *Cloth and Clothing in Medieval Europe: Essays in Memory of Professor E. M. Carus-Wilson* (London: Heinemann Educational Books, 1983), 368–73.

Turnau, Irene, 'The organization of the European textile industry from the thirteenth to the eighteenth century', *Journal of European Economic History*, 17, 3 (1988).

Twigger Holroyd, Amy and Emma Shercliff, *Stitching Together: Good Practice Guidelines. Advice for Facilitators of Participatory Textile Making Workshops and Projects* (Bournemouth: Stitching Together, 2020).

Ulrich, Laurel Thatcher, *The Age of Homespun* (New York: Knopf, 2001).

Unsworth, Rebecca, 'Hands deep in history: Pockets in men and women's dress in western Europe, c. 1480–1630', *Costume*, 51, 2 (2017), 148–70.

Van der Wee, Herman, 'The western European woollen industries, 1500–1750', in D. Jenkins (ed.), *The Cambridge History of Western Textiles* (Cambridge: Cambridge University Press, 2003), vol. 1, 397–472.

Veale, Elspeth M., *The English Fur Trade in the Later Middle Ages* (London: London Record Society, 2003).

Vecellio, Cesare, *De gli habiti antichi et moderni di diverse parti del mondo* (Venice: Damian Zanaro, 1590).

Vecellio, Cesare, *The Clothing of the Renaissance World: Europe, Asia, Africa, the Americas: Cesare Vecellio's Habiti Antichi et Moderni*, ed. and trans. Margaret F. Rosenthal and Ann Rosalind Jones (London: Thames and Hudson, 2008).

Venborg Pedersen, M. 'Material culture: A source material to everyday life in early modern Eiderstedt, Schleswig', *Kieler Blätter zur Volkskunde*, 35 (2003), 89–104.

Venturelli, Paola, 'Segreti di Leonardo da Vinci per ottenere "perle grosse"', *Arte Lombarda*, 132, 2 (2001), 42–7.

Vianello, Francesco, 'Rural manufactures and patterns of economic specialization: Cases from the Venetian mainland', in Paola Lanaro (ed.), *At the Centre of the Old World: Trade and Manufacturing in Venice and the Venetian Mainland, 1400–1800* (Toronto: Centre for Reformation and Renaissance Studies, 2006), 343–66.

Vigarello, Georges, *Concepts of Cleanliness: Changing Attitudes in France Since the Middle Ages* (Cambridge: Cambridge University Press, 1988).

Wadsworth, Alfred P. and Julia de Lacy Mann, *The Cotton Trade and Industrial Lancashire: 1600–1780* (Manchester: Manchester University Press, 1931).

Walker, Jonathan, 'Bravi and Venetian nobles, c. 1550–1650', *Studi veneziani*, 36 (1998), 85–114.

Walker, Jonathan, 'Honour and the culture of male Venetian nobles, c. 1500–1650' (PhD thesis, University of Cambridge, 1998).

Warsh, Molly A., *American Baroque: Pearls and the Nature of Empire, 1492–1700* (Chapel Hill: University of North Carolina Press, 2018).

Waters, Laurie, 'A new interpretation of certain bobbin lace patterns in Le Pompe, 1559', *Textiles and Politics: Textile Society of America 13th Biennial Symposium Proceedings*, Washington, DC, September 18 – September 22, 2012, https://digitalcommons.unl.edu/tsaconf/754/.

Weatherill, Lorna, *Consumer Behaviour and Material Culture in Britain, 1660–1760* (London: Routledge, 1996).

Weatherill, Lorna, *The Account Book of Richard Latham, 1724–1767* (Oxford: Oxford University Press, 1990).

Weinmann, Arno, *Reval 1646 bis 1672: vom Frieden von Brömsebro bis zum Beginn der selbständigen Regierung Karls XI* (Bonn: Kulturstiftung der Deutschen Vertriebenen, 1991).

Welch, Evelyn 'Art on the edge: Hair and hands in Renaissance Italy', *Renaissance Studies*, 23, 3 (2008), 241–68.

Welch, Evelyn, 'Introduction', in Evelyn Welch (ed.), *Fashioning the Early Modern: Dress, Textiles, and Innovation in Europe, 1500–1800* (Oxford: Oxford University Press, 2016).

Welch, Evelyn, 'New, old and second-hand culture: The case of the Renaissance sleeve', in *Revaluing Renaissance Art* (Farnham: Ashgate, 2000), 101–19.

Welch, Evelyn, 'Scented buttons and perfumed gloves: Smelling things in Renaissance Italy', in Bella Mirabella (ed.), *Ornamentalism: The Art of Renaissance Accessories* (Ann Arbor: University of Michigan Press, 2011), 13–39.

Welch, Evelyn, *Shopping in the Renaissance: Consumer Cultures in Italy, 1400–1600* (New Haven: Yale University Press, 2005).

Wells, Frederick A., *The British Hosiery and Knitwear Industry: Its History and Organisation* (New York: Barnes & Noble, 1972).

Whittle, Jane, 'A critique of approaches to "domestic work": women, work and the pre-industrial economy', *Past & Present*, 243 (2019), 35–70.

Whittle, Jane and Elizabeth Griffiths, *Consumption and Gender in the Early Seventeenth-Century Household: The World of Alice Le Strange* (Oxford: Oxford University Press, 2012).

Wickham, Chris, 'How did the feudal economy work? The economic logic of medieval societies', *Past & Present*, 251 (2021), 3–40.

Wilkinson-Latham, Robert, *Phaidon Guide to Antique Weapons and Armour* (New Jersey: Prentice Hall, 1981).

Wilson, Bronwen, 'Foggie diverse di vestire de' Turchi: Turkish costume illustration and cultural translation', *The Journal of Medieval and Early Modern Studies*, 37, 1 (2007), 97–139.

Wilson, Charles, 'Cloth production and international competition in the seventeenth century', *Economic History Review*, 13 (1960), 209–21.

Worlidge, John, *Systema Agriculturae: The Mystery of Husbandry Discovered* (London, 1675).

Wunder, Amanda, 'Spanish fashion and sumptuary legislation from the thirteenth to the eighteenth century,' in Giorgio Riello and Ulinka Rublack (eds), *The Right to Dress: Sumptuary Laws in a Global Perspective c. 1200–1800* (Cambridge: Cambridge University Press, 2019), 243–72.

Young, Arthur, *A Six Months Tour through the North of England*, 4 vols (London, 1771).

Young, Arthur, *A Tour in Ireland: With General Observations on the State of that Kingdom*, second edition, 2 vols (London, 1780).

Zander-Seidel, Jutta, 'Nicht nur tierisch warm: Pelz in Kleidung und Mode', in Georg Ulrich Großmann (ed.), *Vom Ansehen der Tiere* (Nuremberg: Germanisches Nationalmuseum, 2009), 118–31.

Zecchin, Paolo, 'La nascita delle conterie veneziane', *Journal of Glass Studies*, 47 (2005), 77–92.

Index

Note: page numbers in *italic* refer to illustrations